Textile Mechanics

Textile Mechanics

Sandeep Roy

RANDOM

RANDOM PUBLICATIONS

NEW DELHI (INDIA)

Textile Mechanics

ISBN 978-93-5111-647-9

Published in 2015 in India by

RANDOM PUBLICATIONS

4376-A/4B, Gali Murari Lal, Ansari Road
New Delhi-110 002
Phone : +9111-43580356, 011-23289044, 011-43142548
e-mail: sales@randompublications.com,
info@randompublications.com, randomexports@gmail.com

Reprinted 2019

Type Setting by : Friends Media, Delhi-110089
Digitally Printed at : Replika Press Pvt. Ltd.

Preface

In a machine, drives are required to transmit power from electric motor to various elements and from one element to others so that the various elements could perform their desired operations. Between any two elements of a machine, the one which drives the other is called 'Driver'; and the other one is termed 'Driven'. The power/motion is transmitted by means of various drives such as belts, chains and gears. The machine designer has the options of selecting the nature/type of drives. Different drives have certain advantages and limitations.

Textile mechanics set up, adjust and maintain machines used in the textile, clothing and footwear industries. A knitting mechanic prepares patterning equipment that controls automatic knitting machines to produce a variety of knitted fabric and clothing, including hosiery, jumpers, shirts and underwear. A sewing machine mechanic assembles, adjusts, lubricates and repairs industrial sewing machines. A spinning mechanic studies specifications to determine settings for machines that use materials ranging from raw fibre through to finished thread such as wool and cotton. They then install and align gears, arrange yarns on the machines and adjust machines to produce different types and thicknesses of yarn.

I would like to thank my team for standing beside me throughout my career and writing this book. My special thanks go to "Random Publications" who have published the book.

– Sandeep Roy

Contents

1

Textile Mechanics and Marketing Trends

This is a marketing research on the Textiles industry and can include information on the background, market structure, definitions, competitors, trends and developments of textiles and is related to other apparel, clothing, yarn, thread and fibre.

MARKET TRENDS

Global market trends in the textile industry increasingly are being determined by the transition to a quota-free textile trade environment. The quota system that has been in place since the early 1960s ended on December 31, 2004. This situation is driving a realignment of markets and a restructuring of the industry that are well under way. The response of textile manufacturers to intensifying global competition has been threefold. First, industry product mixes are shifting among countries. There is ongoing rationalization of production to emphasize products in which the manufacturer has the greatest competitive advantages and de-emphasize products that are more vulnerable to foreign competition.

In the United States and other developed countries, this implies a de-emphasis on textile products consumed in the apparel industry and a greater emphasis on other products, such as home furnishings and industrial textiles. Second, companies continue to invest in new, more productive plant and equipment. Textile production is no longer a low-technology, labour-intensive industry enterprise. Textile firms around the world are under constant pressure to become more efficient through technology upgrades. Third and perhaps most important, manufacturing is expanding into countries close to home markets. This is being undertaken primarily to support production sharing arrangements for apparel, sometimes in partnership with customers, suppliers, or other textile manufacturers.

Textiles refer to yarns, threads and wools that can be spun, woven, tufted, tied and otherwise used to manufacture cloth. Production of textiles has been altered almost beyond recognition by mass-production and the introduction of modern manufacturing techniques.

MARKET STRUCTURE

Economies of scale in textile manufacturing are significant and create entry barriers in the market. The textile industry is more capital intensive than the clothing industry and is highly automated. It consists of spinning, weaving and finishing functions in integrated plants. For example, the approximate cost of a new fibre plant is $100 million depending on various factors. Costs of raw materials are volatile and can account for 50-60per cent of the cost of the finished products. For purposes of hedging against supply problems, many manufacturers are backward integrated into chemical intermediaries. Forward integration into apparel and product manufacturing is also seen in industry.

In 1995, the World Trade Organization (WTO), in the Agreement on Textiles and Clothing (ATC), agreed that all quotas on textiles and clothing would disappear between WTO member countries on January 1, 2005. The expiration of ATC marked the end of quotas, limiting textile and clothing trade between the WTO members. Developing countries such as China, India and Pakistan had been the most restricted by the quotas. While India and China are likely to emerge as winners, the main losers after quota will be quota-restricted countries that had enjoyed the benefits and protection for more than 40 years. The post quota market has changed with producers already affected by significant changes in retailing. Large retailers are now buying up independent brands to give consumers more value and enhance their shopping experience. Manufacturers in developed countries are more likely to adapt by relocating operations to production centers in low wage countries. Some of the major textile industries can be divided as follows:

- Awnings, textile
- Blankets
- Bags or sacks, textile
- Blinds, textile
- Canvas goods
- Cordage
- Elasticized fabrics
- Fabrics, textile
- Felt (except floor coverings)
- Glass fibre fabrics
- Household linen
- Lace
- Narrow fabrics
- Netting
- Rope (except wire rope)

- Sailcloth
- Sewing thread
- Soft furnishings
- String
- Tarpaulins
- Tents
- Thread
- Towels
- Trimmings, textile
- Yarns

For U.S. textile producers, the most important regional expansion and partnering are taking place with Mexico.

Broad market forces, including Asian pricing pressures, have accelerated the flow of textile capital into Mexico.

INDUSTRY DEFINITIONS

- *Acrylic*: Manufactured fibre derived from polyacrylonitrile. Its major properties include a soft, wool-like hand, machine washable and dryable, excellent colour retention. Solution-dyed versions have excellent resistance to sunlight and chlorine degradation.
- *Anti-dumping duty*: An extra duty imposed on an imported product by an importing country (or group of countries, as in the case of the EU) to compensate for the dumping of goods by a foreign supplier.
- *Aramid*: The generic name for a special group of synthetic fibres (aromatic polyamide) having high strength; examples are "Kevlar" and "Twaron".
- *Calico*: Tightly woven cotton type fabric with an all-over print, usually a small floral pattern on a contrasting background colour. Common end-uses include dresses, aprons, and quilts.
- *Countervailing duty*: An extra duty imposed on an imported product by an importing country (or group of countries, as in the case of the EU) to compensate for subsidies deemed to be illegal which are given to the manufacturer of the product in the exporting country.
- *Delocalization*: The geographical move of a production unit to a low cost country. (Note that the term is increasingly being used to describe all forms of shifts in production, including foreign sourcing and subcontracting.)
- *Industrial textiles*: A category of technical textiles used as part of an industrial process, or incorporated into final products.
- *Lycra*: DuPont's brand name for its elastane or spandex fibre.
- *PET*: Polyethylene terephthalate, the most common form of polyester.
- *Polyester*: A manufactured fibre introduced in the early 1950s, and is second only to cotton in worldwide use. Polyester has high strength (although somewhat lower than nylon), excellent resiliency, and high

abrasion resistance. Low absorbency allows the fibre to dry quickly.

- *Quota*: A quantitative restraint imposed by an importing country on an exporting country - or established by agreement between the trading partners - which is designed to limit shipments of a product from the exporting to the importing country.
- *Spandex fibre*: The generic name used in the USA to denote elastane fibre. A manufactured elastomeric fibre that can be repeatedly stretched over 500per cent without breaking, and will still recover to its original length.
- *Technical textiles*: Textile materials and products manufactured primarily for their technical performance and functional properties rather than their aesthetic or decorative characteristics. End uses include aerospace, industrial, marine, medical, military, safety and transport textiles, and geotextiles.
- *Unit production systems*: An advanced apparel manufacturing system in which a single garment is progressed through a sequence of operations. Using a unit production system, a garment is automatically transported via a computer-controlled overhead hanging system, which has been ergonomically designed to reduce the amount of handling of the garment (see also progressive bundle system).
- *Yarn*: A continuous strand of textile fibres created when a cluster of individual fibres are twisted together. These long yarns are used to create fabrics, either by knitting or weaving.
- *Yarn Numbering System*: Systems for sizing yarn fall into two basic types. The yarn number is based on the length of yarn needed to make up a specified weight. The finer the yarn, the higher the number. Cotton, wool and linen are numbered with such systems. The yarn number is based on the mass of a specified length of yarn. The finer the yarn, the lower the number. Silk, synthetic fibres and jute are numbered with such systems.
- *Textile HS' codes*: Harmonized System Codes (HS Codes) are a government, regulated system used to classify products and their corresponding tariffs. The Harmonized System is an internationally developed and implemented commodity-description and coding system, on which the tariffs of most countries of the world (including the NAFTA countries) are based.
- *Manufactured Fibres*: It is important to understand that all manufactured fibres are not alike. Each fibre has a unique composition and physical properties.]

  The U. S. Federal Trade Commission has established generic names and definitions for manufactured fibres, including acetate, acrylic, lyocell, modacrylic, nylon, polyester, polypropylene (olefin), rayon, and spandex.

SUSTAINABILITY

Sustain means "to maintain" or " to uphold" and with regard to industrial processes sustainability means establishing hose principles and practices which can help to maintain the equilibrium of nature in other words to avoid causing irreversible damage to the earth's natural resources. Moving to a greater degree of sustainability in our industrial processes and systems requires that we achieve a better balance between the social, economic and environmental aspects of textile production.

- A sustainable product is one that is manufactured in a way that respects the social elements of fair trade and human rights of the people involved in the whole of the manufacturing chain.
- A sustainable product is one that is manufactured in such a way that it has the lowest possible adverse effect on the environment *e.g.* by making the most efficient use of resources such as water and energy, and which goes the extra mile to recover raw materials, *e.g.* by the recycling of as much water as possible or by recovering the heat from wastewater discharges.
- But equally important, a sustainable product is one which can compete effectively in the global marketplace against less sustainable products. *i.e.* which offers value benefits to the consumer, and where the economic returns from its success are fairly distributed back along the supply chain.

In order to achieve the above agenda, it is important to work at each and every aspect of textile production where natural resources, energy and chemicals are consumed and emissions to air, water and land arise.

SUSTAINABLE TEXTILE PROCESSING

As we have already noted the textile industry is one of the most polluting industry sectors. A vast range and quantity of chemicals is used at every stage and the after- effects in terms of wastewater treatment and air pollution are critical to manage.

There is a strong need to establish more sustainable textile processing measures in the industry. The main objective of these measures should be to minimize and eliminate the most harmful inputs and the most polluting outputs and to reduce the level of chemical residues left on the textile.

But it would be wrong to focus solely on the chemical inputs and ignore the consumption of energy and water which are the primary impacts of the textile processing industry. In summary a sustainable approach covers the following points:

- Minimum use of resources (water and energy)
- Minimum chemical consumption
- Minimum pollution load

- Toxic chemicals eliminated from supply chain
- Harmful chemical residues eliminated from final textile

But in order to minimize the usage, it is important to measure the inputs and in order to eliminate the most harmful chemicals, it is important to know and understand what is being used. Uncontrolled or unknown inputs lead to unmanaged use of resources and uncontrolled outputs. The measurement of the parameters shown below is essential to establish an understanding of the input-output balance of the textile processing operation.

The measurement and control of these inputs and outputs can lead to:

- Improved resource productivity
- Improved eco-efficiency
- Improved cost efficiency And
- Improved customer satisfaction
- Improved brand reputation

Once the understanding about the basics of processes and chemicals involved with the supply chain is established and awareness about the inputs is there, then control over the output of the production process can be achieved.

COLOUR COMMUNICATION

Finally we should note another important element that is often overlooked in seeking to improve the sustainability of the textile and clothing supply chain and that is accurate and timely colour communication. We will return to this subject later in the series but for now a few tips for improving colour communication:

- Communicate your colour accurately using physical and/or digital colour standards. Accurate communication of the colour you require will reduce lab dip rejection rates and result in better right first time bulk dyeing.
- Communicate electronically when feasible: avoid wasting time due to delayed communication
- Do not set unachievable colour standards- avoid wasting dye and chemicals in trying to achieve heavy depths or brilliant shades on certain fibres.

CONCLUSION

Sustainable design, informed selection of dyes and chemicals, accurate colour communication, and controlled colouration using Best Available Technology are key elements in raising the standard of sustainability in the textile supply chain.

CREATING A GLOBAL VISION FOR SUSTAINABLE TEXTILES

Certification, such as eco-labels, plays a major role in giving credible assurance to retailers and end consumers that products comply with standards based on social, ecological and environmental standards. Of the 309 eco-labels identified world wide, 41 cover textiles (Ecolabelling, 2008) and some 9000 textile and clothing manufacturing companies have been certified. Organic Exchange Fibre Report (2008/09) estimated a 54per cent increase in *cultivation* of organic cotton from the previous year, but *production* of organic cotton only 0.959per cent of conventional cotton, *i.e.* the growth in eco-labelled textiles is not reflected in consumer demand, raising questions about the impact eco-labelled or 'sustainable' textiles. A number of issues may impede the spread of eco-labelled textiles through the supply chain: costs and time required to achieve, use and renew the eco-label, recession and potential loss of competitive advantages. This paper will present the findings from in depth interviews examining the decision making around buying and sourcing of eco-labelled fibre, fabrics or textile products. The seven companies located both in India and the UK, spanned the supply chain, from fibre to product: textile manufacturers, eco-parameter testing labs, Certification Company and retailer. The aim of the research was to understand and investigate the marketing strategies for sustainable textile products. Our goal was to understand how designers, manufacturers and retailers may collaborate to deliver eco-labelled textiles attractive to the end consumer and we conclude by reflecting on potential implications for the supply chain integration

ECO-LABELS

Eco-labelling is becoming a differentiating factor on a worldwide scale in retail markets for textile and apparel purchase. Consumers are becoming increasingly concerned with the adverse impacts of industrial pollution on the environment and their health, resulting mounting pressure on textile, fashion industry to adopt more eco-friendly, chemicals and manufacturing processes. Environmental concerns raised by production systems have been recognized since the late 1960's and attempts to move towards more sustainable and environmentally friendly approaches have been through a range of regulatory measures from green taxes to strict bans. One approach acquiring increasing importance is that of 'environmental labelling' or 'eco-labelling', which, according to Piotrowski and Kratz (2005) differ in that *environmental labelling* is broad and covers a range of labels and declarations of environmental performance and focus on consumption rather than the production of a given product; *e.g.* recyclable material while *eco-labels* are a sub-group of environmental labelling and convey environmental information about a product to the consumer and communicate that the environmental impacts are reduced over the entire life cycle of a product without specifying the production practices. In brief an ecolabel –

- Identifies the overall environmental preferences of a product;
- Provides information on environment related product qualities;

- Are tools for consumers to identify environmentally safe product;
- Enables manufacturers to use ecofriendly raw material and ingredients;
- Is an additional product quality which can be used as a marketing tool;
- Can be issued by private or public body;
- Causes less stress on the environment
- Enables to earn premium on products.

SIGNIFICANCE OF ECO-LABELS

The eco-label has a role in the Integrated Product Policy (IPP) which aims to minimise the environmental degradation caused by any of the phases of a product's life cycle (tangible or intangible, such as service), *e.g.* manufacture, development, use or disposal (European Commission, 2008). All phases of a product life cycle are examined with the objective of improving their environmental performance. This approach requires all participants in this process to be engaged: eg, designers, industry, marketers, retailers and consumers. The US EPA (1994) defined the following five factors for measuring effectiveness of an eco-label, the first four of which serve to support the last:

- Consumer awareness of labels
- Consumer acceptance of labels (credibility and understanding)
- Changes in consumer behaviour
- Changes in manufacturer behaviour
- Net environmental gains

TYPES OF ECO-LABELS

Eco-labels may be voluntary or mandatory. Mandatory labelling is always third party labelling (*i.e.* an independent body is required to attest to required standards having been achieved), voluntary programmers may be established by firms or business associations as well as third party. Currently, there are no eco-labels in textiles and clothing enforced by mandatory rules. Eco-labels are normally issued either by government supported or private enterprises once it has been proved that the product of the applicant has met the criteria.,

- Government: Blue Angel (Germany), Eco Mark (Japan), Environmental Choice (Canada), White Swan (Nordic Countries), EU, Eco-Mark (India), Green Label (Singapore)
- Private: Eco-tex, Oeko-Tex (textiles and clothing) (Germany). Green Seal (United States),

The criteria for granting eco-labels are mostly based on the "cradle-to-grave" approach, *i.e.* the life-cycle analysis of the product and assessment of its impact on the environment from processing of raw materials, production, distribution, consumption and maintenance, (*i.e.* washing, ironing, dry-cleaning) and finally disposal of the product. A 'Cradle to Cradle' certification programme assesses the sustainability of product ingredients for human and environmental health, as well as their recyclability or compostability making it easier at the design stage to create ecologically-intelligent products through choosing materials that meet key sustainability criteria for material health and material reutilisation (Braungart and McDonough, 2008). Differences between various eco-labelling schemes confuse public understanding of eco-labels: some are based on detailed analysis of the environmental impacts while others analyse only certain stages of the life-cycle. Voluntary labels are classified according to International Standards Organisation (ISO) (Baumann, 2007). ISO is the world's largest non-governmental organization that develops and publishes International Standards and is a network of the national standards institutes of 163 countries. There are now eco-labelling schemes both in developed and developing countries and the ISO has classified the existing environmental labels into three typologies –*Type I, II and III*, specifying preferential principles and procedures for each one of them (beyond the remit of this paper to detail). Many other prominent international trade and environmental organisations deal with issues related to eco-labelling, *e.g.*: the United Nations, the World Trade Organisation through its International Trade Centre and Committee on Trade and Environment, the US Environmental Protection Agency, as well as the Organisation for Economic Co-operation and Development (Naumann, 2001).

DEVELOPING AN ECO-LABEL

This is complex and complicated but can be generalised into four broad phases:

- Selection of a product category by a labelling board through suggestions from industry, environmentalist, consumers, and other interested parties. De Man et al (1997) proposed that four levels of actors function throughout the industry:
  - *Primary economic*: Production /consumption decision-makers (producers, importers, consumers).
  - *Secondary economic*: Influence the decision making of primary actors.
  - *Governmental and administrative*: set the framework for the actors.
  - *Others*: Try to influence the behaviour of all actors to improve status quo.
- Life-cycle analysis to assess environmental impact of products in chosen category and examine the material and energy inputs for manufacture and use of a product and the solid, liquid, and gaseous waste generated at each stage of lifecycle, *e.g.* raw material, production, distribution, packing use and disposal.

- Criteria and thresholds for the award of an eco-label set taking into consideration technical feasibility and environmental impacts in different media like air, water and soil against one another. Different eco-labels have differing methodologies, *e.g.*, Oeko-tex 100 examines harmful residues on the product, while GOTS tends to look environmental as well as residual parameters.
- The product category and criteria is reviewed and refined. Interested parties including industry and environmental and consumer groups are asked for their inputs, although they are often already included much earlier on in the process.

ACQUIRING AN ECO-LABEL

Atilgan, (2007) identified fourteen steps to obtaining the EU Flower eco-label for a Turkish textile manufacturing firm In brief, the factory administration decides to apply for eco-label and use it. At the performance control stage, the selected ecolabelling organisation's prohibited chemicals are identified. Detailed monitoring of the chemicals and quality control system is necessary at each stage of production to prove that the products introduced to the certificating institute are appropriate. At this stage, honesty is necessary because the institute is authorised to make tests any time it likes. Failing in these tests results is cancellation of the eco-label certificate (Atilgan, 2007). As well as the time and complexity of obtaining the eco-label, the most significant issue is that of cost, which, for the EU eco-label is set at 0.15per cent of the annual turnover of the eco-labelled product, costing up to □1,300 for registration (*i.e.* to apply for the label), □25,000 per year for the use of the label, with a reduction of 25per cent for SMEs (buyusa.gov 2009, Rubik and Frankl, 2005). Atilgan (2007) indicated that the costs of using eco-labelled production made the finished product between 12-15per cent more expensive to make, depressing interest in their use by manufacturers and retailers. The costs become even greater down the chain. A UK corporate wear supplier estimated the costs of using eco-labelled fabrics (such as Teijin fabrics, from EcoCircle closed loop system of fibre processing) as placing a 57per cent premium on their final product (Sinha and Hussey, 2009).

ECO-LABELS AND CERTIFICATION

Regardless of the costs / benefits, Atilgan (2007) urged the Turkish government and industry to become engaged with eco-labelling as he felt that the next area of purchase and trading selection appeared to be based on the criteria set by the eco-labelling bodies. Furthermore he suggested that these costs may be mitigated through the use of smaller amounts of high quality products, optimising the production techniques *e.g.* by controlling all recipes and procedures, and identifying problem areas. Given the premiums on using eco-labelled textiles, commercial buyers of such products require assurances that goods comply with standards set by the eco-label organisation. This is particularly relevant for organic fibres where the market has flooded by products described in vague terms such as 'Green' and 'Eco' (Rundgren, 1999). To differentiate between sustainable and traditional textiles, third party standards are desirable. The unverified, market-based self-labelling (without outside monitoring) approach for textile ecolabelling, invites fraud due to lack of third-party verification (Moore et al., 2009). Retailers and consumers are starting to demand labels backed by solid third-party certifications to give confidence to all the marketing claims. Commercial buyers therefore may require certification of the product. Like ISO Type I labels, certification schemes are voluntary and provide information on the environmental impacts of a company's production methods and processes (PPMs), ie, impacts of the *entire* activity not just a particular *product* (Rotherham, 1999). Certification provides a comprehensive system for ensuring that certain standards of organic production and processing are met. The system includes-

- Developing rules or standards (standard setting).
- Verifying and evaluating performance against those standards (inspection).
- Recognizing procedures which successfully meet the standards (certification).

SCOPE AND TRANSACTION CERTIFICATES

Most companies only look apply for certification if it is required by the customer. From the literature reviewed to date, shows the flow of the certification across the supply chain. There are two kinds of certificates: *Scope* - issued to the selling company and stating its name, the products and production facility inspected and certified in accordance with standards. *Transaction*: only issued (to the seller) for the sale of products if the applicant has the scope certificate. For example, a retail buyer may require a scope certificate from chosen garment manufacturer to prove that their manufacturing facility is certified and products are being produced in accordance with standards. If the manufacturer is not certified, they contact the certification body for the certification. If the garment manufacturer does not manufacture textiles, then the retail buyer would need to source the required certified raw material to process into fabric. Then they will contact the factories from where they wish to buy the fabric, however that facility must also be certified and to prove it, processor need to provide scope certificate to garment manufacturer, if there is no certificate then they need to contact the certification body to certify their facility. If the ordered garment has embroidery and printing, then either the garment manufacturer should acquire certification for those facilities under their scope, or suggest to the embroiderer or printer to acquire certification themselves through using certified threads, dyes or chemicals. To obtain a certificate can take between 30-45 days, depending on the standards, factory condition, number of factories under one application understanding of standards by applicant, and changes within the factory with respect to compliance with standards. This process continues 'backwards' along the supply chain till supplying of certified (in this case organic) cotton to spinner.

CERTIFICATION FLOW

If all the points are certified the process would take around a minimum of 30-45 working days to issue a certificate from certification body. If during inspection any non-compliance occurs the process would be lengthened as each certification takes about 30-45 days. A further complication is that the entire chain requires consistency in certification bodies used: *i.e.*, if the garment manufacturer is required to produce to GOTS standards, the processors must also have GOTS certification-a different certification will not suffice as they each have their own analysis processes for certification. In brief, certification is a complex, time consuming and costly process however it creates the transparency within supply chain. For each step in the supply of a product for the chain, the transaction certificate provides transparency within the supply chain and it gives assurance to buyer about the compliance of products with standards. The certification body issues the transaction certificate for each application made by their certified clients and is delivered to the buyer to give assurance and to confirm the product is manufactured accordance with standards.

- Demand for Factory certification
- Receiving certified product from supplier
- Contacting certification body for certification
- ontacting ecological testing laboratory for required test reports

COST AND BENEFITS OF CERTIFICATION

An important factor in the success of any eco-labels is its ability to cover its certification costs and therefore stay in business. According to EPA (1998), the ease with which Programmes will be able to cover costs varies depending on two questions:

- Can the Programme charge enough in application, testing, audit and other fees to cover its costs; and
- Can the Programme subsidize its environmental labelling activities from other Programme activities?

Cost of certification can be high in relation to the value of the product and thus can become prohibitive. This is especially true for textiles because of the number of process are involved from production to consumer. It may be more expensive for companies in developing countries to obtain labels and certifications, "due to factors such as the lack of existing management structures

(EMS), the novelty of EMS, insufficient infrastructure, and high auditing costs if companies have to rely on international consultants and certification companies" (Rotherham, 1999). In addition to capital costs, the absence of necessary knowledge and skills and a lack of mutual recognition between different national Programmes can further disadvantage some countries (UNCTAD 1997).

ECO-LABELLED SUSTAINABLE TEXTILE PRODUCTS:

'Sustainability' is the ability to maintain an activity indefinitely over time. A sustainable activity is therefore one that does not exhaust the resources on which it depends. The concept of sustainability gained worldwide recognition following a report in 1987 by the World Commission on Environment and Development (WCED). This report entitled *Our Common Future*, defined sustainability development as (Performance Apparel Market, 2009) –

- "Development that meets the needs of the present without compromising the ability of future generations to meet their own needs."

Based on above definition, it's possible to define the term 'sustainable textile product' (STP) as –

- "A product which is manufactured with the help of services and related products in responds to basic need and bring a better quality of the life with use of minimum amount of natural resources and toxic chemicals during the production, minimum waste emission to the environment over the life cycle of product keeping environmental and social factors in mind throughout the supply chain".

Sustainable textile or apparel are, therefore:

- Safe for human and physical environment;
- Made from renewable materials;
- Produced while making the most efficient use of resources such as water and energy;
- Manufactured by people employed in decent working environment;
- Capable of being washed at low temperature using environmentally friendly laundering agents; and
- Capable of being returned safely to the environment at the end of their useful life (Performance Apparel Market, 2009).

CHALLENGES FOR ECO-LABELLING

There are many challenges for the eco-labelling, the most serious of which are: misleading or fraudulent to uninformative claims, unfair competition and protectionism and lack of stringency or standardisation in the process or mechanisms of eco-labelling. The objective of certification is to gain access to the market for environmentally sustainable products (Rundgren, 1999), and the certification process should help as data collected in the process of certification can be very useful for market planning as well as for extension and research, moreover, improves the 'image of product and increases its credibility and visibility (Rundgren, 1999), Auriol and Schilizzi's (2003) studies have shown that the costlier the certification process, the fewer firms able to afford certification, ie, cost becomes a major factor in deciding market structure, potentially leading to monopoly and ultimately to no certification at all. Influencing factors on consumers' willingness to buy environmentally friendly products have been identified and categorised as: demographics, knowledge, values, attitudes and behaviour (Laroche et al., 2001). However, price has been found to be one of the most decisive factors in determining when consumers actually purchase apparel products. Consumers' willingness to pay and purchase cloths made from sustainable raw material like organic cotton is a complex issue (Gam et al., 2010). Empirical testing has shown, however, that many consumers are willing to pay a premium for eco-labeled products (Imkamp, 2000; Loureiro et al., 2002; Makatouni, 2002; Moon et al., 2002) and that they do purchase such products (Lathrop and Centner, 1998; Teisl et al., 2002). According to Gam et al., (2010) study in one of US, only 35per cent (27 out of 84) were willing to pay more for OCC and only 10.7per cent (9 out of 84) were willing to accept more than a 10per cent increase in price for OCC. In contrast, they found that 52per cent of survey participants would pay a 50per cent price premium and 25per cent of the participants would pay 100per cent more for an OCC over conventional cotton clothing. From above results it's clear that, willingness to pay for STP like product made from organic cotton varies place to place, country to country. Differences in testing and certification methods have created difficulties in the application of an eco-label to a particular product category. For example, should the label represent an overall assessment of a product's environmental burden over its entire life cycle, or some subset of it? What techniques can be used to measure environmental impact? Who determines what specific environmental impacts are the most important? And what criteria are appropriate in rating impacts? Moreover, the consumer is unable to verify the claims made by the eco-label. An analysis of ecological labelling process by Lavallee and Plouffe (2004) concluded that 'cradle-tograve' analysis for ecolabeled products and services is not always, in fact, respected, and that at the present time ecolabel delivery criteria are not sufficiently stringent or standardised leading to confusion in the marketplace, making it difficult for companies to identify stakeholder preferences and for justified environmental claims to be considered credible (Rotherham, 1999; EPA 1998).

THE RESEARCH QUESTION AND METHODOLOGY:

The research undertook to understand the issues within and across the textile supply chain that come to bear upon the growth of eco-labelled sustainable textiles products. While the fear of losing market share is a motivating factor, it should be stressed that market impact of eco-labelled textiles products are only one indication of an ecolabelling programme's success. The effectiveness of an eco-label ultimately depends on the extent to which consumers perceive, recognise and act on the information it conveys. The next section describes the research methodology adopted and presents some of the most pertinent results. Multiple in-depth interviews were conducted using semi-structured interview schedules designed as open ended questions. The decision was made to focus the study on organic cotton (one of a range of sustainable fibres) as this (and its products) is produced in large quantity, consumers are much more aware of it, retailers as well as consumers have accepted it and year by year demand is growing throughout world. 9 in-depth interviews were conducted with company employees who ranged from director of development, director of Marketing, Company Owner/ Director, etc. As indicated, the companies were selected on the basis that they had at least one following production facility in order to complete as much as possible the supply chain: spinning, weaving, knitting, wet processing, stitching/garmenting, retailing, testing, authorization of certification and authorization to issue eco-label. Companies interviewed are listed in table 1; the majority of the companies were in India as this is a centre of organics cotton production.

RESULTS

As mentioned earlier, the research tried to understand the issues within and across the textile supply chain that come to bear upon the growth of eco-labelled sustainable textiles products in particular and in developing a sustainable textiles industry in general. From the interviews, the following issues arose:

Sustainable textiles products (STP) are needed: All companies interviewed agreed that STPs are needed. According to textile manufactures companies A and B, STP is a holistic approach and it can be achieved through recycle, reduce and reuse processes. They felt it very important to note that all naturally grown products are not organic or sustainable; for example, organic cotton. All naturally grown cotton is not organic; it might be genetically modified organic cotton. Also, there is no assurance that the land doesn't have any traces of harmful fertilizer, pesticides. According to dyes and chemicals manufacturers companies D and F, STP is a mindset (company D) and its "the product which has manufactured by taking care of all the three elements of sustainability that is: social, economic and environmental sustainability and product design is fashionable to sell and sustain in the market" (company F). STP`s are needed to take care of our eco-system, so that this planet can 'sustain' the lives and livelihood 6 billion inhabitants. From certifiers (company G), STP`s are those products which are manufactured from fibre cultivated by natural or organic method considering social and environmental impact, understanding soil fertility and animal welfare. Further, Mr A (Company H) added that "producing sustainable textile is one of the ways to overcome the Global warming" and Ms. AB (company I) suggested product made from 'organic cotton' are the best example of STP.

'WHY' AND 'HOW' TO BECOME SUSTAINABLE TEXTILE MANUFACTURER

A number of methods are available to enter the sustainability arena; those raised by the interviewed companies included the following:

- All the companies interviewed were either certified or were in the process of receiving certification, ie there was felt to be a general endorsement of the idea of eco-labelling.

BOLSTER THE CLAIMS WITH INDEPENDENT VERIFICATION

Third-party verification of environmental credentials can often bring legitimacy to sustainability. Many of the most successful eco-labels are those that have been backed by issues-led organisations, for example GOTS certification for textile products made from organic cotton. Third-party verification can range in scope from qualitative assurance of general claims to detailed verification of all stages of a full life-cycle product assessment. Given the generally low levels of consumer trust in big business, some degree of external verification is an essential component of any credible environmental claim. According to Mr. H (company G) and Mr. A (company H), the claims made by the manufacturer or retailers are cross checked by certification of responsible eco-labelling body by testing the goods which can be picked from market store to confirm whether claims are right or wrong. In case of dyes and chemical, customer requires proof from third party initiating business. Therefore many companies use the test reports as a marketing tool to prove the product and company integrity. Within this study, it was the analysis, researcher has observed that, all the segments of supply chain apply for certification or verification as a buyer requirement.

EDUCATE, ENABLE, AND ENCOURAGE

Educating, enabling, and encouraging people to act towards sustainability is key for the success of any eco-label and STP as consumers' usage and disposal patterns liberate CO2 and so there should be programmes to educate them. Methods to do this range from placing trust worthy eco-label with required info on it, through various media, and through regulations. Once the consumer gets educated then may be encouraged to prefer buying more sustainable products by linking them with promotions and reward schemes (carefully and consistently in accordance with principles of sustainable consumption). Product information can now be shared via many more 'touch points': at point of sale; in retailer magazines, leaflets and web sites; through road shows, help lines and education packs. Enlisting employees to promote sustainability is another method as employees are key players of any manufacturing facility. Investing in the skills of employees is part of a sustainable and responsible human resource management (Brito et al., 2008). To educate their employees, company A celebrates days like Green Day, Earth day in factory to increase the awareness of sustainability. Company B have organised seminars for their employees, their suppliers, buyers, contractors etc before starting production of organic cotton where they have invited expertise from industry on subjects like eco-friendly dyes and chemicals, GOTS, ethical production practices etc.

REDUCE, REUSE, AND RECYCLE:

To become sustainable, everyone, from manufacturer to consumer should think of 'Reduce', 'Reuse' and 'Recycle'. Textile waste in landfill contributes to the formation of leachate as it decomposes, (which has the potential to contaminate groundwater), methane gas (a major cause of greenhouse gases contributing to global warming) and ammonia (highly toxic for land, water and air) (Productivity Commission 2006). The companies interviewed engaged currently in the following activities to try to reduce waste and reuse material:

Company A –

- By utilizing solar energy company is running around 20 computers within factory,
- From cutting waste, manufactured 5000 bags and sold at local super market subsidiary price
- Currently working on yarn manufactured from solar energy,
- Reusing treated effluent water for gardening and washing purpose (specifically washing of printing screens).

Company B –

- Have reverse osmosis (RO) plant for purification of processing effluent, after purification of that water it utilised for washing, gardening etc.
- Have started recycling of old paper cones (used in spinning for packing) and making new ones.

Company I –

- Offering shopping bags to their customers made of potato starch *i.e.* 100per cent biodegradable.

DISCUSSION AND CONCLUSION

The proliferation of voluntary certification and labelling schemes for environmentally and socially responsible production is often seen as a driven by companies and consumers. Consumers are heavily involved in environmental pollution because of their buying behaviour and consumption of textiles and significant associations were found between environmental shopping attitudes and behaviour and willingness to pay more for organic cotton products and observed that consumers with a greater environmental awareness demand more environmentally friendly merchandise (Fraj and Martinez, 2006). Eco-labels are not simple to understand as they appear and so may not be as appropriate marketing tools as suggested by the government policies. For example, GOTS and OE standards labelling guide are both standards applicable for products made from organic cotton on which retailers, manufacturers can use respective logo on their tags. Under both the standards it's mandatory to mention the percentage of organic cotton on the label and if the product is made from 100per cent organically cotton, manufacturer or retailer can use the statement "Organic" and "Made with 100per cent organically grown cotton" respectively. However, GOTS is based on social, technical and environmental areas while Organic Exchange 100 Standard (OE 100) is for tracking and documenting the purchase, handling, and use of 100per cent certified organically farmed cotton fibres (or organicinconversion cotton fibre) in yarns, fabrics and finished goods. The significance of this difference that, while purchasing the product a consumer will check only the organic content and assume that product is eco-friendly or sustainable. However, this is not a completely accurate picture as an OE standard does not look into social or harmful dyes and chemicals or about the environment related issues and an OE 100 logoed garment may be made from 100per cent organically grown cotton but then finished with harmful dyes and chemicals, printed with non-eco-friendly printing technique like solvent based printing. Therefore a large problem in marketing communication is using eco-labels is the lack of common definition or general understanding exists for what constitutes environmentally friendly clothing and eco-labelling. Eco-labels backed by solid third party certification give confidence to consumers about the genuineness of product. Certification bodies are the key player for the growth of eco-labelled STP who confirms the product, process and manufactures integrity with respect to sustainability. In this study, across the whole supply chain, most of the people had positive thought about requirement of certification. However, cost of certification can be high in relation to the value of the product and thus can become prohibitive. This is especially true for textiles because of the number of process are involved from production to consumer. To overcome the certification and labelling cost, Ibanez and Grolleau (2007) suggested "carrot" approach - the firm who preserve the environment appreciate them by subsidizing a recognized labelling and "stick" approach - increase of labelling cost for polluting firms by enforcing stricter labelling guidelines and severe punishment in case of deceptive use of environmental claims. According to Getz and Shreck (2006), despite much analysis of third-party certification, little is known about how certification is enabled or enacted at the point of production. The insights of those few who have explored some of the political and social effects of certification at the point of production are worthy of further examination. In this study researcher has analysed the 'how' certification process works within the organic textile-clothing industry and 'why' it is required.

Having reviewed the process of certification and how individual actors of supply chain get involved into this very complex, time consuming and sometime costly process of certification. Therefore we suggest that manufacturers take a vertically integrated approach through networking or developing two or three processing facilities to overcome cost and time of certification, reduce handling, transportation CO2 emission and time to retailers' shelves. Companies such as MandS, HandM, Zara, who are bringing STP's to the mass market must be operating in this manner. Moreover we suggest (table 3), a fourth to the traditional three pillars of sustainable development; "technical" by which the first three may be achieved.

- Demand for Factory certification
- Receiving certified product from supplier
- Contacting certification body for certification
- Contacting ecological testing laboratory for required test reports

SD PILLARS WHY HOW

- Economic Competitiveness
- Process and product innovation; Process
- and product substitution.
- Environmental Clean out puts Reduce, Reuse and Recycle
- Social Social fairness Better human resource management
- Technical Eco-friendly inputs Use of certified products

RECOMMENDATIONS FOR FURTHER RESEARCH AND INDUSTRY

There are numerous possibilities for future research in eco-labelled STP. For this research manufactures were from India, results may vary from country to country. However, based on general understanding, the study results indicate the following areas of further research:

- Most of the research is carried out to identify the consumers' willingness to pay for sustainable, organic or fair tread products. However, research has been not carried out about
  - How much consumer knows about eco-labels, and eco-textile standards and
  - The shopping behaviour and attitudes of STP consumers have not yet been analysed through actual purchase data.

- Investigate the health claims dictated by the lack of direct categorical data available on environmental benefits claims versus health benefits claims on labels (Nimon et al., 1999). Are there are any serious health benefits of STP especially Organic Cotton to human being?
- Why are eco-standards not mandatory throughout the textiles industry?

LIMITATIONS TO THE RESEARCH

This study's limitations need to be taken into account when considering its contributions to theory and findings. This being a one year period of study, time and cost were main limitations and affected the sampling, since it was impossible to visit entire population due to budget restrictions, large dispersal of locations.

It was also not possible to include either the logistics sector (also responsible for CO2 emission); government authorities' whose views may put some light on government policies and approach towards eco-labelled STP or the manufacturers who are not certified by any of the standards to understand why they are not willing to get certify. The results have not been validated, again due to lack of time. Bias may arisen, some participants tend to express views that are consistent with organic cotton, GOTS, OE standards only and try not to present themselves negatively. In addition, the interviewees may have also been unable to disclose some information due to the privacy issues with various companies. Information collected may be prone to some inaccuracy as a result of less than accurate recall, lack of information, or discomfort with self-disclosure.

MARKET METRICS

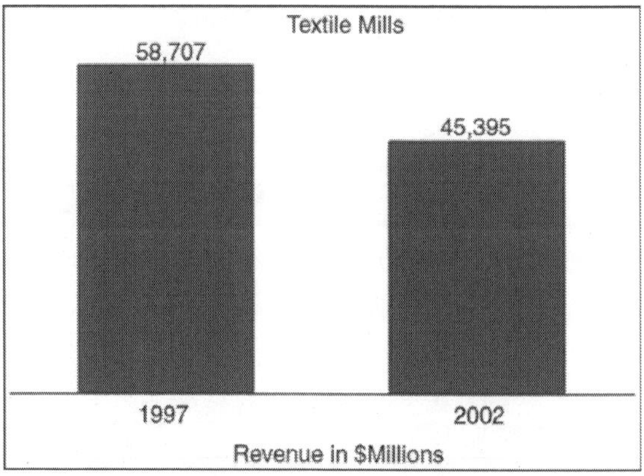

Fig. U.S.

The textile industry is currently more than US $400 billion and continues to expand. The globalization of the textile trade has increased the outsourcing of production.

A variety of fabrics are used worldwide in different applications such as apparel, household textiles and furnishings, medical equipment, industrial and technical products. Recent studies have highlighted that fabric weaving alone consumes around 28 million tons of fibre annually. It is predicted that global production will grow by 25per cent between 2002 and 2010, to reach more than 35 million tons and Asia is one of the key regions for growth.

The United States ships raw materials to other countries in which the products are made. The cost of labour is cheaper and provides retail companies the ability to buy and sell at lower costs. This increases competitiveness and has led to several textile companies having to shut down operations. At present, there is a minor trade war developing between the United States and China in regards to textiles and apparel. Chinese exports of textiles to the United States increased by 320 percent between January 2002 and November 2003 while prices of Chinese textiles and apparel declined by 55 percent. The growth in imports of textiles to U.S. during 1995 to 2002 was about 9per cent annually in dollar terms.

INDUSTRY PLAYERS

These are major players in this market, but not an exhaustive list of all key firms. Revenues, Net Income and Market Capitalization are expressed in US$ Millions.

Ticker	Company	Location	Revenue	Net Income	Profit Marginper cent	Market Cap
VFC	VF CP	Greensboro, North Carolina	6860	590	7.81per cent	8040
LIZ	LIZ Claiborne	New york, New york	4950	149	2.75per cent	2410
HBI	Hanesbrands	Winston-Salem, North Carolina	2250	74	3.28per cent	2770
Rl	Ralaph Lauren Polo CP	New york, New york	4540	387	8.52per cent	6870
ZQK	Quiksilver	Huntington Beach, California	2362	93	3.94per cent	1280

Research and development and design are important in industrial textiles where the material technology provides competitive advantage. In industrial countries, notably the United States, an increasing share of the textile sector produces household appliances and other industrial fabrics such as for the furniture and car industries. This is a more RandD intensive segment of the industry and subject to less frequent changes in patterns, material and colours. Only about a third of US textile production was used for clothing in the late 1990s.

- Mohawk Industries, Inc., is the market leader both in terms of market capitalization and annual sales in 2006. It engages in the production and sale of floor covering products for residential and commercial applications in the United States and Europe. It operates in three segments: Mohawk, Dal-Tile, and Unilin. The Mohawk segment designs, manufactures, and distributes floor covering products, including carpets, rugs, ceramic tiles, hardwood, resilient, and laminates. It markets its products through independent floor covering retailers, home centers, mass merchandisers, department stores, commercial dealers, and commercial end users, as well as through private labeling Programmes.

- Albany International Corp. produces paper machine clothing and doors worldwide. It operates in three segments: Paper Machine Clothing, Applied Technologies, and Albany Door Systems. Paper Machine Clothing segment includes fabrics and belts used in the manufacture of paper and paperboard. This segment offers forming, pressing, and dryer fabrics, and process belts.
- Polymer Group, Inc. manufactures and markets nonwoven and oriented polyolefin products. Its Nonwovens division provides nonwoven materials, which are used in diapers, training pants, feminine sanitary protection, adult incontinence, baby wipes, and consumer wiping products. Its products are used in medical applications, such as wound care sponges and dressings, and disposable surgical packs, as well as for apparel, including operating room gowns and drapes, face masks, and shoe covers; and in industrial applications.
- Interface, Inc. engages in the design, production, and sale of modular carpet products in Americas, Europe, and Asia-Pacific. It also manufactures broadloom carpet, panel fabrics, and upholstery fabrics for the commercial interiors industry.
- International Textile Group, Inc., together with its subsidiaries, operates as a textile manufacturer. The company produces automotive airbag fabrics and airbag cushions for automotive airbag modules. International Textile Group also manufactures and sells various technical and value added fabrics for use in fire service apparel, ballistics materials, filtration, military fabrics, and outdoor awnings and covers. The company also produces denim, cotton, synthetic and worsted apparel fabrics for use in bottom garments, such as pants, skirts, and shorts. It manufactures fabrics for military dress uniforms and battle dress uniforms for the U.S. Government and government contractors.
- Xerium Technologies, Inc. manufactures and supplies consumable products used in the production of paper clothing and roll covers.
- The Dixie Group, Inc. engages in the manufacture, marketing, and sale of carpets and rugs to residential and commercial customers in the United States.

TRENDS AND RECENT DEVELOPMENTS

To sharpen their competitive edge, many U.S. textile companies are (1) restructuring their operations through mergers, acquisitions, and divestitures, (2) concentrating on more profitable niche markets and product diversification, (3) improving productivity through new capital investment, (4) building strategic partnerships with retail and apparel companies, and (5) using quick response and other communication-oriented techniques to enhance their overall flexibility and competitiveness. U.S. textile companies have been acting to improve their competitive positions through reorganization, restructuring, and downsizing. By merging, textile companies gain market share and access to capital. Through acquisitions and vertical integration, textile companies have been able to achieve economies of scale and expand the range of their services.

Another trend is for U.S. mills and converters to move slowly away from staple fabrics for the apparel industry and towards niche products and diversification into multiple sewn products markets. Technical fibres for sportswear and active-wear have experienced tremendous growth. Other profitable growth opportunities in non-apparel fabric markets include automotive, medical, fire safety, and telecommunications (fibre-optics). The U.S. textile industry is a technologically advanced industry, achieving such advancement through restructuring and investment in modern machinery. Spinning, weaving, and knitting processes are highly modernized in many companies, and this has allowed them to reduce the share of labour in production costs and increase quality and speed. There has been a substantial job loss in the U.S. textile industry, a trend that has been ongoing for almost 30 years. This job loss can be attributed partly to imports, but like many other industries in the United States, the textile industry has been able to increase its production through new technologies while decreasing the number of employees. Overall textile production has increased more than 20 percent in fifteen years even though industry employment has dropped over 13 percent. The textile industry continues to invest in machinery and technology, spending upward of $2 billion a year to maintain modern manufacturing facilities. Continued investment in modern technology, including information technology, is very important. Textile production is very capital intensive, and up-to-date technology is essential to meet the increasingly rigorous demand for high-quality products.

Proximity to major markets has assumed increasing economic significance and tariffs are restraining trade due to the fact that products cross borders several times. Developing countries are catching up with China in terms of unit labour costs in the textile and clothing sector.

Future trends include multifunctional textiles, intelligent textiles, eco-textiles, e-textiles and customized textiles in the textile-apparel industry. Nano-Tex is the company manufacturing "enhanced textiles" which are stain resistant transparent fabrics manufactured using nanotechnology.

2

Textile Manufacturing by Pre-Industrial Methods

Textile manufacturing is one of the oldest human activities. The oldest known textiles date back to about 5000 B.C. In order to make textiles, the first requirement is a source of fibre from which a yarn can be made, primarily by spinning. The yarn is processed by knitting or weaving to create cloth. The machine used for weaving is the loom. Cloth is finished by what are described as wet processes to become fabric. The fabric may be dyed, printed or decorated by embroidering with coloured yarns.

The three main types of fibres are natural vegetable fibres (such as cotton, linen, jute and hemp), man-made fibres (made by industrial processes) and protein based fibres (such as wool, silk).

Almost all commercial textiles are produced by industrial methods. Textiles are still produced by pre-industrial processes in village communities in Asia, Africa and South America, as an artisan craft and a hobby in Europe and North America.

YARN FORMATION

WOOL

Wool is a protein based fibre, being the coat of a sheep. The wool is removed by shearing.

Wool is the textile fibre obtained from sheep and certain other animals, including cashmere from goats, mohair from goats, qiviut from muskoxen, vicuña, alpaca, and camel from animals in the camel family, and angora from rabbits.

Wool has several qualities that distinguish it from hair or fur: it is crimped, it is elastic, and it grows in staples (clusters).

Characteristics

Wool's scaling and crimp make it easier to spin the fleece by helping the individual fibres attach to each other, so that they stay together. Because of the crimp, wool fabrics have a greater bulk than other textiles, and retain air, which causes the product to retain heat. Insulation also works both ways; Bedouins and Tuaregs use wool clothes to keep the heat out.

The amount of crimp corresponds to the fineness of the wool fibres. A fine wool like Merino may have up to 100 crimps per inch, while the coarser wools like karakul may have as few as 1 to 2. Hair, by contrast, has little if any scale and no crimp, and little ability to bind into yarn. On sheep, the hair part of the fleece is called kemp. The relative amounts of kemp to wool vary from breed to breed, and make some fleeces more desirable for spinning, felting, or carding into batts for quilts or other insulating products.

Wool fibres are hygroscopic, meaning they readily absorb moisture. Wool fibres are hollow. Wool can absorb moisture almost one-third of its own weight. Wool absorbs sound like many other fabrics. Wool is generally a creamy white colour, although some breeds of sheep produce natural colours such as black, brown, silver, and random mixes.

Wool ignites at a higher temperature than cotton and some synthetic fibres. It has lower rate of flame spread, low heat release, low heat of combustion, and does not melt or drip; it forms a char which is insulating and self-extinguishing, and contributes less to toxic gases and smoke than other flooring products, when used in carpets. Wool carpets are specified for high safety environments, such as trains and aircraft. Wool is usually specified for garments for fire-fighters, soldiers, and others in occupations where they are exposed to the likelihood of fire.

Wool is resistant to static electricity, as the moisture retained within the fabric conducts electricity. This is why wool garments are much less likely to spark or cling to the body. The use of wool car seat covers or carpets reduces the risk of a shock when a person touches a grounded object. Wool is considered by the medical profession to be hypoallergenic.

SHEEP SHEARING

Sheep shearing, shearing or clipping is the process by which the woolen fleece of a sheep is cut off. The person who removes the sheep's wool is called a *shearer*. Typically each adult sheep is shorn once each year (a sheep may be said to have been "shorn" or "sheared", depending upon dialect). The annual shearing most often occurs in a shearing shed, a facility especially designed to process often hundreds and sometimes more than 3,000 sheep per day.

History

Shearers wear moccasins to protect their feet, grip wooden floors well, and absorb sweat.

Up until the 1870s squatters washed their sheep in nearby creeks prior to shearing. Later some expensive hot water installations were constructed on some of the larger stations for the washing. Sheep washing in Australia was influenced by the Saxony sheep breeders in Germany who washed their sheep and by the Spanish practice of washing the wool after shearing. There were three main reasons for the custom in Australia:

- The English manufacturers demanded that Australian woolgrowers provide their fleeces free from vegetable matter, burrs, soil, etc.
- The dirty fleeces were hard to shear and demanded that the metal blade shears be sharpened more often.
- Wool in Australia was carted by bullock team or horse teams and charged by weight. Washed wool was lighter and did not cost as much to transport.

The practice of washing the wool rather than the sheep evolved from the fact that hotter water could be used to wash the wool, than that used to wash the sheep. When the practice of selling wool in the grease occurred in the 1890s, wool washing became obsolete.

Australia and New Zealand had to discard the old methods of wool harvesting and evolve more efficient systems to cope with the huge numbers of sheep involved. After 1888 machine shearing was introduced, reducing second cuts and shearing time. By 1915 most large sheep station sheds in Australia had installed machines, driven by steam or later by internal combustion engines.

Shearing tables were invented in the 1950s and have not proved popular, although some are still used for crutching.

MODERN SHEARING

Today, large flocks of sheep are shorn by professional shearing teams working eight hour days, most often in spring, by machine shearing. These contractor teams will consist of shearers, shed hands and a cook (in the more isolated areas). The shed staff working hours and wages are regulated by industry awards. A working day starts at 7:30 AM and the day is divided into four "runs" of two hours each. "Smoko" breaks of a half hour each are at 9:30 AM and again at 3:00 PM.

The lunch break is taken at 12:00 PM for one hour. Most shearers are paid on a piece rate, *i.e.*, per sheep. Shearers who "tally" more than 200 sheep per day are known as "gun shearers". Typical mass shearing of sheep today follows a well-defined workflow: remove the wool, throw the fleece onto the wool table, skirt, roll and class the fleece, place it in the appropriate wool bin, press and store the wool until it is transported.

In 1984 Australia became the last country in the world to permit the use of wide combs, due to previous Australian Workers Union rules. Although they were rare in sheds, women now take a large part in the shearing industry by working as pressers, wool rollers, rouse about, wool classers and also shearing, too.

REMOVING THE WOOL

A sheep is caught by the shearer from the catching pen and taken to his "stand" on the shearing board. It is then shorn using mechanical hand piece (see *Shearing devices* below). The wool is removed by following an efficient set of movements, devised by Godfrey Bowen in c. 1950, (the *Bowen Technique*) or the *Tally-Hi* method. In 1963 the Tally-hi shearing system was developed by the Australian Wool Corporation and promoted using synchronized shearing demonstrations. Sheep struggle less using the Tally-Hi method, reducing strain on the shearer and there is a saving of about 30 seconds shearing each sheep. The shearer begins by removing the sheep's belly wool, which is separated from the main fleece by a rouseabout, while the sheep is still being shorn. A professional or "gun" shearer typically removes a fleece without badly marking or cutting the sheep in two to three minutes, depending on the size and condition of the sheep, or less than two in elite competitive shearing. The shorn sheep is moved from the board via a chute in the floor, or wall, to a counting out pen, efficiently removing it from the shed.

The CSIRO in Australia has developed a non-mechanical method of shearing sheep using an injected protein that creates a natural break in the wool fibres. After fitting a retaining net to enclose the wool, sheep are injected with the protein. When the net is removed after a week, the fleece has separated and is removed by hand. In some breeds a similar process occurs naturally (see below).

SKIRTING THE FLEECE

Once the entire fleece has been removed from the sheep, the fleece is *thrown*, clean side down, on to a wool table by a shed hand (commonly known in New Zealand and Australian sheds as a *rouseabout* or *roustie*). The wool table top consists of slats spaced approximately 12 cm apart. This enables short pieces of wool, the *locks* and other debris, to gather beneath the table separately from the fleece.

The fleece is then *skirted* by one or more wool rollers to remove the sweat fribs and other less desirable parts of the fleece. The removed pieces largely consist of shorter, seeded, burry or dusty wool etc. which is still useful in the industry. As such they are placed in separate containers and sold along with fleece wool. Other items removed from the fleece on the table, such as faeces, skin fragments or twigs and leaves, are discarded a short distance from the wool table so as not to contaminate the wool and fleece.

WOOL CLASSING

Following the skirting of the fleece, it is folded, rolled and examined for its quality in a process known as wool classing, which is performed by a registered and qualified wool classer. Based on its type, the fleece is placed into the relevant wool bin ready to be pressed (mechanically compressed) when there is sufficient wool to make a wool bale.

SHEARING DEVICES

Blade shears

Blade shears consist of two blades arranged similarly to scissors except that the hinge is at the end farthest from the point (not in the middle). The cutting edges pass each other as the shearer squeezes them together and shear the wool close to the animal's skin. Blade shears are still used today but in a more limited way. Blade shears leave some wool on a sheep and this is more suitable for cold climates where the sheep needs some protection from the elements. For those areas where no powered-machinery is available blade shears are the only option. Blades are more commonly used to shear stud rams.

Machine shears

Machine shears, known as hand pieces, operate in a similar manner to human hair clippers in that a power-driven toothed blade, known as a cutter, is driven back and forth over the surface of a comb and the wool is cut from the animal. The original machine shears were powered by a fixed hand-crank linked to the hand piece by a shaft with only two universal joints, which afforded a very limited range of motion. Later models have more joints to allow easier positioning of the hand piece on the animal. Electric motors on each stand have generally replaced overhead gear for driving the hand pieces. The jointed arm is replaced in many instances with a flexible shaft. Smaller motors allowed the production of shears in which the motor is in the hand piece; these are generally not used by professional shearers as the weight and heat of the motor becomes bothersome with long use.

SHEARING IN AUSTRALASIAN CULTURE

A culture has evolved out of the practice of sheep shearing, especially in post-colonial Australia and New Zealand. *Shearing the Rams*, a painting by Australian impressionist painter Tom Roberts is considered to be iconic of the livestock-growing culture or "life on the land" in Australia.

For an inversion, Michael Leunig's *Ramming the Shears* can be seen as a sign of the shifts in Australian culture, and the extent to which the dominant rural culture is being eroded by an increasingly urban population.

The expression that Australia's wealth rode on *the sheep's back* in parts of the twentieth century no longer has the currency it once had.

In 2001, Mandy Francis of Hardy's Bay, NSW constructed a black butt seat for the Street Furniture Project at Walcha, New South Wales. This seat was inspired by the combs, cutters, wool tables and grating associated with the craft and industry of shearing.

During the long weekend in June 2010, 111 machine shearers and 78 blade shearers shore 6,000 Merino ewes and 178 rams at the historic 72 stand *North Tuppal* station. Along with the shearers there were 107 wool handlers and penners-up and more than 10,000 visitors to witness this event in the restored shed. Over this weekend the scene in Tom Robert's *Shearing of the Rams* was twice re-enacted for the visitors.

Many stations across Australia no longer carry sheep due to lower wool prices, drought and other disasters, but their shearing sheds remain, in a wide variety of materials and styles, and have been the subject of books and documentation for heritage authorities. Some farmers are reluctant to remove either the equipment or the sheds, and many unused sheds remain intact.

Contests

Sheep shearing and wool handling competitions are held regularly in parts of the world, particularly Ireland, the UK, South Africa, New Zealand and Australia. As sheep shearing is an arduous task, speed shearers, for all types of equipment and sheep, are usually very fit and well trained. In Wales a sheep shearing contest is one of the events of the Royal Welsh Show, the country's premier agricultural show held near Builth Wells.

The world's largest sheep shearing and wool handling contest, the Golden Shears, is held in the Wairarapa district, New Zealand.

SCOURING

Wool straight off a sheep, known as "greasy wool" or "wool in the grease", contains a high level of valuable lanolin, as well as dirt, dead skin, sweat residue, pesticide, and vegetable matter. Before the wool can be used for commercial purposes, it must be scoured, a process of cleaning the greasy wool. Scouring may be as simple as a bath in warm water, or as complicated as an industrial process using detergent and alkali, and specialized equipment. In commercial wool, vegetable matter is often removed by chemical carbonization. In less processed wools, vegetable matter may be removed by hand, and some of the lanolin left intact through use of gentler detergents. This semi-grease wool can be worked into yarn and knitted into particularly water-resistant mittens or sweaters, such as those of the Aran Island fishermen. Lanolin removed from wool is widely used in cosmetic products such as hand creams.

QUALITY

The quality of wool is determined by the following factors, fibre diameter, crimp, yield, colour, and staple strength. Fibre diameter is the single most important wool characteristic determining quality and price.

Merino wool is typically 3-5 inches in length and is very fine (between 12-24 microns). The finest and most valuable wool comes from Merino hoggets. Wool taken from sheep produced for meat is typically more coarse, and has fibres that are 1.5 to 6 inches in length. Damage or breaks in the wool can occur if the sheep is stressed while it is growing its fleece, resulting in a thin spot where the fleece is likely to break.

Wool is also separated into grades based on the measurement of the wool's diameter in microns and also its style. These grades may vary depending on the breed or purpose of the wool. For example:

- <15.5 - Ultrafine Merino
- 15.6-18.5 - Superfine Merino
- 18.6-20 - Fine Merino
- 20.1-23 - Medium Merino
- 23< - Strong Merino
- Comeback: 21-26 microns, white, 90–180 mm long
- Fine crossbred: 27-31 microns, Corriedales etc.
- Medium crossbred: 32–35 microns
- Downs: 23-34 microns, typically lacks luster and brightness. Examples, Aussiedown, Dorset Horn, Suffolk etc.
- Coarse crossbred: 36> microns
- Carpet wools: 35-45 microns

Any wool finer than 25 microns can be used for garments, while coarser grades are used for outerwear or rugs. The finer the wool, the softer it is, while coarser grades are more durable and less prone to pilling.

The finest Australian and New Zealand Merino wools are known as 1PP which is the industry benchmark of excellence for Merino wool that is 16.9 micron and finer. This style represents the top level of fineness, character, colour, and style as determined on the basis of a series of parameters in accordance with the original dictates of British Wool as applied today by the Australian Wool Exchange (AWEX) Council. Only a few dozen of the millions of bales auctioned every year can be classified and marked 1PP.

HISTORY

As the raw material has been readily available since the widespread domestication of sheep - and of goats, another major provider of wool - the use of felted or woven wool for clothing and other fabrics characterizes some of the earliest civilizations. Prior to invention of shears - probably in the Iron Age - the wool was plucked out by hand or by bronze combs. The oldest known European wool textile, ca. 1500 BCE, was preserved in a Danish bog. Wool fibres from wild goats found in a prehistoric cave in the Republic of Georgia as far back 34,000 BCE suggest that wool fabrics were made even earlier than this.

In Roman times, wool, linen, and leather clothed the European population; the cotton of India was a curiosity that only naturalists had heard of; and silk, imported along the Silk Road from China, was an extravagant luxury. Pliny the

Elder records in his Natural History that the reputation for producing the finest wool was enjoyed by Tarentum, where selective breeding had produced sheep with a superior fleece, but which required special care.

In medieval times, as trade connections expanded, the Champagne fairs revolved around the production of wool cloth in small centers such as Proving; the network that the sequence of annual fairs developed meant that the woollens of Provins might find their way to Naples, Sicily, Cyprus, Majorca, Spain, and even Constantinople. The wool trade developed into serious business, the generator of capital. In the thirteenth century, the wool trade was the economic engine of the Low Countries and of Central Italy; by the end of the following century Italy predominated, though in the 16th century Italian production turned to silk. Both pre-industries were based on English raw wool exports - rivaled only by the sheepwalks of Castile, developed from the fifteenth century - which were a significant source of income to the English crown, which from 1275 imposed an export tax on wool called the "Great Custom". The importance of wool to the English economy can be shown by the fact that since the 14th Century, the presiding officer of the House of Lords has sat on the "Woolsack", a chair stuffed with wool.

Economies of scale were instituted in the Cistercian houses, which had accumulated great tracts of land during the twelfth and early thirteenth centuries, when land prices were low and labour still scarce.

Raw wool was baled and shipped from North Sea ports to the textile cities of Flanders, notably Ypres and Ghent, where it was dyed and worked up as cloth. At the time of the Black Death, English textile industries accounted for about 10per cent of English wool production; the English textile trade grew during the fifteenth century, to the point where export of wool was discouraged. Over the centuries, various British laws controlled the wool trade or required the use of wool even in burials.

The smuggling of wool out of the country, known as owling, was at one time punishable by the cutting off of a hand. After the Restoration, fine English woollens began to compete with silks in the international market, partly aided by the Navigation Acts; in 1699 English crown forbade its American colonies to trade wool with anyone but England herself.

A great deal of the value of woolen textiles was in the dyeing and finishing of the woven product.

In each of the centers of the textile trade, the manufacturing process came to be subdivided into a collection of trades, overseen by an entrepreneur in a system called by the English the "putting-out" system, or "cottage industry", and the *Verlagssystem* by the Germans.

In this system of producing wool cloth, until recently perpetuated in the production of Harris tweeds, the entrepreneur provides the raw materials and an advance, the remainder being paid upon delivery of the product. Written contracts bound the artisans to specified terms. Fernand Braudel traces the appearance of the system in the thirteenth-century economic boom, quoting a document of 1275 The system effectively by-passed the guilds' restrictions.

Before the flowering of the Renaissance, the Medici and other great banking houses of Florence had built their wealth and banking system on their textile industry based on wool, overseen by the Arte della Lana, the wool guild: wool textile interests guided Florentine policies. Francesco Datini, the "merchant of Prato", established in 1383 an *Arte della Lana* for that small Tuscan city. The sheepwalks of Castile shaped the landscape and the fortunes of the *meseta* that lies in the heart of the Iberian peninsula; in the sixteenth century, a unified Spain allowed export of Merino lambs only with royal permission. The German wool market - based on sheep of Spanish origin - did not overtake British wool until comparatively late. Australia's colonial economy was based on sheep raising, and the Australian wool trade eventually overtook that of the Germans by 1845, furnishing wool for Bradford, which developed as the heart of industrialized woollens production. A World War I era poster sponsored by the United States Department of Agriculture encouraging children to raise sheep to provide needed war supplies.

Due to decreasing demand with increased use of synthetic fibres, wool production is much less than what it was in the past. The collapse in the price of wool began in late 1966 with a 40per cent drop; with occasional interruptions, the price has tended down. The result has been sharply reduced production and movement of resources into production of other commodities, in the case of sheep growers, to production of meat.

Superwash wool (or washable wool) technology first appeared in the early 1970s to produce wool that has been specially treated so that it is machine washable and may be tumble-dried. This wool is produced using an acid bath that removes the "scales" from the fibre, or by coating the fibre with a polymer that prevents the scales from attaching to each other and causing shrinkage. This process results in a fibre that holds longevity and durability over synthetic materials, while retaining its shape.

In December 2004, a bale of the world's finest wool, averaging 11.8 micron, sold for $3,000 per kilogram at auction in Melbourne, Victoria. This fleece wool tested with an average yield of 74.5per cent, 68 mm long, and had 40 newtons per kilotex strength. The result was $AUD279,000 for the bale. The finest bale of wool ever auctioned sold for a seasonal record of 269,000 cents per kilo during June 2008. This bale was produced by the Hillcreston Pinehill Partnership and measured 11.6 microns, 72.1per cent yield and had a 43 Newtons per kilotex strength measurement. The bale realized $247,480 and was exported to India.

During 2007 a new wool suit was developed and sold in Japan that can be washed in the shower, and dries off ready to wear within hours with no ironing required. The suit was developed using Australian Merino wool and it enables woven products made from wool, such as suits, trousers and skirts, to be cleaned using a domestic shower at home.

In December 2006 the General Assembly of the United Nations proclaimed 2009 to be the International Year of Natural Fibres, so to raise the profile of wool and other natural fibres.

PRODUCTION

Global wool production is approximately 1.3 million tonnes per year, of which 60per cent goes into apparel. Australia is the leading producer of wool which is mostly from Merino sheep. New Zealand is the second-largest producer of wool, and the largest producer of crossbred wool. China is the third-largest producer of wool. Breeds such as Lincoln, Romney, Tukidale, Drysdale and Elliotdale produce coarser fibres, and wool from these sheep is usually used for making carpets.

In the United States, Texas, New Mexico and Colourado have large commercial sheep flocks and their mainstay is the Rambouillet (or French Merino). There is also a thriving home-flock contingent of small-scale farmers who raise small hobby flocks of specialty sheep for the hand spinning market. These small-scale farmers offer a wide selection of fleece.

Global woolclip (total amount of wool shorn) 2004/2005

- Australia: 25per cent of global woolclip (475 million kg greasy, 2004/2005)
- China: 18per cent
- New Zealand: 11per cent
- Argentina: 3per cent
- Turkey: 2per cent
- Iran: 2per cent
- United Kingdom: 2per cent
- India: 2per cent
- Sudan: 2per cent
- South Africa: 1per cent
- United States: 0.77per cent

Organic wool is becoming more and more popular. This wool is very limited in supply and much of it comes from New Zealand and Australia. It is becoming easier to find in clothing and other products, but these products often carry a higher price. Wool is environmentally preferable (as compared to petroleum-based Nylon or Polypropylene) as a material for carpets as well, in particular when combined with a natural binding and the use of formaldehyde-free glues.

Animal rights groups have noted issues with the production of wool, such as Mulesing.

MARKETING

About 85per cent of wool sold in Australia is sold by open cry auction. Sale by Sample is a method in which a mechanical claw takes a sample from each bale in a line or lot of wool. These grab samples are bulked, objectively measured, and a sample of not less than 4 kg is displayed in a box for the buyer to examine. The Australian Wool Exchange (AWEX) conducts sales primarily in Sydney, Melbourne, Newcastle, and Fremantle. There are about 80 brokers and agents throughout Australia.

About 7per cent of Australian wool is sold by private treaty on farms or to local wool-handling facilities. This option gives wool growers benefit from reduced transport, warehousing, and selling costs. This method is preferred for small lots or mixed butts in order to make savings on reclassing and testing.

About 5per cent of Australian wool is sold over the internet on an electronic offer board. This option gives wool growers the ability to set firm price targets, reoffer passed in wool and offer lots to the market quickly and efficiently. This method works well for tested lots as buyers use these results to make a purchase. 97per cent of wool is sold without sample inspection however as of dec 2009, 59per cent of wool listed had been passed in from auction. Growers through certain brokers can allocate their wool to a sale and what price their wool will be reserved at.

Sale by tender can achieve considerable cost savings on wool clips large enough to make it worthwhile for potential buyers to submit tenders. Some marketing firms sell wool on a consignment basis, obtaining a fixed percentage as commission.

Forward selling: Some buyers offer a secure price for forward delivery of wool based on estimated measurements or the results of previous clips. Prices are quoted at current market rates and are locked in for the season. Premiums and discounts are added to cover variations in micron, yield, tensile strength, etc., which are confirmed by actual test results when available.

Another method of selling wool includes sales direct to wool mills. The British Wool Marketing Board operates a central marketing system for UK fleece wool with the aim of achieving the best possible net returns for farmers.

Less than half of New Zealand's wool is sold at auction, while around 45per cent for farmers sell wool directly to private buyers and end-users. Some businesses in New Zealand like Blue House Yarns have turned to selling organic wool, a new trend on wool production.

United States sheep producers' market wool with private or cooperative wool warehouses, but wool pools are common in many states. In some cases, wool is pooled in a local market area but sold through a wool warehouse. Wool offered with objective measurement test results is preferred. Imported apparel wool and carpet wool goes directly to central markets where it is handled by the large merchants and manufacturers.

USES

In addition to clothing, wool has been used for blankets, horse rugs, saddle cloths, carpeting, felt, wool insulation (also see links) and upholstery. Wool felt covers piano hammers, and it is used to absorb odors and noise in heavy machinery and stereo speakers. Ancient Greeks lined their helmets with felt, and Roman legionnaires used breastplates made of wool felt.

Wool has also been traditionally used to cover cloth diapers. Wool fibre exteriors are hydrophobic (repel water) and the interior of the wool fibre is hygroscopic (attracts water); this makes a wool garment able to cover a wet diaper while inhibiting wicking, so outer garments remain dry. Wool felted and treated with lanolin is water resistant, air permeable, and slightly antibacterial, so it resists the buildup of odour. Some modern cloth diapers use felted wool fabric for covers, and there are several modern commercial knitting patterns for wool diaper covers.

Initial studies of woollen underwear have found it prevented heat and sweat rashes because it more readily absorbs the moisture than other fibres.

Merino wool has been used in baby sleep products such as swaddle baby wrap blankets and infant sleeping bags.

Wool is an animal protein, and as such it can be used as a soil fertilizer, being a slow release source of nitrogen and ready made amino acids.

VIRGIN WOOL

Wool spun for the first time is called Virgin wool.

SHODDY OR RECYCLED WOOL:

It is made by cutting or tearing apart existing wool fabric and respinning the resulting fibres. As this process makes the wool fibres shorter, the remanufactured fabric is inferior to the original. The recycled wool may be mixed with raw wool, wool noil, or another fibre such as cotton to increase the average fibre length. Such yarns are typically used as weft yarns with a cotton warp. This process was invented in the Heavy Woollen District of West Yorkshire and created a micro-economy in this area for many years.

Ragg is a sturdy wool fibre made into yarn and used in many rugged applications like gloves.

Worsted is a strong, long-staple, combed wool yarn with a hard surface.

Woollen is a soft, short-staple, carded wool yarn typically used for knitting. In traditional weaving, woollen weft yarn (for softness and warmth) is frequently combined with a worsted warp yarn for strength on the loom.

EVENTS

A buyer of Merino wool, Ermenegildo Zegna, has offered awards for Australian wool producers. In 1963, the first Ermenegildo Zegna Perpetual Trophy was presented in Tasmania for growers of "Superfine skirted Merino fleece". In 1980, a national award, the Ermenegildo Zegna Trophy for Extrafine Wool Production, was launched. In 2004, this award became known as the Ermenegildo Zegna Unprotected Wool Trophy. In 1998, an Ermenegildo Zegna

Protected Wool Trophy was launched for fleece from sheep coated for around nine months of the year.

In 2002, the Ermenegildo Zegna Vellus Aureum Trophy was launched for wool that is 13.9 micron and finer. Wool from Australia, New Zealand, Argentina, and South Africa may enter, and a winner is named from each country. In April 2008, New Zealand won the Ermenegildo Zegna Vellus Aureum Trophy for the first time with a fleece that measured 10.8 microns. This contest awards the winning fleece weight with the same weight in gold as a prize, hence the name.

In 2010 an ultra-fine, 10 micron fleece, from Windradeen, near Pyramul, New South Wales set a new world record in the fineness of wool fleeces when it won the Ermenegildo Zegna Vellus Aureum International Trophy.

Since 2000, Loro Piana has awarded a cup for the world's finest bale of wool that produces just enough fabric for 50 tailor-made suits. The prize is awarded to an Australian or New Zealand wool grower who produces the year's finest bale.

The New England Merino Field days which display local studs, wool, and sheep are held during January, every two years (in even numbered years) around the Walcha, New South Wales district. The Annual Wool Fashion Awards, which showcase the use of Merino wool by fashion designers, are hosted by the city of Armidale, New South Wales in March each year.

This event encourages young and established fashion designers to display their talents. During each May, Armidale hosts the annual New England Wool Expo to display wool fashions, handicrafts, demonstrations, shearing competitions, yard dog trials, and more.

In July, the annual Australian Sheep and Wool Show is held in Bendigo, Victoria. This is the largest sheep and wool show in the world, with goats and alpacas as well as woolcraft competitions and displays, fleece competitions, sheepdog trials, shearing, and wool handling.

The largest competition in the world for objectively-measured fleeces is the Australian Fleece Competition, which is held annually at Bendigo. In 2008, there were 475 entries from all states of Australia with first and second prizes going to the Northern Tablelands, New South Wales fleeces.

WOOL CLASSING

Wool classing is an occupation for which people are trained to produce uniform, predictable, low risk lines of wool. This is carried out by examining the characteristics of the wool in its raw state. The characteristics which a wool classer would examine are:

Breed of the sheep: Shedding breeds will increase the risk of medulated and/or pigmented fibres. Any sheep likely to have dark fibres should be shorn last to avoid contamination.

The age of the sheep will have a bearing on the fibre diameter and value of wool, too.

Chemical usage: Ensure that all rules have been followed.

Brands, seedy jowls and *shanks*: Must be removed from fleeces and broken.

Stain: Must be removed from bellies and fleeces and identified in a separate line.

Wool crimp: The number of bends per unit length along the wool fibre approximately indicates spinning capacity of the wool. fibres with a fine crimp have many bends and usually have a small diameter. Such fibre can be spun into fine yarns, with great lengths of yarn for a given weight of wool, and greater market value. Fine fibres may be utilised in the production of fine garments such as men's suits whereas the coarser fibres may be used for the production of carpet and other sturdy products. Crimp is measured in crimps per inch or crimps per centimetre. Average diameter or mean fibre diameter is measured in micrometres (microns). For generations, English wool-handlers categorized wool along the above lines estimating spinning capacity by eye and touch. This spread worldwide as the Bradford system.

Wool Strength (also known as *tensile strength*) determines wool's ability to withstand processing. Weaker wools produce more waste in carding and spinning. Weaker wools may be used for production of felt, or combined with other fibres, etc.

Wool colour: Indicates whether wool is able to be dyed in light shades. Colour may be graded depending upon the natural colour, impurities and various stains present. Severely stained wool decreases prices dramatically. However, it is difficult to assess colour accurately without proper measurement, since some stains will wash out in the processing, whereas others are quite persistent.

The fleece is skirted to remove excess frib, seed and burr etc to leave the fleece as reasonably even as possible in good respects. The parts of wool taken from a sheep are graded separately.

The fleece forming the bulk of the yield is placed with other fleece wool as the main line, other pieces such as the neck, belly and skirtings (inferior wool from edges) are sold for such purposes where the shorter wools are required (for example: fillings, carpets, insulation).

Whilst in some places crimp may determine which grade the fleece will be placed into, this subjective assessment is not always reliable and processors prefer that wools are measured objectively by qualified laboratories. Some of the superfine wool growers do in shed wool testing, but this can only be used as a guide.

This enables wool classers to place wool into lines of a consistent quality. A shedhand, known as a wool presser, places the wool into approved wool packs in a wool press to produce a bale of wool that must meet regulations concerning its fastenings, length, weight and branding if it is to be sold at auction in Australasia. All Merino fleece wool sold at auction in Australia is objectively measured for fibre diameter, yield (including the amount of vegetable matter), staple length, staple strength and sometimes colour.

Classers are also responsible for ensuring that a pre-shearing check is made to ensure that the wool and sheep areas are free of possible contaminants. They are to supervise shed staff during shearing and train any inexperienced hands. At the end of shearing classers have to provide full documentation concerning the clip.

ROOING

In some primitive sheep (for example in many Shetlands), there is a natural break in the growth of the wool in spring. By late spring this causes the fleece to begin to peel away from the body, and it may then be plucked by hand without cutting – this is known as *rooing*. Individual sheep may reach this stage at slightly different times.

Shearing can be done with use of hand-shears or powered shears. Professional sheep shearers can shear a sheep in under a minute, without nicking the sheep.

The fleece is removed in one piece. Second cuts can be made but produce only short fibres, which are more difficult to spin. Primitive breeds, like the Scottish Soay sheep have to be plucked, not sheared, as the kemps are still longer than the soft fleece, (a process called rooing).

SKIRTING

Skirting is disposing of all wool that is unsuitable for spinning. Recovering can be attempted. It can also be done at the same time as carding.

CLEANING

The wool is cleaned. At this point the fleece is full of lanolin and often contains extraneous vegetable matter, such as sticks, twigs, burrs and straw. These may all be removed, though lanolin may be left in the wool till after the spinning, a technique known as spinning 'in the grease'. Indeed if the fabric is to be water repellent, lanolin is not removed at any stage.

Washing the wool at this stage can be a tedious process. Some people wash it a small handful at a time very carefully, and then set it out to dry on a table in the sun. Others will wash the whole fleece. Lanolin is removed by soaking the fleece in very hot water. If the fleece gets agitated, it will become felt, and then spinning is impossible. Felting, when done on purpose (with needles, chemicals, or simply rubbing the fibres against each other), can be used to create garments.

CARDING OR COMBING

It is possible to spin directly from a clean fleece, but it is much easier to spin a carded fleece. Carding by hand yields a rolag, a loose woollen roll of fibres. Using a drum carder yields a bat, which is a mat of fibres in a flat, rectangular shape. Carding mills return the fleece in a roving, which is a stretched bat; it is very long and often the thickness of a wrist.A pencil roving is a roving thinned to the width of a pencil. It can used for knitting without any spinning, or for apprentices.

One good-sized fleece may take weeks to card with a drum-carder, or an eternity by hand. If the fleece is sent to a carding mill, it must be washed before it is carded. Most mills offer washing the wool as a service, with extra fees if the wool is exceptionally dirty. Fibres can be purchased pre-carded. Combing is another method to align the fibres parallel to the yarn, and thus is good for spinning a worsted yarn, whereas the rolag from handcards produces a woolen yarn.

SPINNING

Hand spinning can be done by using a spindle or the spinning wheel. Spinning turns the carded wool fibres into yarn which can then be directly woven, knitted (flat or circular), crocheted, or by other means turned into fabric or a garment.

The spinning wheel collects the yarn on a bobbin. A woollen yarn is lightly spun so it is airey, and is a good insulator and suitable for knitting, while a worsted yarn is spun tight to exclude air, and has greater strength and is suited to weaving..

Once the bobbin is full, the spinner either puts on a new bobbin, or forms a skein, or balls the yarn. A skein is a coil of yarn twisted into a loose knot. Yarn is skeined using a niddy-noddy or other type of skein -winder. Yarn is rarely balled directly after spinning, it will be stored in skein form, and transferred to a ball only if needed. Knitting from a skein, is difficult as the yarn forms knots, in this case it is best to ball. Yarn to be plied is left on the bobbin.

A skein is either formed on a niddy noddy or some other type of skein winder. Traditionally niddy-noddys looked like an uppercase "i", with the bottom half rotated 90 degrees. Now spinning wheel manufactures also make niddy-noddys that attach onto the spinning wheel for faster skein winding.

PLYING

Plying yarn is when one takes a strand of spun yarn (one strand is often called a single) and spins it together with other strands in order to make a thicker yarn.

Regular plying consists of taking two or more singles and twisting them together, the against their twist. This can be done on a spinning wheel or on a spindle. If the yarn was spun clockwise (which is called a "Z" twist), to ply, the wheel must spin counter-clockwise (an "S" twist). This is the most common way. When plying from bobbins a device called a lazy kate is often used to hold them.

Most spinners (who use spinning wheels) ply from bobbins. This is easier than plying from balls because there is less chance for the yarn to become tangled and knotted if it is simply unwound from the bobbins. So that the bobbins can unwind freely, they are put in a device called a lazy kate, or sometimes simply *kate*. The simplest lazy kate consists of wooden bars with a metal rod running between them. Most hold between three and four bobbins. The bobbin sits on the metal rod. Other lazy kates are built with devices that create an adjustable amount of tension, so that if the yarn is jerked, a whole bunch of yarn is not wound off, then wound up again in the opposite direction. Some spinning wheels come with a built in lazy kate.

Navajo plying consists of making large loops, similar to crocheting.A loop about 8 inches long is made on the leader the end on the leader. (A leader is the string left on the bobbin to spin off.) The three strands together are spun in the opposite direction. When a third of the loop remains, a new loop is created and the spinning continues. The process is repeated until the yarn is all plied. The advantage of this method is that only one single is needed and if the single is already dyed this technique allows it to be plied without ruining the colour scheme. This technique also allows the spinner to try to match up thick and thin spots in the yarn, thus making for a smoother end product.

WASHING

If the lanolin is unwanted, and has not already been washed out, this is done now. The skein is tied in six points and steeped overnight in detergent, it is rinsed and air-dried, and re-skinned.

unless the lanolin is to be left in the cloth as a water repellent. When washing a skein it works well to let the wool soak in soapy water overnight, and rinse the soap out in the morning. Dishwashing detergents are commonly used, and a special laundry detergent designed for washing wool is not required. The dishwashing detergent works and does not harm the wool. After washing, let the wool dry (air drying works best). Once it is dry, or just a bit damp, one can stretch it out a bit on a niddy-noddy. Putting the wool back on the niddy-noddy makes for a nicer looking finished skein. Before taking a skein and washing it, the skein must be tied up loosely in about six places. If the skein is not tied up, it will be very hard to unravel when done washing.

FLAX

The preparations for spinning is similar across most plant fibres, including Flax and Hemp. Flax is the fibre used to create linen. Cotton is handled differently since it uses the fruit of the plant and not the stalk.

HARVESTING

Flax is pulled out of the ground about a month after the initial blooming when the lower part of the plant begins to turn yellow, and when the most forward of the seeds are found in a soft state. It is pulled in handfuls and several handfuls are tied together with slip knot into a 'beet'. The string is tightened as the stalks dry. The seed heads are removed and the seeds collected, by threshing and winnowing.

RETTING

Threshing and dressing flax at the Roscheider Hof Open Air Museum

Retting is the process of rotting away the inner stalk, leaving the outer fibres intact. A standing pool of warm water is needed, into which the beets are submerged. An acid is produced when retting, and it would corrode a metal container.

At 80 °F (27 °C), the retting process takes 4 or 5 days, it takes longer takes longer when colder.

When the retting is complete the bundles feel soft and slimy, The process can be overdone, and the fibres rot too.

DRESSING THE FLAX

Dressing is removing the fibres from the straw and cleaning it enough to be spun. The flax is broken, scutched and hackled in this step.

Breaking The process of breaking breaks up the straw into short segments. The beets are untied and fed between the beater of the breaking machine, the set of wooden blades that mesh together when the upper jaw is lowered. Scutching In order to remove some of the straw from the fibre a wooden scutching knife is scaped down the fibres while they hang vertically.

Heckling Fibre is pulled through various different sized heckling combs. A Heckling comb is a bed of sharp, long-tapered, tempered, polished steel pins driven into wooden blocks at regular spacing. A good progression is from 4 pins per square inch, to 12, to 25 to 48 to 80.

The first three will remove the straw, and the last two will split and polish the fibres. Some of the finer stuff that comes off in the last heckles can be carded like wool and spun.

It will produce a coarser yarn than the fibres pulled through the heckles because it will still contain some straw.

SPINNING THE FLAX

Flax can either be spun from a distaff, or from the spinner's lap. Spinners keep their fingers wet when spinning, to prevent forming fuzzy thread. Usually singles are spun with an "S" twist.

After flax is spun it is washed in a pot of boiling water for a couple of hours to set the twist and reduce fuzziness.

Many handspinners, will buy a roving of flax. This roving is spun in the same manner as above. The rovings may come with very long fibres (4 to 8 inches), or much shorter fibres (2 to 3 inches).

SILK

Silk is a natural protein fibre, some forms of which can be woven into textiles. The best-known type of silk is obtained from the cocoons of the larvae of the mulberry silkworm *Bombyx mori* reared in captivity (sericulture). The shimmering appearance of silk is due to the triangular prism-like structure of the silk fibre, which allows silk cloth to refract incoming light at different angles, thus producing different colours.

Silks are produced by several other insects, but only the silk of moth caterpillars has been used for textile manufacturing. There has been some research into other silks, which differ at the molecular level. Silks are mainly produced by the larvae of insects undergoing complete metamorphosis, but also by some adult insects such as webspinners. Silk production is especially common in the Hymenoptera (bees, wasps, and ants), and is sometimes used in nest construction. Other types of arthropod produce silk, most notably various arachnids such as spiders (see spider silk).

HISTORY

Woven silk textile from tomb no 1. at Mawangdui in Changsha, Hunan province, China, from the Western Han Dynasty, 2nd century BC

WILD SILK

A variety of wild silks, produced by caterpillars other than the mulberry silkworm have been known and used in China, South Asia, and Europe since ancient times. However, the scale of production was always far smaller than that of cultivated silks.

They differ from the domesticated varieties in colour and texture, and cocoons gathered in the wild usually have been damaged by the emerging moth before the cocoons are gathered, so the silk thread that makes up the cocoon has been torn into shorter lengths. Commercially reared silkworm pupae are killed by dipping them in boiling water before the adult moths emerge, or by piercing them with a needle, allowing the whole cocoon to be unravelled as one continuous thread. This permits a much stronger cloth to be woven from the silk. Wild silks also tend to be more difficult to dye than silk from the cultivated silkworm.

CHINA

Silk fabric was first developed in ancient China, with some of the earliest examples found as early as 3500 BC. Legend gives credit for developing silk to a Chinese empress, Lei Zu (Hsi-Ling-Shih, Lei-Tzu). Silks were originally reserved for the Kings of China for their own use and gifts to others, but spread gradually through Chinese culture and trade both geographically and socially, and then to many regions of Asia. Silk rapidly became a popular luxury fabric in the many areas accessible to Chinese merchants because of its texture and luster.

Silk was in great demand, and became a staple of pre-industrial international trade. In July 2007, archeologists discovered intricately woven and dyed silk textiles in a tomb in Jiangxi province, dated to the Eastern Zhou Dynasty roughly 2,500 years ago. Although historians have suspected a long history of a formative textile industry in ancient China, this find of silk textiles employing "complicated techniques" of weaving and dyeing provides direct and concrete evidence for silks dating before the Mawangdui-discovery and other silks dating to the Han Dynasty (202 BC-220 AD).

The first evidence of the silk trade is the finding of silk in the hair of an Egyptian mummy of the 21st dynasty, c.1070 BC. Ultimately the silk trade reached as far as the Indian subcontinent, the Middle East, Europe, and North Africa. This trade was so extensive that the major set of trade routes between Europe and Asia has become known as the Silk Road. The highest development was in China.

The Emperors of China strove to keep knowledge of sericulture secret to maintain the Chinese monopoly. Nonetheless sericulture reached Korea around 200 BC, about the first half of the 1st century AD had reached ancient Khotan, and by AD 300 the practice had been established in India.

THAILAND

Silk is produced, year round, in Thailand by two types of silkworms, the cultured Bombycidae and wild Saturniidae. Most production is after the rice harvest in the southern and northeast parts of the country.

Women traditionally weave silk on hand looms, and pass the skill on to their daughters as weaving is considered to be a sign of maturity and eligibility for marriage. Thai silk textiles often use complicated patterns in various colours and styles. Most regions of Thailand have their own typical silks. A single thread filament is too thin to use on its own so women combine many threads to produce a thicker, usable fibre. They do this by hand-reeling the threads onto a wooden spindle to produce a uniform strand of raw silk. The process takes around 40 hours to produce a half kilogram of Thai silk.

Many local operations use a reeling machine for this task, but some silk threads are still hand-reeled. The difference is that hand-reeled threads produce three grades of silk: two fine grades that are ideal for lightweight fabrics, and a thick grade for heavier material.

The silk fabric is soaked in extremely cold water and bleached before dyeing to remove the natural yellow colouring of Thai silk yarn.

To do this, skeins of silk thread are immersed in large tubs of hydrogen peroxide. Once washed and dried, the silk is woven on a traditional hand operated loom.

INDIA

Silk, known as "Paat" in Eastern India, *Pattu* in southern parts of India and *Resham* in Hindi/Urdu, has a long history in India. Recent archaeological discoveries in Harappa and Chanhu-daro suggest that sericulture, employing wild silk threads from native silkworm species, existed in South Asia during the time of the Indus Valley Civilization, roughly contemporaneous with the earliest known silk use in China. Silk is widely produced today. India is the second largest producer of silk after China. A majority of the silk in India is produced in Karnataka State, particularly in Mysore and the North Bangalore regions of Muddenahalli, Kanivenarayanapura, and Doddaballapur. India is also the largest consumer of silk in the world. The tradition of wearing silk sarees in marriages by the brides is followed in southern parts of India. Silk is worn by people as a symbol of royalty while attending functions and during festivals. Historically silk was used by the upper classes, while cotton was used by the poorer classes. Today silk is mainly produced in Bhoodhan Pochampally (also known as Silk City), Kanchipuram, Dharmavaram, Mysore, etc. in South India and Banaras in the North for manufacturing garments and sarees. "Murshidabad silk", famous from historical times, is mainly produced in Malda and Murshidabad district of West Bengal and woven with hand looms in Birbhum and Murshidabad district. Another place famous for production of silk is Bhagalpur. The silk from Pochampally is particularly well-known for its classic designs and enduring quality. The silk is traditionally hand-woven and hand-dyed and usually also has silver threads woven into the cloth. Most of this silk is used to make sarees. The sarees usually are very expensive and vibrant in colour. Garments made from silk form an integral part of Indian weddings and other celebrations. In the northeastern state of Assam, three different types of silk are produced, collectively called Assam silk: Muga, Eri and Pat silk. Muga, the golden silk, and Eri are produced by silkworms that are native only to Assam. The heritage of silk rearing and weaving is very old and continues today especially with the production of Muga and Pat *riha* and *mekhela chador*, the three-piece silk sarees woven with traditional motifs. *Mysore Silk Sarees*, which are known for their soft texture, last many years if carefully maintained.

ANCIENT MEDITERRANEAN

In the Odyssey, 19.233, when Odysseus, while pretending to be someone else, is questioned by Penelope about her husband's clothing, he says that he wore a shirt "gleaming like the skin of a dried onion" (varies with translations, literal translation here) which could refer to the lustrous quality of silk fabric. The Roman Empire knew of and traded in silk. During the reign of emperor Tiberius, sumptuary laws were passed that forbade men from wearing silk garments, but these proved ineffectual. Despite the popularity of silk, the secret of silk-making only reached Europe around AD 550, via the Byzantine Empire.

Legend has it those monks working for the emperor Justinian I smuggled silkworm eggs to Constantinople in hollow canes from China. All top-quality looms and weavers were located inside the Palace complex in Constantinople and the cloth produced was used in imperial robes or in diplomacy, as gifts to foreign dignitaries. The remainder was sold at very high prices.

MIDDLE EAST

In Islamic teachings, Muslim men are forbidden to wear silk. Many religious jurists believe the reasoning behind the prohibition lies in avoiding clothing for men that can be considered feminine or extravagant. There are disputes regarding the amount of silk a fabric can consist of (*i.e.*, whether a small decorative silk piece on a cotton caftan is permissible or not) for it to be lawful for men to wear but the dominant opinion of most Muslim scholars is that the wearing of silk for men is forbidden.

Despite injunctions against silk for men, silk has retained its popularity in the Islamic world because of its permissibility for women. The Muslim Moors brought silk with them to Spain during their conquest of the Iberian Peninsula.

MEDIEVAL AND MODERN EUROPE

Venetian merchants traded extensively in silk and encouraged silk growers to settle in Italy. By the 13th century, Italian silk was a significant source of trade. Since that period, the silk worked in the province of Como has been the most valuable silk in the world. The wealth of Florence was largely built on textiles, both wool and silk and other cities like Lucca also grew rich on the trade. Italian silk was so popular in Europe that Francis I of France invited Armenian silk makers to France to create a French silk industry, especially in Lyon. Mass emigration (especially of Huguenots) during periods of religious dispute had seriously damaged French industry and introduced these various textile industries, including silk, to other countries.

James I attempted to establish silk production in England, purchasing and planting 100,000 mulberry trees, some on land adjacent to Hampton Court Palace, but they were of a species unsuited to the silk worms, and the attempt failed. British enterprise also established silk filature in Cyprus in 1928. In England in the mid 20th Century, silk was produced at Lullingstone Castle in Kent. Silkworms were raised and reeled under the direction of Zoe Lady Hart Dyke. Production started elsewhere later.

In Italy, the Stazione Bacologica Sperimentale was founded in Padua in 1871 to research sericulture. In the late 19th century, China, Japan, and Italy were the major producers of silk.

The most important cities for silk production in Italy were Como and Meldola (Forlì). In medieval times, it was common for silk to be used to make elaborate casings for bananas and other fruits.

Silk was expensive in Medieval Europe and used only by the rich. Italian merchants like Giovanni Arnolfini became hugely wealthy trading it to the Courts of Northern Europe.

NORTH AMERICA

James I of England introduced silk-growing to the American colonies around 1619, ostensibly to discourage tobacco planting. The Shakers in Kentucky adopted the practice as did a cottage industry in New England. In the 19th century a new attempt at a silk industry began with European-born workers in Paterson, New Jersey, and the city became a US silk center, although Japanese imports were still more important.

World War II interrupted the silk trade from Japan. Silk prices increased dramatically, and US industry began to look for substitutes, which led to the use of synthetics such as nylon. Synthetic silks have also been made from local, a type of cellulose fibre, and are often difficult to distinguish from real silk (see spider silk for more on synthetic silks).

PROPERTIES

Physical Properties

Silk fibres from the *Bombyx mori* silkworm have a triangular cross section with rounded corners, 5-10 ìm wide. The fibroin-heavy chain is composed mostly of beta-sheets, due to a 59-mer amino acid repeat sequence with some variations. The flat surfaces of the fibrils reflect light at many angles, giving silk a natural shine. The cross-section from other silkworms can vary in shape and diameter: crescent-like for *Anaphe* and elongated wedge for *tussah*. Silkworm fibres are naturally extruded from two silkworm glands as a pair of primary filaments (brin), which are stuck together, with sericin proteins that act like glue, to form a bave. Bave diameters for tussah silk can reach 65 ìm. See cited reference for cross-sectional SEM photographs. Silk has a smooth, soft texture that is not slippery, unlike many synthetic fibres.

Silk is one of the strongest natural fibres but loses up to 20per cent of its strength when wet. It has a good moisture regain of 11per cent. Its elasticity is moderate to poor: if elongated even a small amount, it remains stretched. It can be weakened if exposed to too much sunlight. It may also be attacked by insects, especially if left dirty.

Silk is a poor conductor of electricity and thus susceptible to static cling.

Unwashed silk chiffon may shrink up to 8per cent due to a relaxation of the fibre macrostructure. So silk should either be pre-washed prior to garment construction, or dry cleaned. Dry cleaning may still shrink the chiffon up to 4per cent. Occasionally, this shrinkage can be reversed by a gentle steaming with a press cloth. There is almost no gradual shrinkage nor shrinkage due to molecular-level deformation.

Natural and synthetic silk is known to manifest piezoelectric properties in proteins, probably due to its molecular structure.

Silkworm silk was used as the standard for the denier, a measurement of linear density in fibres. Silkworm silk therefore has a linear density of approximately 1 den, or 1.1 dtex.

Comparison of silk fibres	Linear Density(dtex)	Diameter (mm)	Coeff. Variation
Moth: Bombyx mori	1.17	12.9	24.8per cent
Spider: Argiope aurentia	0.14	3.57	14.8per cent

Chemical Properties

Silk emitted by the silkworm consists of two main proteins, sericin and fibroin, fibroin being the structural center of the silk, and serecin being the sticky material surrounding it. Fibroin is made up of the amino acids Gly-Ser-Gly-Ala-Gly-Ala and forms beta pleated sheets. Hydrogen bonds form between chains, and side chains form above and below the plane of the hydrogen bond network.

The high proportion (50per cent) of glycine, which is a small amino acid, allows tight packing and the fibres are strong and resistant to stretching. The tensile strength is due to the many interseeded hydrogen bonds. Since the protein forms a beta sheet, when stretched the force is applied to these strong bonds and they do not break. Silk is resistant to most mineral acids, except for sulfuric acid, which dissolves it. It is yellowed by perspiration.

USES

Silk's absorbency makes it comfortable to wear in warm weather and while active. Its low conductivity keeps warm air close to the skin during cold weather. It is often used for clothing such as shirts, ties, blouses, formal dresses, high fashion clothes, lingerie, pyjamas, robes, dress suits, sun dresses and kimonos.

Silk's attractive luster and drape makes it suitable for many furnishing applications. It is used for upholstery, wall coverings, window treatments (if blended with another fibre), rugs, bedding and wall hangings

While on the decline now, due to artificial fibres, silk has had many industrial and commercial uses; parachutes, bicycle tires, comforter filling and artillery gunpowder bags.

A special manufacturing process removes the outer irritant sericin coating of the silk, which makes it suitable as non-absorbable surgical sutures.

This process has also recently led to the introduction of specialist silk underclothing for children and adults with eczema where it can significantly reduce itch.

PRODUCTION

The cultivation of silk is called sericulture. Over 30 countries produce silk, the major ones are China (54per cent) and India (14per cent).

To produce 1 kg of silk, 104 kg of mulberry leaves must be eaten by 3000 silkworms. It takes about 5000 silkworms to make a pure silk kimono.

Table. Top Ten Cocoons (Reelable) Producers — 2005.

Country	Production (Int $1000)	Footnote	Production (1000 kg)	Footnote
People's Republic of China	978,013	C	290,003	F
India	259,679	C	77,000	F
Uzbekistan	57,332	C	17,000	F
Brazil	37,097	C	11,000	F
Iran	20,235	C	6,000	F
Thailand	16,862	C	5,000	F
Vietnam	10,117	C	3,000	F
Democratic People's Republic of Korea	5,059	C	1,500	F
Romania	3,372	C	1,000	F
Japan	2,023	C	600	F

CULTIVATION

Silk moths lay eggs on specially prepared paper. The eggs hatch and the caterpillars (silkworms) are fed fresh mulberry leaves. After about 35 days and 4 moltings, the caterpillars are 10,000 times heavier than when hatched and are ready to begin spinning a cocoon. A straw frame is placed over the tray of caterpillars, and each caterpillar begins spinning a cocoon by moving its head in a pattern. Two glands produce liquid silk and force it through openings in the head called spinnerets.

Liquid silk is coated in sericin, a water-soluble protective gum, and solidifies on contact with the air. Within 2–3 days, the caterpillar spins about 1 mile of filament and is completely encased in a cocoon. The silk farmers then kill most caterpillars by heat, leaving some to metamorphose into moths to breed the next generation of caterpillars.

Harvested cocoons are then soaked in boiling water to soften the sericin holding the silk fibres together in a cocoon shape. The fibres are then unwound to produce a continuous thread. Since a single thread is too fine and fragile for commercial use, anywhere from three to ten strands are spun together to form a single thread of silk.

ANIMAL RIGHTS

As the process of harvesting the silk from the cocoon kills the larvae, sericulture has been criticized in the early 21st century by animal rights activists, especially since artificial silks are available. Mohandas Gandhi was also critical of silk production based on the Ahimsa philosophy "not to hurt any living thing." This led to Gandhi's promotion of cotton spinning machines, an example of which can be seen at the Gandhi Institute. He also promoted *Ahimsa silk*, wild silk made from the cocoons of wild and semi-wild silk moths. Ahimsa silk is promoted in parts of Southern India for those who prefer not to wear silk produced by killing silkworms.

COTTON

Cotton is a soft, fluffy staple fibre that grows in a boll, or protective capsule, around the seeds of cotton plants of the genus *Gossypium*. The plant is a shrub native to tropical and subtropical regions around the world, including the Americas, Africa, India, and Pakistan. The fibre most often is spun into yarn or thread and used to make a soft, breathable textile, which is the most widely used natural-fibre cloth in clothing today. The English name derives from the Arabic *(al) qutn*, which began to be used circa 1400. The botanical purpose of cotton fibre is to aid in seed dispersal.

HISTORY

According to the Foods and Nutrition Encyclopaedia, the earliest cultivation of cotton discovered thus far in the Americas occurred in Mexico, some 8,000 years ago. The indigenous species was *Gossypium hirsutum*, which is today the most widely planted species of cotton in the world, constituting about 89.9per cent of all production worldwide. The greatest diversity of wild cotton species is found in Mexico, followed by Australia and Africa.

Cotton was first cultivated in the Old World 7,000 years ago (5th–4th millennia BC), by the inhabitants of the Indus Valley Civilization, which covered a huge swath of the northwestern part of the Indian subcontinent, comprising today parts of eastern Pakistan and northwestern India. The Indus cotton industry was well developed and some methods used in cotton spinning and fabrication continued to be used until the modern industrialization of India. Well before the Common Era, the use of cotton textiles had spread from India to the Mediterranean and beyond.

Greeks and the Arabs were apparently ignorant about cotton until the Wars of Alexander the Great, as his contemporary Megasthenes told Seleucus I Nicator of "there being trees on which wool grows" in "Indica".

According to The Columbia Encyclopaedia, Sixth Edition:

Cotton has been spun, woven, and dyed since prehistoric times. It clothed the people of ancient India, Egypt, and China. Hundreds of years before the Christian era, cotton textiles were woven in India with matchless skill, and their use spread to the Mediterranean countries. In the first century, Arab traders brought fine muslin and calico to Italy and Spain. The Moors introduced the cultivation of cotton into Spain in the 9th century. Fustians and dimities were woven there and in the 14th century in Venice and Milan, at first with a linen warp. Little cotton cloth was imported to England before the 15th century, although small amounts were obtained chiefly for candlewicks. By the 17th century, the East India Company was bringing rare fabrics from India. Native Americans skillfully spun and wove cotton into fine garments and dyed tapestries. Cotton fabrics found in Peruvian tombs are said to belong to a pre-Inca culture.

In Iran (Persia), the history of cotton dates back to the Achaemenid era (5th century BC); however, there are few sources about the planting of cotton in pre-Islamic Iran. The planting of cotton was common in Merv, Ray and Pars of Iran. In the poems of Persian poets, especially Ferdowsi's Shahname, there are many references to cotton ("panbe" in Persian). Marco Polo (13th century) refers to the major products of Persia, including cotton. John Chardin, a famous French traveler of 17th century, who had visited the Safavid Persia, has approved the vast cotton farms of Persia.

In Peru, cultivation of the indigenous cotton species *Gossypium barbadense* was the backbone of the development of coastal cultures, such as the Norte Chico, Moche and Nazca. Cotton was grown upriver, made into nets and traded with fishing villages along the coast for large supplies of fish. The Spanish who came to Mexico and Peru in the early 16th century found the people growing cotton and wearing clothing made of it.

During the late medieval period, cotton became known as an imported fibre in northern Europe, without any knowledge of how it was derived, other than that it was a plant; noting its similarities to wool, people in the region could only imagine that cotton must be produced by plant-borne sheep. John Mandeville, writing in 1350, stated as fact the now-preposterous belief: "There grew there [India] a wonderful tree which bore tiny lambs on the endes of its branches. These branches were so pliable that they bent down to allow the lambs to feed when they are hungrie [*sic*]." (See Vegetable Lamb of Tartary.) This aspect is retained in the name for cotton in many European languages, such as German *Baumwolle*, which translates as "tree wool" (*Baum* means "tree"; *Wolle* means "wool"). By the end of the 16th century, cotton was cultivated throughout the warmer regions in Asia and the Americas.

India's cotton-processing sector gradually declined during British expansion in India and the establishment of colonial rule during the late 18th and early 19th centuries. This was largely due to aggressive colonialist mercantile policies of the British East India Company, which made cotton processing and manufacturing workshops in India uncompetitive. Indian markets were increasingly forced to supply only raw cotton and were forced, by British-imposed law, to purchase manufactured textiles from Britain.

INDUSTRIAL REVOLUTION IN BRITAIN

The advent of the Industrial Revolution in Britain provided a great boost to cotton manufacture, as textiles emerged as Britain's leading export. In 1738, Lewis Paul and John Wyatt, of Birmingham, England, patented the roller spinning machine, and the flyer-and-bobbin system for drawing cotton to a more even thickness using two sets of rollers that traveled at different speeds. Later, the invention of the spinning jenny in 1764 and Richard Arkwright's spinning frame (based on the roller spinning machine) in 1769 enabled British weavers to produce cotton yarn and cloth at much higher rates. From the late 18th century onwards, the British city of Manchester acquired the nickname *"Cottonopolis"* due to the cotton industry's omnipresence within the city, and Manchester's role as the heart of the global cotton trade. Production capacity in Britain and the United States was further improved by the invention of the cotton gin by the American Eli Whitney in 1793. Improving technology and increasing control of world markets allowed British traders to develop a commercial chain in which raw cotton fibres were (at first) purchased from colonial plantations, processed into cotton cloth in the mills of Lancashire, and then exported on British ships to captive colonial markets in West Africa, India, and China (via Shanghai and Hong Kong).

By the 1840s, India was no longer capable of supplying the vast quantities of cotton fibres needed by mechanized British factories, while shipping bulky, low-price cotton from India to Britain was time-consuming and expensive. This, coupled with the emergence of American cotton as a superior type (due to the longer, stronger fibres of the two domesticated native American species, *Gossypium hirsutum* and *Gossypium barbadense*), encouraged British traders to purchase cotton from plantations in the United States and the Caribbean. By the mid 19th century, "King Cotton" had become the backbone of the southern American economy. In the United States, cultivating and harvesting cotton became the leading occupation of slaves.

During the American Civil War, American cotton exports slumped due to a Union blockade on Southern ports, also because of a strategic decision by the Confederate government to cut exports, hoping to force Britain to recognize the Confederacy or enter the war, prompting the main purchasers of cotton, Britain and France to turn to Egyptian cotton. British and French traders invested heavily in cotton plantations and the Egyptian government of Viceroy Isma'il took out substantial loans from European bankers and stock exchanges. After the American Civil War ended in 1865, British and French traders abandoned Egyptian cotton and returned to cheap American exports, sending Egypt into a deficit spiral that led to the country declaring bankruptcy in 1876, a key factor behind Egypt's annexation by the British Empire in 1882.

During this time, cotton cultivation in the British Empire, especially India, greatly increased to replace the lost production of the American South. Through tariffs and other restrictions, the British government discouraged the production of cotton cloth in India; rather, the raw fibre was sent to England for processing. The Indian patriot Mahatma Gandhi described the process:

- English people buy Indian cotton in the field, picked by Indian labour at seven cents a day, through an optional monopoly.
- This cotton is shipped on British ships, a three-week journey across the Indian Ocean, down the Red Sea, across the Mediterranean, through Gibraltar, across the Bay of Biscay and the Atlantic Ocean

to London. One hundred per cent profit on this freight is regarded as small.

- The cotton is turned into cloth in Lancashire. You pay shilling wages instead of Indian pennies to your workers. The English worker not only has the advantage of better wages, but the steel companies of England get the profit of building the factories and machines. Wages; profits; all these are spent in England.
- The finished product is sent back to India at European shipping rates, once again on British ships. The captains, officers, sailors of these ships, whose wages must be paid, are English. The only Indians who profit are a few lascars who do the dirty work on the boats for a few cents a day.
- The cloth is finally sold back to the kings and landlords of India who got the money to buy this expensive cloth out of the poor peasants of India who worked at seven cents a day. (Fisher 1932 pp 154–156)

In the United States, Southern cotton provided capital for the continuing development of the North. The cotton produced by enslaved African Americans not only helped the South, but also enriched Northern merchants. Much of the Southern cotton was transshipped through the northern ports.

Cotton remained a key crop in the Southern economy after emancipation and the end of the Civil War in 1865. Across the South, sharecropping evolved, in which free black farmers and landless white farmers worked on white-owned cotton plantations of the wealthy in return for a share of the profits. Cotton plantations required vast labour forces to hand-pick cotton, and it was not until the 1950s that reliable harvesting machinery was introduced into the South (prior to this, cotton-harvesting machinery had been too clumsy to pick cotton without shredding the fibres). During the early 20th century, employment in the cotton industry fell, as machines began to replace laborers, and the South's rural labour force dwindled during the First and Second World Wars. Today, cotton remains a major export of the southern United States, and a majority of the world's annual cotton crop is of the long-staple American variety.

TANGÜIS COTTON

In 1901, Peru's cotton industry suffered because of a fungus plague caused by a plant disease known as "cotton wilt" or, more correctly, "fusarium wilt", caused by the fungus *Fusarium vasinfectum*. The plant disease, which spread throughout Peru, entered plant's roots and worked its way up the stem until the plant was completely dried up. Fermín Tangüis, a Puerto Rican agriculturist who lived in Peru, studied some species of the plant that were affected by the disease to a lesser extent and experimented in germination with the seeds of various cotton plants. In 1911, after 10 years of experimenting and failures, Tangüis was able to develop a seed which produced a superior cotton plant resistant to the disease. The seeds produced a plant that had a 40per cent longer (between 29 mm and 33 mm) and thicker fibre that did not break easily and required little water. The Tangüis cotton, as it became known, is the variety which is preferred by the Peruvian national textile industry. It constituted 75per cent of all the Peruvian cotton production, both for domestic use and apparel exports. The Tangüis cotton crop was estimated at 225,000 bales that year.

CULTIVATION

Successful cultivation of cotton requires a long frost-free period, plenty of sunshine, and a moderate rainfall, usually from 600 to 1200 mm (24 to 48 inches). Soils usually need to be fairly heavy, although the level of nutrients does not need to be exceptional. In general, these conditions are met within the seasonally dry tropics and subtropics in the Northern and Southern hemispheres, but a large proportion of the cotton grown today is cultivated in areas with less rainfall that obtain the water from irrigation. Production of the crop for a given year usually starts soon after harvesting the preceding autumn. Planting time in spring in the Northern hemisphere varies from the beginning of February to the beginning of June. The area of the United States known as the South Plains is the largest contiguous cotton-growing region in the world. While dryland (non-irrigated) cotton is successfully grown in this region, consistent yields are only produced with heavy reliance on irrigation water drawn from the Ogallala Aquifer. Since cotton is somewhat salt and drought tolerant, this makes it an attractive crop for arid and semiarid regions. As water resources get tighter around the world, economies that rely on it face difficulties and conflict, as well as potential environmental problems. For example, improper cropping and irrigation practices have led to desertification in areas of Uzbekistan, where cotton is a major export. In the days of the Soviet Union, the Aral Sea was tapped for agricultural irrigation, largely of cotton, and now salination is widespread.

GENETIC MODIFICATION

Genetically modified (GM) cotton was developed to reduce the heavy reliance on pesticides. The bacterium *Bacillus thuringiensis* (Bt) naturally produces a chemical harmful only to a small fraction of insects, most notably the larvae of moths and butterflies, beetles, and flies, and harmless to other forms of life. The gene coding for BT toxin has been inserted into cotton, causing cotton to produce this natural insecticide in its tissues. In many regions, the main pests in commercial cotton are lepidopteron larvae, which are killed by the BT protein in the transgenic cotton they eat. This eliminates the need to use large amounts of broad-spectrum insecticides to kill lepidopteron pests (some of which have developed pyrethroid resistance). This spares natural insect predators in the farm ecology and further contributes to no insecticide pest management.

BT cotton is ineffective against many cotton pests, however, such as plant bugs, stink bugs, and aphids; depending on circumstances it may still be desirable to use insecticides against these. A 2006 study done by Cornell researchers, the Center for Chinese Agricultural Policy and the Chinese Academy of Science on Bt cotton farming in China found that after seven years these secondary pests that were normally controlled by pesticide had increased, necessitating the use of pesticides at similar levels to non-Bt cotton and causing less profit for farmers because of the extra expense of GM seeds. However a more recent 2009 study by the Chinese Academy of Sciences, Stanford University and Rutgers University refutes this.

They concluded that the GM cotton effectively controlled bollworm. The secondary pests were mostly miridae (plant bugs) whose increase was related to local temperature and rainfall and only continued to increase in half the villages studied. Moreover, the increase in insecticide use for the control of these secondary insects was far smaller than the reduction in total insecticide use due to BT cotton adoption. The International Service for the Acquisition of Agri-biotech Applications (ISAAA) said that, worldwide, GM cotton was planted on an area of 16 million hectares in 2009.

This was 49per cent of the worldwide total area planted in cotton. The U.S. cotton crop was 93per cent GM in 2010 and the Chinese cotton crop was 68per cent GM in 2009.

The initial introduction of GM cotton proved to be a huge success in Australia - the yields were equivalent to the no transgenic varieties and the crop used much less pesticide to produce (85per cent reduction). The subsequent introduction of a second variety of GM cotton led to increases in GM cotton production until 95per cent of the Australian cotton crop was GM in 2009.

Cotton has also been genetically modified for resistance to glyphosate (marketed as Roundup in North America), an inexpensive and highly effective, but broad-spectrum herbicide.

Originally, it was only possible to achieve glyph sate resistance when the plant was young, but with the development of Roundup Ready Flex, it is possible to achieve glyphosate resistance much later in the growing season.

GM cotton acreage in India continues to grow at a rapid rate, increasing from 50,000 hectares in 2002 to 8.4 million hectares in 2009. The total cotton area in India was 9.6 million hectares (the largest in the world or, about 35per cent of world cotton area), so GM cotton was grown on 87per cent of the cotton area in 2009. This makes India the country with the largest area of GM cotton in the world, surpassing China (3.7 million hectares in 2009).

The major reasons for this increase is a combination of increased farm income ($225/ha) and a reduction in pesticide use to control the cotton bollworm.

Cotton has gossypol, a toxin that makes it inedible.

However, scientists have silenced the gene that produces the toxin, making it a potential food crop.

ORGANIC PRODUCTION

Organic cotton is generally understood as cotton, from plants not genetically modified, that is certified to be grown without the use of any synthetic agricultural chemicals, such as fertilizers or pesticides. Its production also promotes and enhances biodiversity and biological cycles.

United States cotton plantations are required to enforce the National Organic Programme (NOP). This institution determines the allowed practices for pest control, growing, fertilizing, and handling of organic crops.As of 2007, 265,517 bales of organic cotton were produced in 24 countries, and worldwide production was growing at a rate of more than 50per cent per year.

PESTS AND WEEDS

The cotton industry relies heavily on chemicals, such as fertilizers and insecticides, although a very small number of farmers are moving towards an organic model of production, and organic cotton products are now available for purchase at limited locations.

These are popular for baby clothes and diapers. Under most definitions, organic products do not use genetic engineering.

Historically, in North America, one of the most economically destructive pests in cotton production has been the boll weevil.

Due to the US Department of Agriculture's highly successful Boll Weevil Eradication Programme (BWEP), this pest has been eliminated from cotton in most of the United States. This Programme, along with the introduction of genetically engineered Bt cotton (which contains a bacterial gene that codes for a plant-produced protein that is toxic to a number of pests such as cotton bollworm and pink bollworm), has allowed a reduction in the use of synthetic insecticides. Other significant global pests of cotton include the pink bollworm, *Pectinophora gossypiella*; the chili thrips, *Scirtothrips dorsalis*; and the cotton seed bug, *Oxycarenus hyalinipennis*.

HARVESTING

Most cotton in the United States, Europe, and Australia is harvested mechanically, either by a cotton picker, a machine that removes the cotton from the boll without damaging the cotton plant, or by a cotton stripper, which strips the entire boll off the plant.

Cotton strippers are used in regions where it is too windy to grow picker varieties of cotton, and usually after application of a chemical defoliant or the natural defoliation that occurs after a freeze.

Cotton is a perennial crop in the tropics, and without defoliation or freezing, the plant will continue to grow. Cotton continues to be picked by hand in developing countries.

COMPETITION FROM SYNTHETIC FIBRES

The era of manufactured fibres began with the development of rayon in France in the 1890s. Rayon is derived from a natural cellulose and cannot be considered synthetic, but requires extensive processing in a manufacturing process, and led the less expensive replacement of more naturally derived materials. A succession of new synthetic fibres were introduced by the chemicals industry in the following decades. Acetate in fibre form was developed in 1924. Nylon, the first fibre synthesized entirely from petrochemicals, was introduced as a sewing thread by DuPont in 1936, followed by DuPont's acrylic in 1944. Some garments were created from fabrics based on these fibres, such as women's hosiery from nylon, but it was not until the introduction of polyester into the fibre marketplace in the early 1950s that the market for cotton came under threat. The rapid uptake of polyester garments in the 1960s caused economic hardship in cotton-exporting economies, especially in Central American countries, such as Nicaragua, where cotton production had boomed tenfold between 1950 and 1965 with the advent of cheap chemical pesticides. Cotton production recovered in the 1970s, but crashed to pre-1960 levels in the early 1990s.

Beginning as a self-help Programme in the mid-1960s, the Cotton Research and Promotion Programme (CRPP) was organized by U.S. cotton producers in response to cotton's steady decline in market share. At that time, producers voted to set up a per-bale assessment system to fund the Programme, with built-in safeguards to protect their investments. With the passage of the Cotton Research and Promotion Act of 1966, the Programme joined forces and began battling synthetic competitors and re-establishing markets for cotton. Today, the success of this Programme has made cotton the best-selling fibre in the U.S. and one of the best-selling fibres in the world.

Administered by the Cotton Board and conducted by Cotton Incorporated, the CRPP works to greatly increase the demand for and profitability of cotton through various research and promotion activities. It is funded by U.S. cotton producers and importers.

USES

Cotton is used to make a number of textile products. These include terrycloth for highly absorbent bath towels and robes; denim for blue jeans; chambray, popularly used in the manufacture of blue work shirts (from which we get the term "blue-collar"); and corduroy, seersucker, and cotton twill. Socks, underwear, and most T-shirts are made from cotton. Bed sheets often are made from cotton. Cotton also is used to make yarn used in crochet and knitting. Fabric also can be made from recycled or recovered cotton that otherwise would be thrown away during the spinning, weaving, or cutting process. While many fabrics are made completely of cotton, some materials blend cotton with other fibres, including rayon and synthetic fibres such as polyester. It can either be used in knitted or woven fabrics, as it can be blended with elastine to make a stretchier thread for knitted fabrics, and apparel such as stretch jeans.

In addition to the textile industry, cotton is used in fishnets, coffee filters, tents, gunpowder (see nitrocellulose), cotton paper, and in bookbinding. The first Chinese paper was made of cotton fibre. Fire hoses were once made of cotton.

The cottonseed which remains after the cotton is ginned is used to produce cottonseed oil, which, after refining, can be consumed by humans like any other vegetable oil. The cottonseed meal that is left generally is fed to ruminant livestock; the gossypol remaining in the meal is toxic to monogastric animals. Cottonseed hulls can be added to dairy cattle rations for roughage. During the American slavery period, cotton root bark was used in folk remedies as an abortifacient, that is, to induce a miscarriage.

Cotton linters are fine, silky fibres which adhere to the seeds of the cotton plant after ginning. These curly fibres typically are less than 1/8 in (3 mm) long. The term also may apply to the longer textile fibre staple lint as well as the shorter fuzzy fibres from some upland species. Linters are traditionally used in the manufacture of paper and as a raw material in the manufacture of cellulose. In the UK, linters are referred to as "cotton wool". This can also be a refined product (*absorbent cotton* in U.S. usage) which has medical, cosmetic and many other practical uses. The first medical use of cotton wool was by Dr. Joseph Sampson Gamgee at the Queen's Hospital (later the General Hospital) in Birmingham, England.

Shiny cotton is a processed version of the fibre that can be made into cloth resembling satin for shirts and suits. However, it is hydrophobic (does not absorb water easily), which makes it unfit for use in bath and dish towels (although examples of these made from shiny cotton are seen).

The term Egyptian cotton refers to the extra long staple cotton grown in Egypt and favoured for the luxury and upmarket brands worldwide. During the U.S. Civil War, with heavy European investments, Egyptian-grown cotton became a major alternate source for British textile mills. Egyptian cotton is more durable and softer than American Pima cotton, which is why it is more expensive. Pima cotton is American cotton that is grown in the southwestern states of the U.S.

INTERNATIONAL TRADE

The largest producers of cotton, currently (2009), are China and India, with annual production of about 34 million bales and 24 million bales, respectively; most of this production is consumed by their respective textile industries. The largest exporters of raw cotton are the United States, with sales of $4.9 billion, and Africa, with sales of $2.1 billion. The total international trade is estimated to be $12 billion. Africa's share of the cotton trade has doubled since 1980. Neither area has a significant domestic textile industry, textile manufacturing having moved to developing nations in Eastern and South Asia such as India and China. In Africa, cotton is grown by numerous small holders. Dunavant Enterprises, based in Memphis, Tennessee, is the leading cotton broker in Africa, with hundreds of purchasing agents. It operates cotton gins in Uganda, Mozambique, and Zambia. In Zambia, it often offers loans for seed and expenses to the 180,000 small farmers who grow cotton for it, as well as advice on farming methods. Cargill also purchases cotton in Africa for export.

The 25,000 cotton growers in the United States are heavily subsidized at the rate of $2 billion per year. The future of these subsidies is uncertain and has led to anticipatory expansion of cotton brokers' operations in Africa. Dunavant expanded in Africa by buying out local operations. This is only possible in former British colonies and Mozambique; former French colonies continue to maintain tight monopolies, inherited from their former colonialist masters, on cotton purchases at low fixed prices.

LEADING PRODUCER COUNTRIES

Table. Top Ten Cotton Producers — 2009 (480-Pound Bales).

People's Republic of China	32.0 million bales
India	23.5 million bales
United States	12.4 million bales
Pakistan	10.8 million bales
Brazil	5.5 million bales
Uzbekistan	4.4 million bales
Australia	1.8 million bales
Turkey	1.7 million bales
Turkmenistan	1.1 million bales
Syria	1.0 million bales

The five leading exporters of cotton in 2009 are (1) the United States, (2) India, (3) Uzbekistan, (4) Pakistan, and (5) Brazil. The largest nonproducing importers are Korea, Russia, Taiwan, Japan, and Hong Kong.

In India, the states of Maharashtra (26.63per cent), Gujarat (17.96per cent) and Andhra Pradesh (13.75per cent) and also Madhya Pradesh are the leading cotton producing states, these states have a predominantly tropical wet and dry climate.

In Pakistan, cotton is grown predominantly in the provinces of Punjab and Sindh. The leading city in cotton production is the Punjabi city of Faisalabad which is also leading in textiles within Pakistan. The Punjab has a tropical wet and dry climate throughout the year therefore enhancing the growth of cotton.

In the United States, the state of Texas led in total production as of 2004, while the state of California had the highest yield per acre.

FAIR TRADE

Cotton is an enormously important commodity throughout the world. However, many farmers in developing countries receive a low price for their produce, or find it difficult to compete with developed countries.

This has led to an international dispute (see United States – Brazil cotton dispute):

On 27 September 2002, Brazil requested consultations with the US regarding prohibited and actionable subsidies provided to US producers, users and/or exporters of upland cotton, as well as legislation, regulations, statutory instruments and amendments thereto providing such subsidies (including export credits), grants, and any other assistance to the US producers, users and exporters of upland cotton. On 8 September 2004, the Panel Report recommended that the United States "withdraw" export credit guarantees and payments to domestic users and exporters, and "take appropriate steps to remove the adverse effects or withdraw" the mandatory price-contingent subsidy measures.

In addition to concerns over subsidies, the cotton industries of some countries are criticized for employing child labour and damaging workers' health by exposure to pesticides used in production. The Environmental Justice Foundation has campaigned against the prevalent use of forced child and adult labour in cotton production in Uzbekistan, the world's third largest cotton exporter. The international production and trade situation has led to "fair trade" cotton clothing and footwear, joining a rapidly growing market for organic clothing, fair fashion or so-called "ethical fashion". The fair trade system was initiated in 2005 with producers from Cameroon, Mali and Senegal.

TRADE

Cotton is bought and sold by investors and price speculators as a tradable commodity on 2 different stock exchanges in the United States of America.

- Cotton futures contracts are traded on the New York Mercantile Exchange (NYMEX) under the ticker symbol TT. They are delivered every year in March, May, July, October, and December.
- Cotton #2 futures contracts are traded on the New York Board of Trade (NYBOT) under the ticker symbol CT. They are delivered every year in March, May, July, October, and December.

CRITICAL TEMPERATURES

- Favourable travel temperature range: below 25°C (77°F)
- Optimum travel temperature: 21°C (70°F)

- Glow temperature: 205°C (401°F)
- Fire point: 210°C (410°F)
- Autoignition temperature: 407°C (765°F)
- Autoignition temperature (for oily cotton): 120°C (248°F)

Cotton dries out, becomes hard and brittle and loses all elasticity at temperatures above 25°C (77°F). Extended exposure to light causes similar problems. A temperature range of 25°C (77°F) to 35°C (95°F) is the optimal range for mold development. At temperatures below 0°C (32°F), rotting of wet cotton stops. Damaged cotton is sometimes stored at these temperatures to prevent further deterioration.

BRITISH STANDARD YARN MEASURES

- 1 thread = 55 inches (about 137 cm)
- 1 skein or rap = 80 threads (120 yards or about 109 m)
- 1 hank = 7 skeins (840 yards or about 768 m)
- 1 spindle = 18 hanks (15,120 yards or about 13.826 km)

FIBRE PROPERTIES

Property	Evaluation
Shape	Fairly uniform in width, 12-20 micrometers; length varies from 1 cm to 6 cm (½ to 2½ inches); typical length is 2.2 cm to 3.3 cm (^! to 1¼ inches).
Luster	high
Tenacity (strength)	
Dry	3.0-5.0 g/d
Wet	3.3-6.0 g/d
Resiliency	low
Density	1.54-1.56 g/cm^3
Moisture absorption raw:conditioned	8.5per centsaturation 15-25per cent mercerized:conditioned 8.5-10.3per cent saturation 15-27per cent+
Dimensional stability	good
Resistance to acids alkali organic solvents sunlight microorganisms insects	damage, weaken fibres resistant; no harmful effects high resistance to most Prolonged exposure weakens fibres.Mildew and rot-producing bacteria damage fibres. Silver fish damage fibres.
Thermal reactions to heat to flame	Decomposes after prolonged exposure to temperatures of 150°C or over.Burns readily.

The chemical composition of cotton is as follows:

- cellulose 91.00per cent
- water 7.85per cent
- protoplasm, pectins 0.55per cent
- waxes, fatty substances 0.40per cent
- mineral salts 0.20per cent

COTTON GENOME

A public genome sequencing effort of cotton was initiated in 2007 by a consortium of public researchers. They agreed on a strategy to sequence the genome of cultivated, tetraploid cotton. "Tetraploid" means that cultivated cotton actually has two separate genomes within its nucleus, referred to as the A and D genomes. The sequencing consortium first agreed to sequence the D-genome relative of cultivated cotton (*G. raimondii*, a wild Central American cotton species) because of its small size and limited number of repetitive elements. It is nearly one-third the number of bases of tetraploid cotton (AD), and each chromosome is only present once. The A genome of *G. arboreum* would be sequenced next. Its genome is roughly twice the size of *G. raimondii*'s. Part of the difference in size between the two genomes is the amplification of *retrotransposons* (GORGE). Once both diploid genomes are assembled, then research could begin sequencing the actual genomes of cultivated cotton varieties. This strategy is out of necessity; if one were to sequence the tetraploid genome without model diploid genomes, the euchromatic DNA sequences of the AD genomes would co-assemble and the repetitive elements of AD genomes would assembly independently into A and D sequences respectively. Then there would be no way to untangle the mess of AD sequences without comparing them to their diploid counterparts.

The public sector effort continues with the goal to create a high-quality, draft genome sequence from reads generated by all sources. The public-sector effort has generated Sanger reads of BACs, fosmids, and plasmids as well as 454 reads. These later types of reads will be instrumental in assembling an initial draft of the D genome. They announced that they would donate their raw reads to the public. This public relations effort gave them some recognition for sequencing the cotton genome. Once the D genome is assembled from all of this raw material, it will undoubtedly assist in the assembly of the AD genomes of cultivated varieties of cotton, but a lot of hard work remains.

Yield! Maximization in the Cotton Spinning Industry (Yarn Manufacturing)

Yieldper cent shows the performance of any industry, it shows efficiency of the industry to convert the raw material into the finished goods. The industry management focuses so much on the maximization of yield so that maximum profit can be earned. Yieldper cent directly relates with profit of the industry, textile cotton spinning industry usually give the yieldper cent up to 84, this can be increased up to 1 to 2per cent, adding million in the profit of cotton spinning industry. This article presents the methodology to enhance the yield per cent of cotton spinning industry.

INTRODUCTION

Raw material consist impurities up to 4-16 per cent(according to type, region and the efficiency of the ginning), the yarn manufacturing processes (Blow-room to carding) are especially designed to remove these impurities. While removing the impurities some good fibres are also removed due to inefficient machine setting or improper machine sequence selection and there are many more causes which decrease the yield per cent. Some of the causes are discussed here which increase / decrease the yield per cent such as

- Raw material selection
- Raw martial process route selection
- Improper machine setting
- Maintenance of Machines
- Inefficient air conditioning plant
- Improper material handling

RAW MATERIAL SELECTION

Raw martial costs about 50-70per cent of the total cost of the product[1]. It is mentioned above that impurities in the cotton raw material in Pakistan ranges between 4 to 16 percent, these impurities must be removed to achieve the required quality product. If the raw material for the yarn is not properly selected than it may cause decrease in the yield per cent. Finer yarn requires more cleaning of the material than the coarser yarn because impurities present in the fine yarn create hinders in further processing and can be more visible and vice versa than coarse yarn.

Finer yarn required fine quality raw cotton which have sufficient strength and the fibre length if the cotton selected for the finer yarn is of not sufficient length and strength then the good fibres may be wasted during the processing in result of that yieldper cent is decreased.

Another main characteristic of raw material which also contribute in decrease of yield per cent is extra moisture in material because if the material has extra moister present in it then it may difficulty to separate the impurities and good fibres which required more efforts, energy, time etc such material sticks with machine parts which may go into the waste and become the cause in yieldper cent decrease.

YUCCA FIBRE

Yucca fibres were at one time widely used throughout Central America for many things. Currently they are mainly used to make twine. Yucca leaves are harvested and then cut to a standard size. The leaves are crushed in between two large rollers producing the fibres which are bundled up and dried in the sun over trellises. The dried fibres are combined into rolags. At this point it is ready to spin. The waste, a pulpy liquid that stinks, can be used as a fertilizer.

RAW MATERIAL PROCESS ROUTE SELECTION

Route selection for processing the raw material is also another important task for the management. In cotton spinning industry different lines of processing are installed in the Blow Room, routes have different beating points and according to amount of impurities present in material and quality requirement the processing route is selected.

Which reduces the wastage of good fibre, resulting increase in (production) yieldper cent. Processing route selection is based on the type of the material process *e.g.* natural and synthetic material cannot be processed on the same route even you are required to manufacturing the pc (Polyester cotton with any ration) for blend yarn you have to process both fibres separately and at any particular point blend them with required ratio.

Now a days technology facilities process and separate blending machines are available to blend the material with any ratio or you can blend the slivers of the different materials at the draw frame.

Material passage at the draw frame is also pre-decided whether material should be double (Breaker and finisher) or triple passage (breaker, intermediate and finisher).

With less passage the production increased by saving the waste per cent *i.e.* 0.5 but the quality may be effected or not if the raw material quality is enough then the intermediate process may be omitted otherwise It should be carried to improve the quality so that final product can be sold.

If after the processing product is not sold then it means whole material and efforts consumed on it are wasted and yieldper cent becomes 0. Therefore it should be kept in mind that product must be sell out.

IMPROPER MACHINE SETTING

Required quality of the product cannot be achieved without the proper setting of machine, any spinning machine setting parameters includes

- Speed
- Gauge
- Top Roller Pressure
- Air pressure

With the above mentioned setting parameters you can manufacture the required quality of yarn within the required time frame.

With the speed you can adjust the draft, production rate, beats per unit time, and many more. With the gauge and pressure (Air and top Roller) setting you can easily process the material without deteriorating its quality and properly removing the impurities to achieve the targeted product and the result of this surely contributes in the increase of yieldper cent. If any above said setting parameter is improperly set then it will deteriorate the quality and ultimately decrease the yieldper cent.

MAINTENANCE OF MACHINES:

There is old saying that machine cannot become older, if properly maintained time to time, if machine is not properly maintained then it will not deliver the required quality and may also waste good fibres and due to this yieldper cent will decrease.

E.g. if card machine is not properly overhauled and its taker-in and cylinder wires are out of order, when the material is process on such a machine then the machine is unable to remove the impurities and also unable to properly open the fibres up to individual fibre stage, then the product (sliver) delivered by the card machine is not match with requirement thus for the required quality most of good material may become useless or you have option to reprocess the material in both cases yieldper cent is ultimately decrease and any owner/ management cants bear this, therefore for improving the yieldper cent machine maintenance should be primary objective well maintained machines produces required/ targeted quality and yield.

INEFFICIENT AIR CONDITIONING PLANT

Air-conditioning plant contributes most of its share in increase / decrease of the yieldper cent, due to A.C plant the yieldper cent can be at maximum or minimum level therefore if plant is properly maintained providing proper suctions into transportation pipes, machines, atmospheric conditions with the requirement of each department then surely yieldper cent can be increase.

The main objectives of the A.C plant is to maintenance required temperature and RHper cent into every department, provide proper suction into return ducts so that the fluf and good fibres from the floor properly collected and have sufficient air filters, so that the dust, other impurities and good fibres can be separated.

Each machine have also separate suction system *e.g.* at ring frame pnewmafil is collected by the internal machine suction.

Pnemafil 100per cent good fibres which and be reprocessed with minimum efforts if these are not collected then went into waste hence causing decrease in yieldper cent. Similarly A.C plant performs its job in every machine to prevent the wastage of good/processable fibre.

THE OBJECTIVES OF THE AIR-CONDITIONING PLANTS INCLUDES

- Control the RHper cent and Temperature in every department
- Collect and Control the fulf
- Properly separate the collected impurities and good fibre by the help of filters

By achieving the above mentioned objectives we can improve the efficiency of man and machine, Quality of yarn and production of plant in result of that we can achieve our target *i.e.* yield per cent maximization

IMPROPER MATERIAL HANDLING:

Improper handling of material before, during and after process decreases the yieldper cent. *E.g.* the slivers cans during transporting from the card to draw frame or from draw to comber or simplex if not properly carried and by chance material may be deteriorate by hand touching or due to other reason can fall on floor because of uneven floor or cane wheels will contribute in yieldper cent shortage.

Handling of material start when the raw material arrived and you have to plane well where to store it and in which conditions it should me stored considering that when transporting to production floor may not be effected / damage/detoriated. Material handling is also required in each department of cotton spinning staring from blow-room as the material is process through each department *e.g.* final product of blow-room is lap or final product of chute feed system is sliver of card therefore the lap/slivers are properly transported and stored so that their quality cannot be effected similarly the final product of each department is well stored, material handling also deals with the proper storage and transportation of the finished product.

PHOTOCHROMIC

Photochromism is the reversible transformation of a chemical species between two forms by the absorption of electromagnetic radiation, where the two forms have different absorption spectra.This can be described as a reversible change of colour upon exposure to light

This property is a boon for scientists doing research on intelligent textiles where they are making use of this property to store data on the surface of textile fabrics and polymer sheets. Whereas the same property of some reactive dyes is a bane for textile processors. The change in shade after dyeing creates unwanted problems in dyeing.

An optical recording medium contains, on a base, one or more dyes and a polymer which forms liquid-crystalline phases. The information is written into the uniformly oriented liquid-crystalline polymer layer, for example by means of a laser. During this procedure, the polymer heats up locally to above a phase transition temperature. By cooling, the resulting change is frozen in the glass state. The information can be erased by applying an electric field and/or heating. The recording material permits high-contrast storage and possesses high sensitivity, good resolution and excellent stability.

There are other chromatic properties called electrochromatism and thermochromatism of dyes that are affected by electric field and heat respectively.

Photochromic colours are plastisol-based inks, which are off-white when not exposed to UV radiation.

It gains colour when exposed to Sun light / UV light. The colour change is "reversible," *i.e.*, the colour will fade again and appear colour less upon removal from UV light / sun light exposure.

These inks are available in various colours. See our colour availability chart for a complete list of available colours.

The uses of Americos Photo chromic colours are:

- On garment to create novel products and promotional items like T-shirts
- On fabric/garment to print company logo / brand name to prevent duplication
- On garment which are used for party wear
- Thermometers and temperature indicators
- Security printing
- Food industry to indicate temperature of packaged food

NANO TEXTILE

WHAT IS NANOTECHNOLOGY?

Nanotechnology is the technical process of working on the nano-scale – each nano-scale molecule is one million times smaller than a grain of sand. Nanotechnology refers to not only the small size of the materials being used, but also how those materials are engineered to perform specific functions.

Traditional coatings make garments feel stiff and clog the weave of the fabric preventing breath ability.

Using nanotechnology, our treatments are small enough to attach to individual fibres, delivering superior performance characteristics without compromising the look, feel or comfort of the fabric.

COOLEST COMFORT

Night before the big job interview. Sleep elusive. Hot, then cold, then hot again. No fever, just frantic. Good thing your sheets are keeping you cool. Wake up to sunshine, fresh outlook.

Feel like a million bucks. Dress to impress. Blow them away with energy and intelligence. Career happily underway.

With Nano Textiles Coolest Comfort, you stay dry and comfortable. An advanced moisture-wicking system helps balance your body temperature to keep you feeling fresh. Keep dry, stay cool, rest easy.

- Balances body temperature
- Enhances comfort
- Retains fabric's natural softness
- Allows fabric to breathe naturally

RESISTS SPILLS

Playing hooky. Breakfast in bed. Perfect white linens. Reading the *Times*. Everything just right. Phone rings. Newspaper falls on breakfast tray. Cranberry juice flies on favourite duvet. Relaxed mood is ruined. Good thing your bed cover isn't.

Conventional methods only coat fabrics superficially. But Nano Textiles uses an innovative process to build spill resistance into the individual fibres, so fabrics hold up under the toughest spills.

You'll never sacrifice softness or durability, though - Nano Textiles technology makes your duvets, drapery, tablecloths, placemats, and even mattress pads stay looking beautiful, longer.

If it could, your bed would thank you.

- Repels liquids
- Outperforms conventional fabric treatments
- Provides long lasting protection
- Extends the life of the fabric
- Retains fabric's natural softness
- Allows fabric to breathe naturally

REPELS AND RELEASES STAINS

10th annual family reunion. More relatives than you can count. Potluck dinner. Plates piled high. Favourite tablecloth "decorated" with Grandma's special spaghetti sauce, and who knows what else. Luckily, your tablecloth is enhanced with an invisible shield.

Nano Textiles Repels and Releases Stains technology actually prevents stains from sinking into fabrics, so spills are never a problem. Easy cleaning and durable protection keeps your finery looking its finest.

- Repels spills
- Helps stains wash out easily
- Provides long lasting protection
- Extends the life of the fabric
- Retains fabric's natural softness
- Allows fabric to breathe naturally

RESISTS STATIC

Cozy evening in. An old movie in front of a roaring fire. Hot popcorn. Great date. Warm blanket- It's not the TV with the bad static, it's the blanket. Dust, lint, and dog hair getting in the way of real electricity. Fast forward. New blanket. Same great date. The right kind of attraction.

Nano Textiles provides permanent static resistance for throws and blankets. It repels hair, lint, and dust while staying soft, comfortable, and beautifully durable.

Let the sparks fly, just not between you and your blankets.

- Provides permanent static protection
- Repels lint, dust, dirt and pet hair
- Enhances appearance and comfort
- Retains fabric's natural softness
- Allows fabric to breathe naturally

ORGANIC CLOTHING: WEAR WITH AWARENESS

When picking cotton clothing, today's ecologically minded consumer has several choices. Once believed to be a pure, natural fabric, today's cotton is dosed with harmful chemicals that pollute our environment. Today, thanks to strong media awareness and a revitalization of old techniques, we have alternatives. Organically grown cotton, naturally coloured cotton and recycled cotton products give us three choices for truly natural clothing. It may cost a little more, but by supporting the natural cotton industry's growth, we're supporting ourselves and the future of our planet.

PROBLEMS WITH CONVENTIONAL COTTON

The use of cotton dates back to the Egyptians some 4,500 year ago. For thousands of years, cotton, grown using natural methods, was the primary source of textiles. Today, cotton production is far from natural. The United States alone dumps 8.5 million tons of pesticides on cotton fields annually. And that's not all. Conventionally grown cotton accounts for nearly 25 percent of the total insecticide use for crops worldwide. This chemical onslaught harms our environment by polluting ground waters and soil, resulting in the death of wildlife and natural habitats.

Most of the cotton garments sold today is made from cotton grown with chemical pesticides, bleached and then coloured with chemical dyes containing toxic heavy metals. About a third of a pound of chemicals are used to make one adult T-shirt; two-thirds of a pound of chemicals can go into a pair of jeans.

THREE ECO-FRIENDLY ALTERNATIVES

Concerned consumers are choosing clothing made from organic cotton, naturally coloured cotton, and recycled sources. Here are details on these three environmentally healthy ways of producing cotton clothing and other textiles.

ORGANICALLY GROWN COTTON

Organic cotton is grown without the use of synthetic chemical fertilizers, pesticides or defoliants. By incorporating farming practices that increase fertility and diverse eco-systems, organic farmers rely on time-honoured techniques. Crop rotation, cover crops, organic fertilizers, integrated pest management, and human labour for weed control are a few of the methods used. To be certified organic, the soil must be free of synthetic pesticides for at least three years. Farmers and processors are required to pass yearly inspections.

With the obvious environmental and health benefits of growing cotton organically, why aren't more farmers doing it? In the past, demand for organic cotton has been limited due to higher production costs. However, in recent years media attention has been strong. Some major industry players such as Levi Strauss, Nike and The Gap are blending organic cotton fibres with conventionally produced cotton. The growing public awareness of the toxicity of conventional farming methods has resulted in an increase in both the demand for organically grown cotton and the amount of acreage planted.

NATURALLY COLOURED COTTON

Until recently, chemical dyeing was counted as the only viable way to colour cotton clothing. This process requires several steps, each of which creates toxic waste. Cotton is often bleached before it is dyed and heavy metal mordants are used to adhere the dye to the fabric. Because dyes have a hard time adhering to cotton, at least half of the chemicals end up as waste water in rivers and in the soil. Even in small amounts these heavy metals are lethal.

In 1982, entomologist Sally Fox reintroduced naturally coloured cotton, eliminating the need for dyeing altogether. Cottons of different colours have always existed in nature. Like our eyes or hair, cotton is genetically encoded with colours ranging from brown to tan. Native peoples have used these wild cottons for weaving and hand spinning for centuries. Because of its short fibres and inherent weakness, however, it was unable to be processed by modern textile machinery and had limited commercial value.

With a Programme of plant breeding, Fox developed a strong, long-fibre, coloured cotton that can be used commercially. Today, coloured cotton is grown on the stem in shades of brown, reddish brown, green and yellow, totally eliminating the need for dyes. In the long run, no dyeing means a savings in our pocket. The cost of one pound of dyed cotton including dyestuff, water, energy costs and toxic waste disposal is 20 to 40 percent higher than that of coloured cotton. And with coloured cotton there are no hidden costs to the environment. In addition coloured cotton offers:

- An eco-friendly solution. Since there is no bleaching or harmful dyestuff involved in manufacturing, no waste water is produced.
- Unlike dyed material, the colour of the fabrics made from coloured cotton actually deepens with washing.
- Coloured cotton is naturally pest and disease tolerant, making it easier to grow organically.
- Coloured cotton is suitable for chemically sensitive people.
- Coloured cotton provides new yarn and fabric design potentials.
- Without the need for abundant water or energy sources, new mill sites have many more choices for location.

In recent years, the demand for coloured cotton has increased. Large companies like Esprit and Levi have launched popular "green lines" of cotton outerwear using dye-free, unbleached, organic green cotton. The demand for these items far exceed the supply. Coloured cotton is now being grown in the United States, Europe and Australia.

RECYCLED COTTON

Another ecological choice when purchasing cotton clothing is Eco Fibre. Eco Fibre is a recycled cotton fabric made from recovered cotton that would otherwise be cast off during the spinning, weaving or cutting process. There are no harsh chemicals used in the processing of this fabric.

The next time you are shopping for cotton clothing, choose garments made from organic cotton, naturally coloured cotton or recycled cotton. By buying from these natural sources, you are supporting the demand for a new earth-friendly textile industry. Your choice makes a difference.

Hemp is naturally one of the most ecologically friendly fabrics and also the oldest. The Columbia History of the World states that the oldest relics of human industry are bits of hemp fabric discovered in tombs dating back to approximately 8,000 BC.

Hemp fibre is one of the strongest and most durable natural textile fibres. Not only is it strong, but it also holds its shape having one of the lowest percent elongation of any natural fibre. In fact, its combination of ruggedness and comfort were utilized by Levi Strauss as a lightweight duck canvas for the very first pair of jeans made in California.

Furthermore hemp has the best ratio of heat capacity of all fibres giving it superior insulation properties. As a fabric, hemp provides all the warmth and softness of other natural textiles but with a superior durability seldom found in other materials. Natural organic hemp fibre 'breathes' and is biodegradable. Hemp blended with other fibres easily incorporate the desirable qualities of both textiles. When combined with the natural strength of hemp, the soft elasticity of cotton or the smooth texture of silk create a whole new genre of fashion design.

A fibre of a hundred uses besides fabrics, hemp is also used in the production of paper. The oldest piece of paper - over 2000 years old - was discovered in China and is made from hemp. Until 1883, between 75per cent and 90per cent of all paper in the world was made with hemp fibre. The Gutenberg bible (15th century), Lewis Carroll's Alice in Wonderland (19th century) and just about everything in between was printed on hemp paper. Thomas Jefferson wrote the early drafts of the Declaration of Independence on hemp paper produced in Holland. Jefferson grew hemp on his plantation as an industrial crop, selling the dried stalk to the U.S. Navy as outfitting material. George Washington also grew hemp, harvesting the fibrous seed for a variety of commercial uses including a skin lotion.

Other uses include feed for animals and for humans in veggie burgers, salad dressings, and pastas. Hemp seed is nutritious and contains more essential fatty acids than any other source, is second only to soybeans in complete protein (but is more digestible by humans), is high in B-vitamins, and is a good source of dietary fibre. Cosmetics manufacturers include hemp oil in makeup, skin lotions, and shampoo. In Europe, hemp is used in household cleaners as a natural alternative to harsher chemicals.

Hemp is a renewable resource which grows more quickly and easily than trees making hemp more cost effective than waiting decades for trees to grow to be used in man-made fibre production such as lyocell and rayon from wood pulps. The bark of the hemp stalk contains bast fibres, which are among the Earth's longest natural soft fibres and are also rich in cellulose. The cellulose and hemi-cellulose in its inner woody core are called hurds. Hemp fibre is longer, stronger, more absorbent and more insulative than cotton fibre.

Hemp produces more pulp per acre than timber on a sustainable basis, and can be used for every quality of paper.

Hemp paper manufacturing can reduce wastewater contamination. Hemp's low lignin content reduces the need for acids used in pulping, and its creamy colour lends itself to environmentally-friendly bleaching instead of harsh chlorine compounds. Less bleaching results in less dioxin and fewer chemical by-products. Hemp fibre paper resists decomposition, and does not yellow with age when an acid-free process is used. Hemp paper more than 1,500 years old has been found. Hemp paper can also be recycled more times than wood-based paper.

According to the Department of Energy, hemp is an excellent biomass fuel producer and the hydrocarbons in hemp can be processed into a wide range of biomass energy sources, from fuel pellets to liquid fuels and gas. Development of bio-fuels could significantly reduce our consumption of fossil fuels and nuclear power.

Hemp can be grown organically easily and hemp is most often grown without herbicides, fungicides or pesticides. Hemp is also a natural weed suppressor due to the fast growth of the plant's canopy.

Eco-friendly hemp can replace most toxic petrochemical products. Research is being done to use hemp in manufacturing biodegradable plastic products: plant-based cellophane, recycled plastic mixed with hemp for injection-molded products, and resins made from the oil are just a few examples.

- Properties of Hemp
- Uses for Hemp
- Hemp Fibres and Fabrics
- Manufacturing Hemp Fabric
- Dyeing and Finishing
- Ecology of Hemp

- Industrial Hemp for renewable energy
- Ramie Fibre - Another Natural fibre:

WELLNESS FINISH WITH VITAMIN E

Functional aspects for clothing textiles, which promote the wearing comfort and well-being have become a striking sales argument. The following article introduces a product idea for "Wellness Textiles". It describes a transfer system consisting of textile/cyclodextrin-vitamin E complex.

Wellness is by long more than just a buzzword for wellbeing, vitality and fitness. It has become a social phenomenon which confers the wish of 'eternal youth' to ageing persons. The term 'wellness' has become a new life philosophy and from the commercial aspect is an important pulse generator and growth motor for many trade brances.

The 'wellness' fever has also reached the field of textile finishing and blew in some 'fresh wind' to this field with innovative product ideas. The following article describes a so-called transfer system by which vitamin E is converted from the textile onto the human skin. For the first time the transfer of vitamin E from the textile material into a skin model could be proven in an experiment.

VITAMIN E (A-TOCOPHEROL)

Vitamin E belongs to the group of lipid-soluble vitamins and is found in nature in many vegetable oils. The chemical term for vitamin E is "a-Tocopherol".(alpha-Tocopherol)

In the cosmetic industry vitamin E is used as antioxidant and active substance among others because of its moisture binding capacity in aliphatic cosmetic creams, lotions, emulsions, body and face oils for dry skin care as well as for decorative cosmetics like lipsticks. Vitamin E is also successfully applied for various skin diseases.

Very much importance is attached to vitamin E in the food industry, too. Vitamin E is an important lipid-soluble antioxidant which has many positive physiological properties besides its vitamin character.

The term 'antioxidant' describes the capability of molecules neutralizing so called radicals. Thus antioxidants are often called scavengers. Radicals are atoms or molecules which have an unpaired electron in their outer shell. Free radicals also emerge by the normal cell breathing as side products and try to snatch away an electron from other structures for means of completing their outer shell. In this way for example the cell membrane can be damaged. Antioxidants and thus also vitamin E 'deactivate' the free radicals by giving off an electron and in this way protect the cells from 'oxidative stress'.

CYCLODEXTRINS

Cyclodextrins are circular molecules produced by the enzymatic decomposition of starch and they belong to the group of oligosaccharides and consist of 6 to 8 glucose units. The interesting point about cyclodextrins is their cylindrical structure and the resulting properties and application possibilities. The polar OH groups of the individual glucose units are on the outside of the cylinder due to their steric arrangement. The outside is hydrophilic, whereas the inside of the cylinder is non polar and thus hydrophobic (resp. lipophilic).

The cavities of the cyclodextrins (host) can take in 'guest molecules' and release them again. The chemist calls this phenomenon host-guest-chemistry. Guests are all those molecules which could fit into the cavity and are non- polar enough to interact with the lipophilic cavity surface.

By this complexation the properties of the locked in molecules change. For example an increase of the water solubility of non- polar organic compounds or decrease of sightly volatile substances are obtained through complexation with cyclodextrins, but also an increase of the stability y of the locked in substances to light, oxygen and heat.

Altogether the complexation through cyclodextrins has been researched only to a small extent, although cyclodextrins have been marketed for some time now for means of odour absorption (antismell finish). Knowledge about which substances can actually be complexed, is still very fragmentary and quite some more fundamental research will be necessary in this field.

CYCLODEXTRIN-VITAMIN-E COMPLEX

Vitamin E being a lipid soluble substance is virtually predestined for complexation with cyclodextrines. Y-cyclodextrin (8 glucose units) has proven to be a particularly suitable 'active substance'. Examinations have shown that a 2:1 complex consisting of Y-cyclodextrin and vitamin E offers many more advantages in terms of stability than a 1:1 complex.

The antioxidative potential of vitamin E is based on the reactivity of the chromanoxyl radical. The main task of the cyclodextrine is to surround the ring system of the vitamin E molecule (benzopyran-6-ol) and continue to stabilize it.

In this stabilized 'packaging', vitamin E is excellently suitable for Wellness finishes of textiles worn close to the body and these products can be purchased as 'NouWell E' from CHT R Beitlich GmbH.

Fixation and proof of CD vitamin E complex on textile surfaces

As the CD vitamin E complex does not show any substantivity, only those processes are to be considered where the product application can be controlled like for example by padding, spraying, coating or printing. For permanent fixation of the complex, reactive polyurethanes have proven to be advantageous for the following reasons:

- Free from formaldehyde
- Soft handle

- Good permanence
- Applicable on all types of fibres.

Fixation of the complex is physical and particularly on CEL and WO chemical (reaction with -OH, -NH2-groups). Fixation of the charged cyclodextrins with reactant crosslinking agents is also be possible, but then only for application on CEL fibres.

The qualitative proof of vitamin E on the textile can be done through a dyeing reaction at which the reductive properties of vitamin E are taken advantage of.

- Dripping on a $FeCl_3$-solution onto the finished textile. In the presence of vitamin E the Fe^{3+}-Ion is reduced to Fe^{2+}.
- Dripping on a dipyridyl solution. Dipyridyl forms with Fe^{2+}ions a red chelate complex.

By means of this Redox reaction vitamin E can be easily and safely seen on the fabric with the restriction that this of course can only be done on white or pastel shaded fabric.

3

Textile Industry

CHARACTERISTICS OF THE MANUAL

In order to promote energy conservation in the textile industry, this manual analyzes in detail the actual conditions of energy consumption in each field. Thus, it has been compiled to serve as reference material for the practical application of these techniques which are always kept available for the engineers. A step-by-step description is included as to how to implement the energy conservation measures, specifically in the area of the dyeing and finishing process, where many small- to medium-sized companies are operating. We strongly hope that this manual will be used as a guide to promote energy conservation in the textile industry and to rationalize the management.

CHARACTERISTICS OF ENERGY CONSUMPTION

TYPES OF ENERGY USED IN THE TEXTILE INDUSTRY

Table. Energy Consumption Share of Each Specialized Technical Field in the Japanese Textile Industry

Unit: Million Yen

Specialized Technical Field	Fuel	Electricity	Total	Share
Fibre production	32,551	21,498	54,049	21.0per cent
Spinning	3,224	44,262	47,480	18.4
Twisting	291	1,660	1,879	0.7
Textured-yarn production	120	1,543	1,663	0.6
Weaving	4,467	24,848	29,315	11.4
Knitting	4,059	11.709	15,.858	6.1
Dyeing	37,661	28,412	66,073	25.0
Clothing manufacturing	8,240	15,420	23,660	9.2
Others	5,959	12,000	17,959	7.0
Total	96,500	161,442	257,942	100.0

Note:

Calculated from the Tabulated Industrial Statistics

In general, energy in the textile industry is mostly used in the forms of: electricity, as a common power source for machinery, cooling and temperature control systems, lighting, office equipment, etc.; oil as a fuel for boilers which generate steam; liquified petroleum gas; coal; and city gas. Table compares the energy consumption shares of various specialized technical fields and it can be seen that energy consumption is relatively high in the fields of dyeing and finishing, fibre production, spinning, weaving and clothing manufacturing. Table summarizes recent trends in the use of various energy sources in the fibre production and dyeing and finishing divisions of the textile industry, where the energy consumption ratio is relatively high.

Technical Field	Fiscal Year	Electricity (10^3 KWh)	Kerosene (KI)	Gas Oil (Ki)	Fuel Oil A (KI)	Fuel Oil B (KI)	Liquefied Petroleum Gas (t)	Coal (t)	City Gas (10^3 m^3)	Total (10^6 Kcal)
Fiber Production	1986	3,839,760	4,067	10	6,182	16	719,547	712	517,177	15,631,924
	1987	889,582	5,587	17	2,221	..	635,941	699	706,402	15,854,930
	1988	3,882,074	3,821	14	12,209	..	622,291	727	788,698	16,012,279
	1989	3,887,386	2,736	3	28,520	..	544,848	656	838,510	15,458,822
	1990	4,066,002	3,571	5	35,503	..	549,098	662	871,265	15,358,554
Dyeing	1986	1,325,692	24,250	400	39,770	11,751	816,942	83,678	7,187	12,103,452
	1987	1,351,241	22,565	406	50,351	11,906	806,477	87,978	30,084	12,322,468
	1988	1,343,959	21,879	442	97,509	12,323	775,313	95,349	37,846	12,610,598
	1989	1,397,260	21,470	471	180,880	10,507	708,562	104,125	37,580	12,706,610
	1990	1,398,127	20,702	614	199,846	10,068	678,942	109,281	32,827	12,724,916

Note:

Calculated from the Annual Report on Textile Statistics. Dyeing Represents the dyeing process including finishing, for both woven and knitted materials.

PROMOTION OF ENERGY CONSERVATION TECHNOLOGIES

While the significance of energy conservation awareness is relatively easily understood at home, when a Programme is introduced into a factory to promote it, its thorough implementation tends to be delayed at an early stage. Therefore, for its actual course of implementation, it is desired to devise company-wide coordinated measures similar to QC activities at factories. Also, in order to promote energy saving measures efficiently, it is found to be effective to separately consider general management techniques for "rational use of energy" and process-specific techniques to be developed in each specialized technical field.

ENERGY CONSERVATION MANAGEMENT TECHNOLOGIES

Organizational Rationalization

Since energy management is relevant to a wide range of departments within a company, it is necessary to enhance the awareness, improve the knowledge and obtain the participation and cooperation of everybody involved in the production process. Therefore, while it is necessary for engineers and technicians with specialized technical knowledge to play a central role in energy conservation efforts, the implementation of an energy conservation Programme itself should not be left to a handful of specialists or specialized sections. Rather, it is desirable to address the task company-wide, for example by setting up an 'Energy Management Committee'.

Improving Efficiency of Electricity Use

Lighting

Due to its nature of operations, the share of lighting in electricity use is relatively high. After the switch from tungsten bulbs to fluorescent lamps achieved considerable electricity savings, electricity-saving fluorescent lamps have been developed and marketed for further improvements, including those capable of reducing electricity use by several percent for the same level of illumination. In general, the effectiveness of illumination is influenced by various factors, such as the intensity of light source, the reflection coefficient and shape of the reflective fitting, the layout of the room to be illuminated, interior finish, colour and the distance from the light source.

Therefore, it is important to re-examine whether the light source is utilized in the most efficient way and take electricity saving measures, if necessary, such as reducing the number of lamps in use by switching from global lighting to local lighting as much as possible.

Electric Motor

The textile industry uses a vast number of relatively small electric motors. Notably, while a conventional machine was driven by a single motor with the generated mechanical power transmitted to various parts of the machine in a collective manner, many modern machines utilize multiple motors with a control board controlling the movement of each motor, which is directly coupled to a machine part to drive it independently from others.

This is also a rationalized feature in terms of energy saving. However, regarding the selection of each motor, emphasis has been placed on mechanical performance, resulting in a motor with an excessive capacity. This leaves considerable room for reexamination from a energy conservation point of view.

Electric Heating

In the textile industry, electric heating has largely been replaced by other methods for some time in order to achieve cost reductions. However, since electric heating only requires a small initial investment as a result of convenience and simplicity in equipment construction, it is still used for small capacity local heating purposes. Therefore, it is desirable to conduct a comparative investigation into alternative heating methods, such as far-infrared radiation heating, high frequency dielectric heating and microwave heating.

Improvements in Efficient Fuel use

Selection of Fuel

As is described before, fuels utilized in the textile industry have already gone through a switch-over from coal to oil.

More recently, efficient energy use is under investigation, including the revival of coal on the way to a further move from oil to liquefied and city gases, while reflecting various fuel prices.

In selecting fuels, those with good flue gas characteristics in addition to high calorific value and ease of combustion are desired, so that air pollution can be prevented as much as possible.

Selection of Boiler

By and large, boilers used in the Japanese textile industry have experienced a change from Lancastrian- or Scotch-type tubular or smoke tube to water-tube boilers.

As a result, boiler efficiency has improved from the conventional 60's to 70's of percentage points to as high as the 90's. Since high performance boilers are prone to a rapid growth of scales inside their water tubes, feed water management becomes important.

Furthermore, these boilers have small amounts of retained water and high evaporation speeds so that many aspects of their operation are automated, including feed water and combustion management.

Improvement in Efficient use of Steam

Piping

The noted feature of steam use in the textile industry is that the amount of steam involved is not so large but the locations where steam is required are widespread so that steam losses due to heat radiation from steam transportation pipes and pressure drops are considerable.

Therefore, for steam transportation over long distances, high pressure and smalldiameter rather than low pressure and large-diameter piping is desired, with pressure reducing valves placed as necessary to regulate the steam pressure at the point of use, thereby curbing heat losses. Also, as pressure losses around bends are great, it is desirable to make their radii large. In order to prevent steam leaks from joints due to the thermal expansion of the pipe, expansion joints should be placed where required.

Furthermore, in order to maintain the temperature inside the valve, tank and treatment tank as well as the piping, it is necessary to install them heat-insulated, using appropriate heat insulating materials, so as to efficiently use steam while preventing heat losses.

Steam Accumulators

Since live steam is often used in dyeing factories, fluctuations in steam use during working hours are large. On the other hand, since high performance water tube boilers and once-through boilers are designed such that water retained inside the boiler is very little, the boiler cannot react to momentary and sudden load changes, while responding to automatically controlled slow load changes is not a problem. In such a case, a steam accumulator can be installed midway through the heat transporting pipe, between the boiler and the heat consuming load, in order to store excess steam when the load is light by transforming it to heated water. This then transforms the heated water back to steam when the load is heavy in order to reinforce supply to the load. This allows the boiler to continuously operate with the average load and is quite advantageous in view of energy saving.

Recycling of Drain

So far, after its heat energy is consumed, steam has been drained off. However, in view of energy saving, it is necessary to collect and recycle the heat energy carried by the drain water.

Utilization of Heat Exchanger

In each production process of the textile industry, the heating and cooling of gases and liquids as media of heat are frequently required. This is done through heat exchange between different fluids, and in order to avoid contamination or chemical reaction due to their direct contact, heat exchangers are used to carry out indirect heating and cooling. It is important to use the right heat exchanger for the intended purpose.

Measuring Instruments and Automatic Control

Energy saving is an operation to grasp the actual situation of energy use in a factory precisely and quantitatively and to carry out improvement measures in order to rationalize and economize on it. While measuring instruments are needed to obtain quantitative data, it will become more and more important to investigate the use of sophisticated measuring instruments based on recent developments in mechanical and electronic engineering, combined with automatic control systems.

ENERGY USE AND RATIONAL USE OF ENERGY IN PROCESS-SPECIFIC TECHNOLOGIES

Progress in production rationalization is achieved through the implementation of a comprehensive set of measures, including energy conservation technologies as the centerpiece measure, along with time management, labour saving, natural resources saving and space saving. It has been frequently pointed out that, along with management techniques described earlier, the improvement and development of process-specific techniques on energy conservation greatly contribute to the rationalization of production. Here, process-specific techniques relating to energy saving are summarized for each specialized technical field.

Fibre Production

Exhibiting relatively large-scale structural forms in the textile industry, this division has already reached a high level of production rationalization; as is well known, it is technologically aiming at diversification into such high valueadded goods as super extra-fine fibre and inorganic functional fibre, commonly referred to as *shingosen.*

In particular, the following techniques relate to energy saving:

- *Raw Material Production Process*: Implementation of energy saving through improvements in the process and reaction conditions
- *Polymerization Process*: Reduction' in polymerization time by means of high efficiency catalysts, polymerization methods, etc.
- *Spinning Process*: Promotion of energy saving through combining the POY (Pre-oriented Yarn: Yarn with some stability with its molecules partially having gone through orientation) and DTY (Draw Textured Yam: false twisted yarn produced while drawing POY yarn) methods and an expanded use in multi-folded spinning yarn.
- *Newly Built Factories*: The factories built during the high growth period have large margins and allowances for production increase so that high losses would result if production decreased. Therefore, suitably sized factories should be constructed.

Spinning

Regarding technological trends in spinning, moves towards high speed and large package size have been investigated in order to achieve labour saving through as much automation as possible. As a result, energy consumption has been gradually increasing. However, in view of price competition with overseas companies, further labour saving as well as energy saving is desired. Table compares a modern and a traditional factory in terms of electricity consumption for each plant/operation.

Namely, it can be seen that a modern factory as a means of achieving production rationalization requires approximately three time as much electricity as a traditional one, with electricity consumption particularly increasing in the air-conditioning plant.

In terms of processing operations, fine spinning, as the main operation of the spinning process, consumes a large amount of electricity. Thus, energy saving measures are required in these fields.

Ring Spinning Operation

For the fine spinning operation, electricity is consumed in driving the spindles, packaging, spinning, drafting, and operating the lifting and cleaning mechanisms. It is desired to curb the increase of electricity consumption as much as possible by setting an optimal condition for each of these electricity usages.

Air-conditioning

Although as an ideal working environment a room temperature less than 30°C is desirable, in cases where the working environment has been drastically improved in most other aspects with work load also reduced, a slightly increased room temperature may be permitted. As has been reported, there was a case where raising the regulated temperature from 30°C to 32°C resulted in a reduction in the electric power demand of a carrier with a contract demand of some 8,000 kW by 190 kW. Also, there are many instances of seasonal switch-over from a damper to a pulley as a means of readjusting the blown air volume; this is in order to recycle the air sucked from the processing machine for each operation through a filter back to the same room, and it is therefore necessary to recheck the locations of fans for suction and returning.

Textured-yarn Production

While synthetic-fibre textured-yam is mostly produced with false twisting machines, its history of rationalization is characterized by challenges for high speed operation. As their operating speeds increased, driving and heat-curing motors and and other peripheral equipment became larger, accompanied by an inevitable increase in electricity consumption. Although this may be acceptable as long as the production improvement resulting from a high speed operation covers the increase in electricity costs, reductions in energy cost would surface as an avoidable urgent task, should a sharp increase in electricity charge occur. It can reasonably be said that the major form of energy consumed in the production of synthetic finished-yarn is electricity. Although the amount of electricity consumed in each piece of equipment varies with factory scale and the type of false twist machine, and therefore cannot be treated in a standardized manner, generally accepted average values may be taken as 3.5 kWh/kg for a single heater system and 5.0 kWh/kg for a double heater systemas one report suggests. Of all the energy consumed in finished-yam production, 70per cent is accounted for by false twist machines. Table shows a breakdown of this energy consumption.

Sixty per cent of all energy consumption by a false twist machine occurs in the heater. Therefore, improvements in the heat insulation of the heater and the lowering of heater temperature may be considered as energy saving measures. Since the latter has implications in the characteristics of the finished-yam, whether or not it is adopted should be examined on such occasions as in the development of a new product.

Since air-conditioning plants are designed based on the conditions applicable at the time of installation, it is desirable that they be re-examined against the present conditions.

Weaving

Rationalization in fabric production is such that while various improvements in machinery aimed at high speed operation and labour saving have been carried out, the amount of energy use per unit of the product has gradually increased. Regarding loom design, high productivity shuttleless looms such as water jet, rapier and gripper types have successfully been introduced, with air jet models put in practice in the production area of industrial fabric material. The amount of energy consumed by each loom during its weaving operation can be estimated from the motor capacity and weaving speed. Conventional shuttle looms are based on the weft-insertion method, incorporating a shuttle zooming to and fro with a large inertia mass (approx. 400) and mounted with extra weft, and they also use energy consuming pirns as an integral part of the machine. For this reason, the shuttleless looms' contribution to energy saving cannot be regarded as too high. On the other hand, as a large amount of energy is consumed in sizing, as one of the preparatory operations for weaving, the introduction of foam and solvent sizing operations are being investigated. Furthermore, long fibre fabrics using nonsizing filaments have been developed, eliminating the sizing process altogether. In a reported example, the introduction of a new heat exchanger into a sizing machine with a very poor sealing capability achieved more than 40per cent of energy saving.

Knitting

The share of energy cost in the total cost of production is not necessarily high for the knitting process. However, of the main production facilities for this process, knitting machines have also been undergoing a shift towards high speed and large capacity and fine gauge features; the current industry trend is for high added-value goods and multi-line, small-volume production based on advanced systems such as computer-controlled pattern making mechanisms. Therefore, a potential tendency for increased energy consumption should be taken into account. As a result, it is desirable to conduct a comprehensive re-examination of the production schedule along with the implementation of actual energy conservation measures in order to reduce or restrain the share of energy cost in the total production cost.

Dyeing and Finishing

It is very important to advance energy conservation in the dyeing and finishing field, which has a high energy consumption share in terms of both the amounts of money and energy used. The dyeing and finishing process consists of many interwoven unit operations, and it is well known that the process generally goes through repeated wet and dry operations. The heat balance of a unit operation can mainly be considered as the difference between the total supplied heat on the one hand and the sum of the heat required by the system and various forms of heat losses on the other. Figure graphically summarizes the major factors in a thorough implementation of energy savings.

Table. Relationship between Production Rationalization Techniques and Energy Savings

Produciton Rationalization Technique		Mechanism	Effect
Time saving	1. 2. 3.	High speed processing of unit operations Reduction in waiting time between unit operations Elimination or merger of unit operations	Reductions in energy use Per unit operations through an improvement in productivity
Labour saving	1. 2.	Implementation of automation Strengthening colourimetric management	Reductions in the frequency of reprocessing through a reduction in the failure rate
Energy saving	1. 2. 3. 4. 5.	Reduction in bath ratio Reduction in treatment Reduction in margin of temperature rise Re-examination of drying method Switch to non--water-based operations	Reductions in energy cost time
Conservation of natural resources	1.	Utilization of continuous bath	Utilization of system's residual heat
Space saving	1.	Construction of modern factories	Improvements in factory-wide energy saving effects

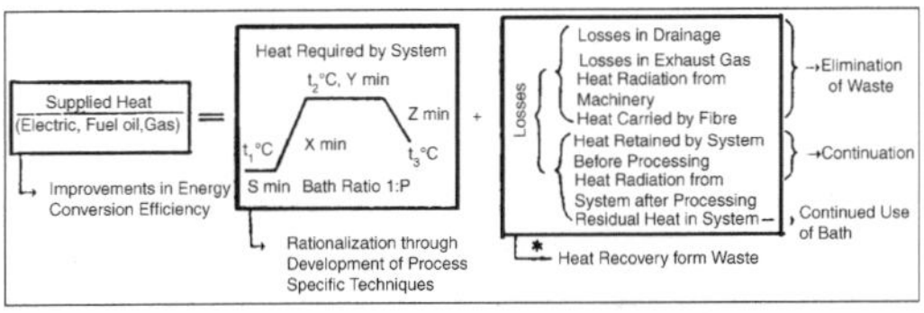

Fig. Heat Balance in Unit Operation

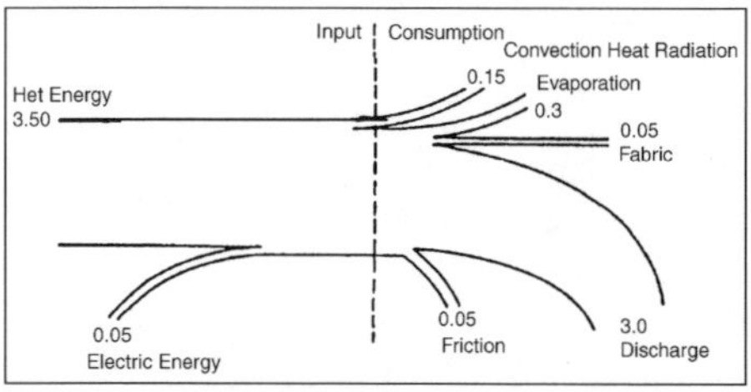

Fig. Shows an Example of Heat Balance in a Continuous Water Cleansing Machine.

This clearly illustrates the importance of the development and utilization of processspecific techniques, apart from the already-described management technologies. Table shows that the implementation of production rationalization eventually relates to energy conservation. The following are brief explanations of typical examples.High Speed Processing of Unit Operations

As the processing machines become faster they also become larger. This means the energy consumption per unit length of time will increase, but generally it will accompany a reduction in energy consumption for the treatment of a unit amount of fabric. Table shows an example of this situation. Therefore, it follows that, as long as the product turnout is maintained, continuous processing with a large machine will be more effective in achieving energy conservation.

Table. Length of Mercerizing Machine and Productivity and Energy Consumptions

Total machine length (m)		38	47	56
Treating speed (m/min)		40	60	80
Treating time (set/m²)		1.5	1	0.75
Product Turnout (m)	1 hour (Operation rate (100%)	2,400	3,600	4,800
	8 hour (Operation rate (58%)	19,200 (16,300)	28,800 (24,500)	38,400 (32,600)
	16 hours (Operation rate 90%)	38,400 (34,500)	57,600 (51,800)	76,800 (69,000)
	24 hours (Operation rate 95%)	57,600 (54,700)	86,400 (82,000)	115,200 (110,000′
Con-Sumption rate	Water (m³)	10.5	14.0	19.0
	Steam (kg)	1,075 (82,691)	1,500 (115,381)	1,850 (142,311)
	Electricity (AC motor (KWh)	21.0	38.0	50.0
	NaOH (30°B′e) (I	288	432	576
Energy and raw Material required For treating m′ of fabric	Water	0.0044	0.0039	0.0040
	Steam	0.4479 (0.0341)	0.4167 (0.0321)	0.3854 (0.0301)
	Electricity	0.0088	0.0106	0.0104
	Amount of Energy (Kcal)	337.2	322.0	302.8
	NaOH	0.12	0.12	0.12

Note:

- The bracketed entries under production trunout show approximate figures which would result from the respective operation rates.
- The bracketed entries under steam show equivalent fuel consumption figures would be needed if a boiler with a evaporation ratio (= evaporation/fuel consumption) of 13 was used.

Elimination or Merger of Unit Operations

The currently employed dyeing techniques are based on unit operations which have been developed and established for use with natural fibre. For this reason, the traditional standard treatment steps are often applied to blended yam fabrics as a matter of principle.

However, through omitting or merging some of the unit operations according to the usage of the product and considering the characteristics of the coexisting synthetic fibres, it becomes possible to achieve energy conservation.

Table. Unit Operations in Preparatory Process of PE/C Blended Fabric and its Processing Characteristics

Combination and Ordering of Operations	Number of Units	Processing Characteristics
Singeing→ Desizing→ Drying → Heat setting → Scouring → Mercerizing → Drying	10	(1) Cumulative Method (2) Strict Control (3) Excessive drying operations
Singeing→ Desizing→ Scouring → Bleaching→ Drying→ Mercerizing→ Drying→ Heat Setting	8	(1) Fewer drying operations (2) Difficulty in controlling scouring and bleaching (3) Great number of wrinkles form processing
Heat setting→ Singeing→ Desizing→ Scouring→ Bleaching→ Drying→ Mercerizing→ Drying	8	(1) Difficulty in desizing (2) Insufficient effectiveness in processing (3) Wrinkles from processing prevented
Singeing→ Desizing→ Scouring → Bleaching→ Drying→ Heat setting→ Mercerizing→ Drying	8	(1) Satisfactory desizing (2) Fewer drying Operations (3) Caution against wrinkles from processing
Singeing→ Mercerizing→ Desizing→ Scouring→ Bleaching→ Drying→ Heat setting	7	(1) Fewest drying operations (2) Difficulty in desizing (3) Small mercerizing effect

Reduction in Processing Bath Ratio

It is easy to understand that a reduction in water use will contribute to energy conservation in the dyeing process which consists of various wet treatment and drying unit operations. It is especially desirable to curb the water consumption because it is linked to the overall water supply cost including that of drainage.

For the reduction of the processing bath ratio, it is necessary to investigate the following measures:

Treatment with Low Bath Ratio

In general, dyeing and finishing methods are classified into the batch and continuous processing methods, and it is recommended to use the latter method where a low bath ratio is desired. However, depending on the details of processing requirements, there are often instances in which the batch method has to be employed. In such cases, batch processing machines which allow lower bath ratios such as the jigger, wince, beam, pad roll and jet flow types should be selected as far as the circumstances permit. Figure shows the relationship between the bath ratio and the production cost for the wince dyeing of a cotton fabric with a reactive dye (bright red, medium shade).

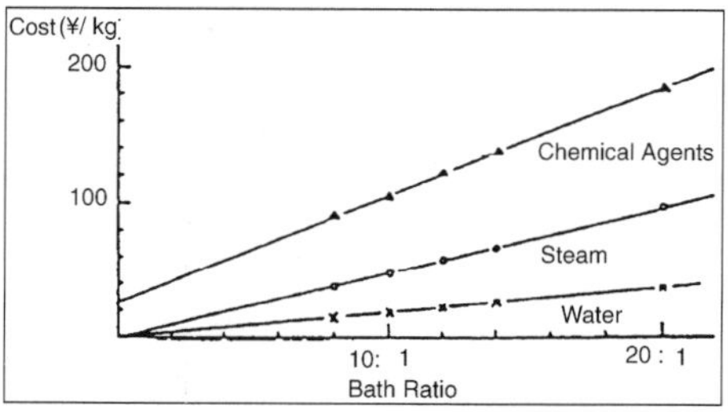

Fig. Relationship between Bath Ratio and Production Cost in Reactive Dyeing

It is easy to see that the bath ratio has a direct influence on the production cost.

Utilization of Low Bath Ratio Processing Equipment

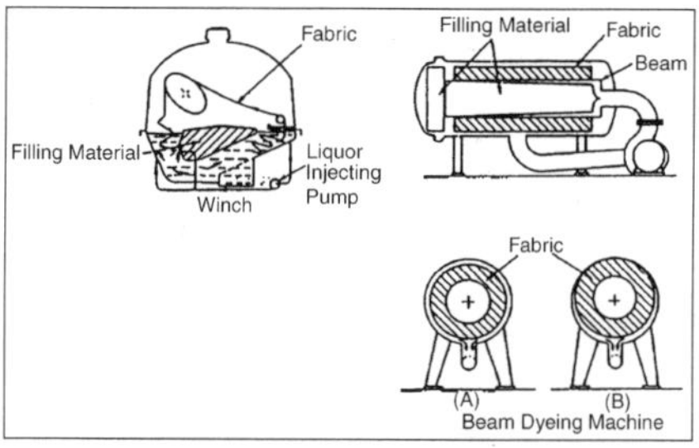

Fig. Example of Low Bath Ratio Operation of Existing Processing Equipment through Insertion of Filling Material

In order to use a lower bath ratio with the existing machinery intact, a method to insert a filling material inside the processing equipment has been proposed. It has been reported that with this method, the bath ratio of a wince could decrease from 25:l to 17:1, and for a beam a reduction was possible from 15:1 to 12.5:1, or even down to as low as less than 10:l where the axis of the beam was made off center with respect to the container body, thus increasing the batch-up volume as shown in (B). More recently, low bath ratio processing machines which are built in with the above mechanisms have been developed and put on the market.

Utilization of Low Add-on Equipment

Several types of processing equipment with a mechanism to uniformly apply the fabric with a minimum amount of liquid necessary in semicontinuous and continuous processing systems are known to be typical examples of energy conservation techniques.

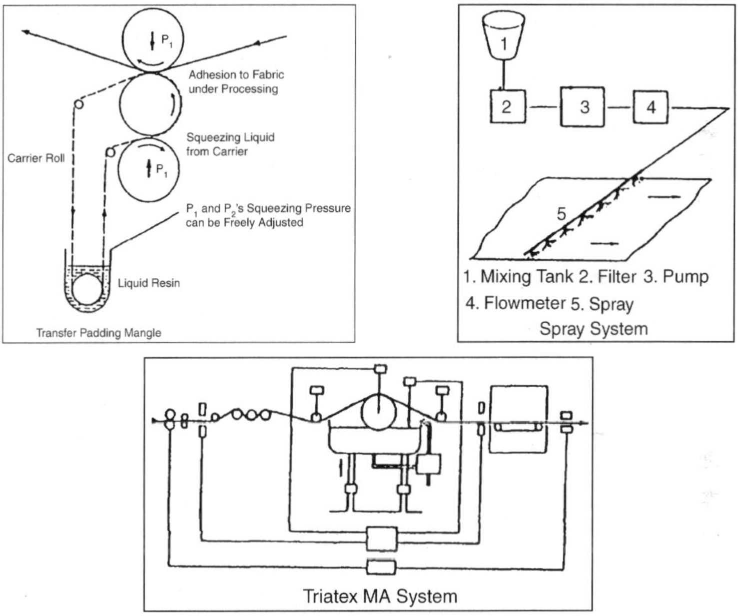

Fig. Typical Low Add-on Machines

Extension of Foam Processing Technique

A typical example of foam processing liquid applying equipment. The foam processing technique is used for the preparatory, dyeing, textile printing and finishing processes, with confirmed effects of promoting energy conservation, but it is desirable to examine details of usage and other practical conditions prior to application.

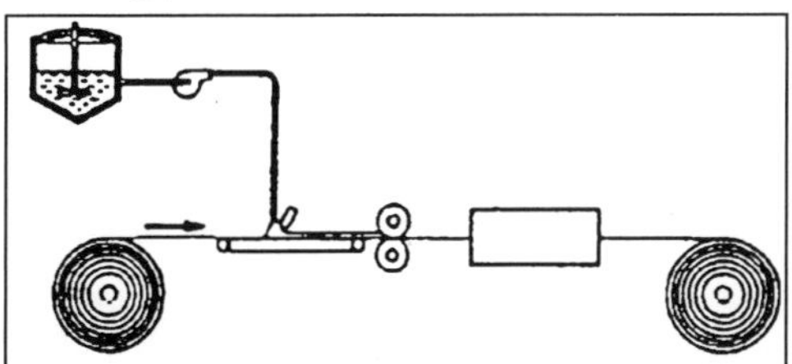

(A) Example of Knife Coat Applying eqipment

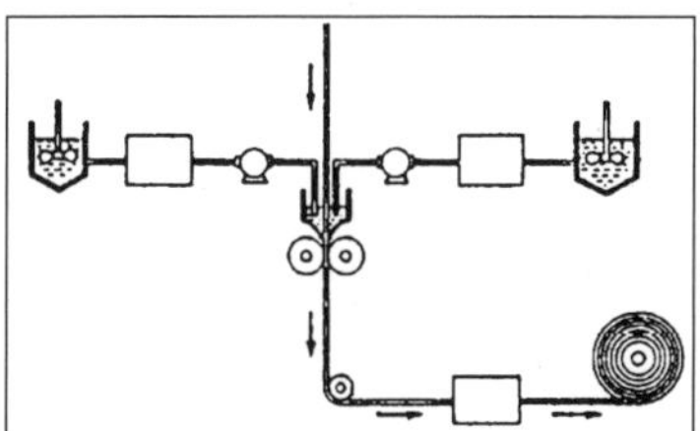

(B) Horizontal Pad Applying Equipment

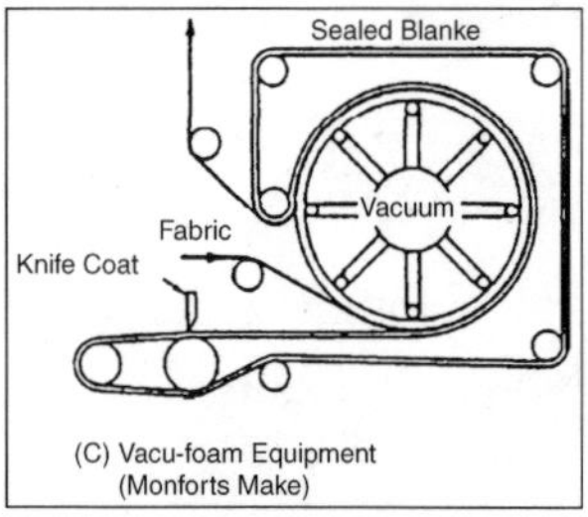

Fig. Typical Foam Applying Equipment

Reduction of Processing Time

As has been pointed out, being a time saving technique aimed at improving productivity, continuous operation with an increase in the size of the processing machine can also further energy conservation. Likewise, for batch processing, the number of technical fields is increasing where the promotion of energy conservation is desired through a reduction in processing time. This tendency becomes more pronounced as the needs of the market become sophisticated. Techniques to accelerate the processing effect with rapid dyeing and plasma treatment are typical examples.

Rapid Dyeing

Rapid dyeing which can drastically reduce the dyeing time and achieve remarkable time savings can also achieve great energy conservation effects when applied to polyester. In order to attain these effects, it is necessary to select dyes with assistants and provide appropriate dyeing equipment. Combined with the foam processing technique, the rapid dyeing technique may also have a potential of leading up to the development of new practical dyeing techniques. Figure shows the position of the dyeing technique in the overall processing bath ratio reduction technique.

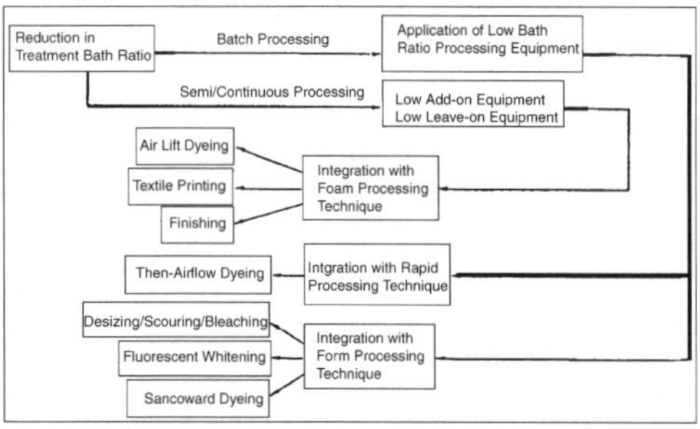

Fig. Situation of Dyeing Technique in Processing Bath Ratio Reduction Technique

Accelerating Techniques for Processing Effects

Aiming at a reduction of processing time, the combined use of a number of new techniques are being studied and it has been reported that processing with plasma, ultrasound, magnetism and radioactive rays accelerates processing effects. Various methods are being investigated to reduce processing time through accelerating processing effects using these techniques in preprocessing, postprocessing, simultaneous processing, etc.

Reduction in Temperature Rise Margin

In many cases, unit operations of the dyeing process are carried out at high temperature. Therefore, to reduce the required margin of temperature rise from heating is very important in view of achieving funda-mental energy conservation along with reductions in processing time. These measures need to be addressed from the following two viewpoints:

Raising Temperature of Inlet Water

If the temperature of the inlet water to be used in the dyeing process becomes relatively higher, the amount of energy to be consumed in raising it to the predetermined value will be reduced. For that purpose, cooperation within a company, or that involving more than one company (using low temperature inlet water at the dyeing factory for cooling purposes, and the high temperature discharge from the cooling system for dyeing purposes) should be investigated, as well as the utilization of natural resources (for example, geothermal and solar energy)

Development and Introduction of Low Temperature Processing Techniques

It is important to continue with technological development aimed specifically at lowering the processing temperature along with raising the inlet water temperature. It would naturally involve the integration of this method with the processing speed acceleration techniques. Low temperature scouring, bleaching, dyeing and curing techniques are some of the practical examples of this.

Re-examination of Drying Method

An important consideration along with the reduction of the processing bath ratio is a re-examination of drying operations. A drying operation is, in principle, inserted after every other unit operation and is an important operation which not only determines drying efficiency as such but also has a direct influence on the morphological stability and texture of the final product. For this reason, various types of drying equipment have been selected and put to practical use, depending on the fibre material and form involved. In view of implementing energy saving measures, it is particularly important to investigate the following three items:

Reduction of Drying Operations in Number

A study of typical preparatory process configurations reveals that drying operations are involved in relatively high numbers, ranging from one to four units or 14per cent to 40per cent of all unit operations in the entire process. Therefore, combinations of unit operations should be sought after such that drying operations between standard unit operations can be eliminated as much as possible. It is especially necessary to cut down on drying operations in preparatory processes which will not directly affect the product's performance or appearance quality. However, an operation which would have come after an eliminated drying operation would have to treat wet fabrics, thereby necessitating special measures that would enable wet on wet treatment.

Improvements in Drying Efficiency

It is desirable to investigate possible improvements in the drying efficiency in terms of efficiencies of both the dewatering and drying steps. While it is well known that the most efficient methods of dewatering and drying are by means of a mangle and a cylinder dryer, methods which have been practiced for a long time, they are also known to have limitations in terms of applicable fibre materials and forms. It is necessary to investigate new drying methods (high frequency drying, microwave heating, far infrared radiation heating, etc.) together with other measures such as utilizing vacuum liquid removal, adding a drying-facilitating organic solvent to the treatment liquid, and combining foam treatment systems with non-foam ones.

Recovery of Heat Energy

Along with active energy saving measures, it is important to carry out the collection and recycling of the energy used in unit operations. The collection of heat energy should start with thoroughly grasping the basic energy balance of each unit operation.

Shift to Solvent Processing

In the dyeing process, although water has been used as the only abundant and cheap resource so far, it is becoming difficult to obtain high quality water in large quantities at a low cost. The worsening of river pollution coinciding with an increase in population density is inevitably creating a situation where the cost of water will gradually increase, including the investment for improvements in waste-water treatment facilities. In addition, although dry-system processing has been contemplated for a long time due to the fact that most energy is consumed in the heating and evaporating operations, it has to date only been applied to a specific area on a limited scale. However, it is a technique which deserves attention as a promising process in the mid- to long-term future. This technique has the following two variations:

Organic Solvent Processing

While the solvents to be used for dyeing processing are categorized into four main groups—halogenated hydrocarbons, petroleum derivatives, aromatics and oxygen-containing solvents—halogenated hydrocarbons are generally recommended as they do not cause a fire or explosion (provided that thorough countermeasures to groundwater pollution are taken). It is well known that in terms of energy conservation, these solvents have an advantage over water-based ones in all of these aspects: specific heat, latent heat for evaporation, heat needed for evaporation and evaporation speed. There are a number of proposals for solvent scouring, solvent dyeing and solvent finishing, including those already put to practical use as a differentiating technique.

Inorganic Solvent Processing

Liquid ammonia is one of the agents being considered for dyeing applications as a inorganic solvent. Of its typical processing techniques, liquid ammonia mercerizing and liquid ammonia dying are given particular attention.

Use of Continuous Bath

While textile manufacturing techniques which promote the conservation of natural resources include grease refinement from raw wool, the collection and recycling of warp sizing agents, and the re-use of alkaline waste liquid arising from the mercerizing process in a scouring bath, in terms of energy conservation the use of continuous baths which utilize the residual heat of the system are particularly important in view of energy conservation. If the continuous use of a processing bath is introduced with the necessary conditions being met, thus allowing only those materials consumed in the dyeing process to be replenished, in particular with the unit operations designed for the batch method, it will greatly contribute to the recycling of heat energy in addition to achieving the conservation of natural resources and the rationalization of countermeasures to water waste. In the dyeing process with a high heat consumption, the use of the continuous bath deserves particular attention as a technique whose practical application is an urgent task to help implement the remaining rationalization measures.

Space Saving

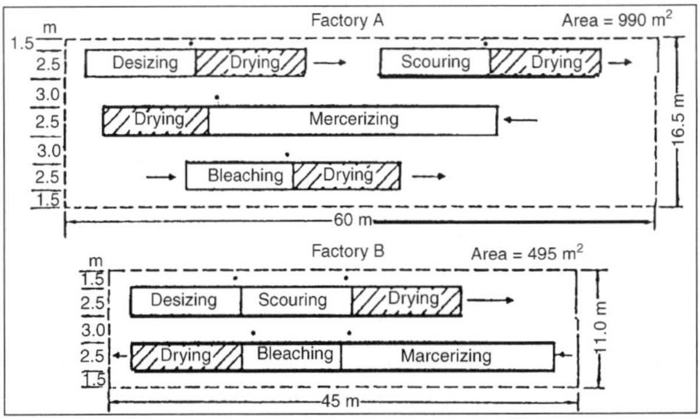

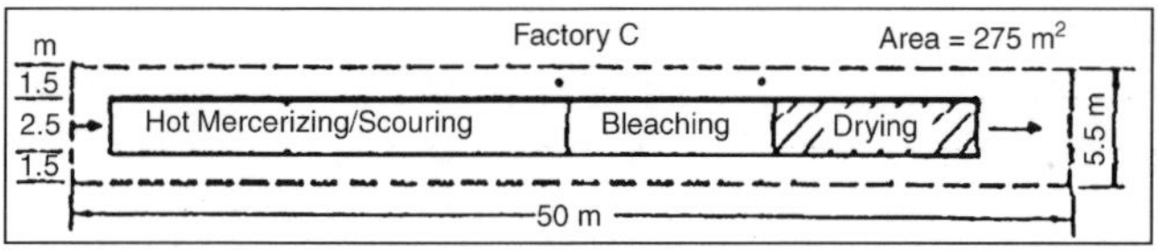

Note:

- *Factory A*: Although it has the same types of machines as Factory B, it has introduced a drying operation for each unit operation, emphasizing flexibility.
- *Factory B*: Considers the production rationalization resulting from continuous processing./
- *Factory C*: It has the same production capability as Factory B, but has reduced the preparatory process related to mercerizing.

Fig. Space Saving Resulting from Introduction of Hot Mercerizing

Table. Compares Operating Conditions of the Same Three Factories

Factory	Investment	Area (m^2)	Workers (person)	Electric Power (kW)	Steam (kg/hour)	Water Supply (1/hour)	Waste Water (1/hour)
A	109%	990 (200 %)	14 (155 %)	242 (114 %)	5,727 (130 %)	14,535 (100 %)	14,535 (100 %)
B	100%	495 (100 %)	10 (100 %0	216 (100 %)	4,386 (100 %)	14,535 (100 %)	14,535 (100 %)
C	67%	275 (55.5 %)	5 (50 %)	114 (52 %)	2,500 (57 %)	7,785 (54 %)	4,785 (54 %)

With an expansion in the practical use of knit mercerizing and ammonia mercerizing, the characteristics of the hot mercerizing technique is also attracting attention. It has been pointed out that, since poor uniformity associated with the traditional mercerizing due to the hydrophobic nature of cotton grey fabric is dissolved in hot mercerizing which uses heated sodium hydroxide, the rationalization of the preparatory process can be greatly advanced. Figure shows an example of comparing some space saving effects achieved by the introduction of hot mercerizing.

At the time a new space-efficient factory is being built, it is possible to incorporate a Programme to introduce such facilities as to be complete with factory-wide energy conservation measures.

Clothing Manufacturing

The energy consumption share of the clothing manufacturing division which consists of large numbers of small-sized companies and their employees in the overall textile industry is not necessarily low but the ratio of energy cost to the total cost is relatively low. However, the energy cost forecast is inevitably a gradual increase under circumstances where the production of high value-added goods is required, along with the implementation of labour saving measures, as a result of the challenging market environment characterized by personalized and diversified consumer needs, high demand for quality goods, short product cycles, etc. Therefore, it is desirable that a comprehensive rationalization Programme be investigated apart from reductions in energy consumption.

STRUCTURES OF TEXTILE MARKETS

In general, it is well known that the developmental process of a textile market moves from a product-oriented stage, where all goods produced are sold to a consumer-oriented stage where the desired goods are only those which will satisfy the consumer's demand, comprising his wants and purchasing power. This is via a mass-consumption stage which is supported by a mass-production system.

The characteristics of textile products demanded in the markets in these different developmental stages may be classified into price-sensitive mass-production type basic clothing, which applies to the earlier stages, and multi-line, small-volume production type fashion clothing which pertains to the last stage and needs to satisfy each individual consumer's taste.

BASIC CLOTHING

Assuming the amount of fabric necessary for people to live a physically comfortable life in a given environment can be derived from the weight of clothes which is enough to keep a certain level of body temperature, given such factors as air temperature, humidity and wind velocity prevalent in the region, basic clothing can be regarded as a cumulative total of such clothes.

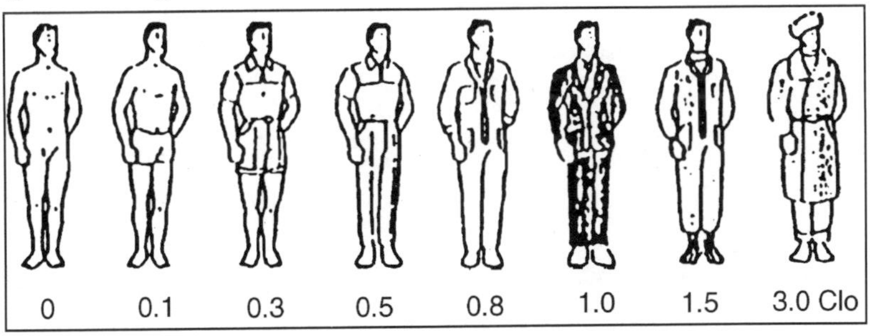

Fig. Typical Relationship between Clothing and Clo Value

Figure Summarizes the Relationship between Some Clo Values and Examples of the Corresponding Clothes.

Table. shows the results of calculations to find the weight of clothes required under average climatic conditions in summer and winter in three Japanese cities.

Although the ultimately required quantity is difficult to determine, the per capita annual textile consumption (kg/person) can be estimated through multiplying the above weight by coefficient a (obtained from the service life of a textile product and its number of units demanded per annum).

Table. Example of Effect of Climatic Conditions on Clo Value and Weight of Clothing.

Region	Season	Temperature (°C)	Humidity (per cent)	Wind Velocity (m/sec)	Clo Value	Required Weight of Clothing (g)
Sapporo	Jan	–4.9	74	2.3	4.1	4005
	Aug	21.3	79	2.7	1.2	1189
Tokyo	Jan	4.7	53	3.1	3.1	3034
	Aug	26.7	75	3.0	0.6	607
Fukuoka	Jan	5.7	67	3.7	3.0	2937
	Aug	27.3	76	2.9	0.6	607

Notes:

- The clo values are shown assuming an average skin temperature of 33.5°C.
- One unit of clo value is 0.155°C/W/m^2.
- The required weight values are calculated with Hanada's formula for female clothing:

$$Y = 0.00103W - 0.025$$

Where

Y: Clo Value

W: Total Weight of Clothing

FASHION CLOTHING

Although it is even more difficult to define the amount of fabric consumed in fashion clothing, a market which has satisfied the demand for minimum basic clothes has a tendency to shift its emphasis towards textile products with stronger fashion overtones and grow rapidly.

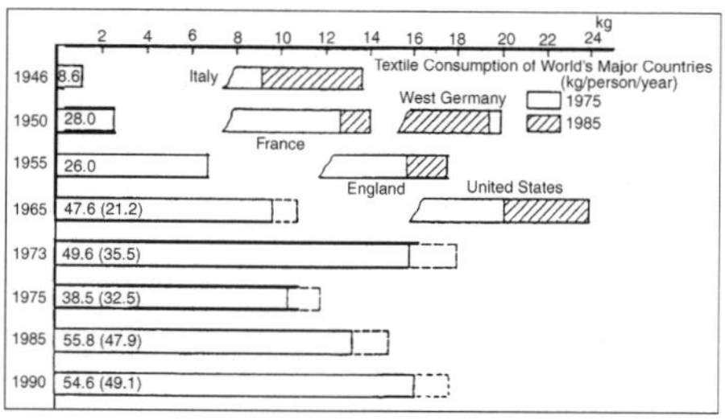

Fig. Changes in Textile Consumption of Japan and World's Major Countries

Notes:

- The figures inside the bar graphs with and without brackets are the shares of chemical and synthetic fibres in the total consumption, respectively.
- The broken-line portions of bar graphs represent the consumption of textile goods to be used as industrial materials.
- The sources are the Annual Statistics on Textile Goods for Japan and FAO data for other countries, respectively.

This is illustrated by Figure, showing chronological changes in textile consumption in Japan and the world's major countries, a graphically expressed correlation between per capita textile consumption and GNP.

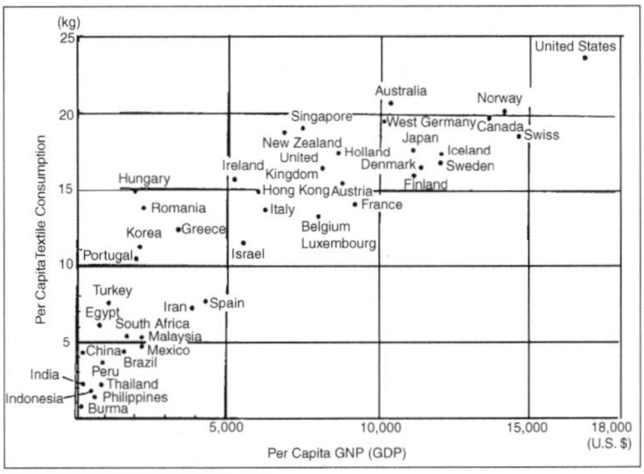

Fig. Correlation between Per Capita Textile Consumption and GNP (GDP).

Notes:

- Data on developed and developing countries are for 1985 and 1984, respectively.
- Entries for Japan, the United States, the United Kingdom, France and Italy reflect per capita GNP, while those for China and other countries are based on per capita income and per capita GDP, respectively.

WORLD'S TOTAL TEXTILE DEMAND AND PRODUCTION BASE DISTRIBUTION

Figure shows the relationship between world population and the total textile demand. Assuming a global environment in which world population will grow from the present 5.4 billion to 10 billion in 2050, and further to 11.6 billion in 2150 when it is expected to reach a static state, the total textile consumption is forecast to double, even using the current figure of per capita annual average textile consumption (8kg/person).

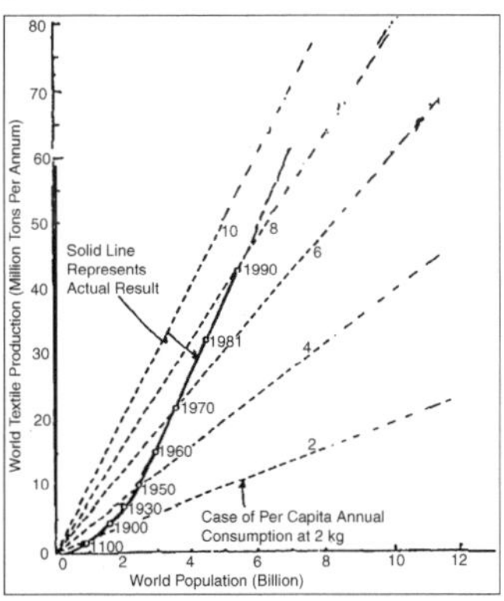

Fig. Relationship between World Population and World Textile Consumption and Production

The textile industry is traditionally regarded as a typical labour intensive industry developed on the basis of abundant labour supply and has a tendency to expand to overseas markets once the domestic demand is satisfied, as illustrated by examples of the established textile industries of many developed countries. For this reason, even when the textile industry of a specific country is to be examined, it is widely recognized that a business strategy taking into consideration the global textile industry setup is very important. Figure shows an example of the schematic representation of such a global setup of textile industries.

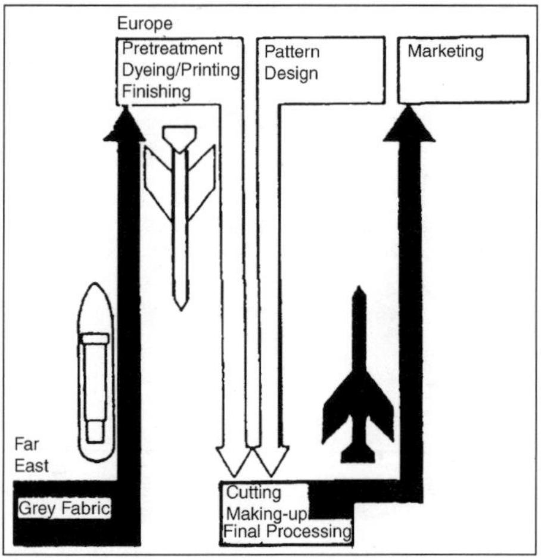

Fig. Example of Schematic Representation of World Textile Industry

CHARACTERISTICS OF TEXTILE MARKET

Most textile consumption takes place in apparel products, and the general tendency is that where further expansion is intended, it will be carried out through the development of industrial applications in which textile products are used as production materials (ranging from fishing nets, tire cords and canvas cloth to geotextile). Therefore, in order to study the textile manufacturing industry, it is generally sufficient to consider the production of apparel goods.

Specialized Techniques necessary for Apparel Goods Production

There are a number of intertwined technical factors involved in various subprocesses or specialized technical fields which make up the overall production process that fabricates and reshapes the raw fibre material into the final textile product to be used by the consumer.

Comparison of Characteristics of Specialized Technical Fields

In Fiscal Year 1989, the textile industry had a share of 4.4per cent in shipment value. 14.8per cent in the number of businesses (20per cent), and

10.4per cent in the number of employees (23.3per cent) in the overall Japanese manufacturing industries, characterized by a gradual decline in relative share, although in terms of absolute shipment value it has actually expanded to 1,053per cent of the Fiscal 1955 level. As has been frequently pointed out throughout its development, the Japanese textile industry has a unique organizational structure consisting of groups of independent companies where all companies in a group belong to one of the specialized technical fields and operate in a horizontal specialization configuration. A comparison and summary of the major indices of these company groups, classifying them in accordance with their specialized technical fields. From these indices, common company characteristics for each group may emerge. Namely, in terms of business size, the fibre production and spinning subdivisions are in contrast with the rest of the textile industry where relatively large numbers of small businesses coexist. In addition, these groups of small companies can only stay in business by relying on the supply of abundant cheap labour, exhibiting a legacy of the textile industry's past as a labour intensive industry, even on its path towards modernization.

Table compares the energy consumption shares of various specialized technical fields and it can be seen that energy consumption is relatively high in the fields of dyeing and finishing, fibre production, spinning, weaving and clothing manufacturing. As for water consumption, the share of the textile industry in the entire manufacturing industries is 5.2per cent for fresh water and 1.1per cent for sea water. Considering the fact that most of this water is used for easy-to-recycle temperature control and cooling purposes, the industry's total water consumption cannot necessarily be regarded as high. However, as is widely accepted, the dyeing and finishing division is placed in a special position in that its water consumption is mainly for processing and washing purposes.

CONCLUSIONS

- There is no panacea for achieving energy conservation in the textile manufacturing industry.
- With the actual implementation of an energy conservation Programme, it is important to grasp the current level of energy consumption and its actual conditions in detail, set goals (energy consumption and corresponding cost), and achieve the goals through a company-wide effort as far as possible.
- In the textile manufacturing industry, it is important to thoroughly understand that, depending on the trend of the market, the company is targeting, consumer requirements for the textile products to be supplied differ, thereby urging the implementation of energy conservation measures which are relevant to the production of the goods that suit the market.

- Therefore, it is necessary to expect that, when multi-line, small-volume production type high value-added goods are produced, energy consumption may increase rather than decrease with production rationalization, in contrast with mass-production type goods.
- When differentiated goods are produced, the share of energy costs in the overall production cost should be given importance rather than energy consumption.
- It is reasonable to consider that ultimately desired energy conservation promoting techniques will depend on the development and practical application of innovative technologies in each specialized technical field.

4

Mechanical Tools of Fabric Construction

INTRODUCTION

Fabric is any piece of cloth. Fabrics can be made or constructed by using a number of techniques as given below:

- Weaving: By interlacing of yarns
- Knitting: By interlocking the loops
- Felting: From matting of the wool fibres
- Bonding: By adhering the fibres to each other
- Knotting: By tying the fabrics.

SOME OF THE IMPORTANT FEATURES OF THESE TECHNIQUES

- *Weaving*: Weaving is the most commonly used method of fabric construction. You must have seen a chatai being made. Weaving is similar to it, where two sets of yarns are interlaced with one another at right angles. Weaving gives a firm fabric. Have you heard of fabrics like poplin, denim and cambric? Yes, these are available in all the cloth shops and you must have used them to make your garments.
- *Knitting*: When you knit a sweater, there is normally one ball of yarn which is interlooped to get a fabric. This technique is called knitting and it gives a lot of stretch and easy-care properties to fabric. Knits are mainly used for hosiery. Knit fabrics are specially useful for garments like underwear, T-shirts, socks etc. knitting is construction of a fabric by forming a yarn into interconnecting loops which hangs one upon other. The fabric may be constructed by machine or hand. The quality of the knitted fabric is affected by the quality of the needles used. The thickness, length and the closeness of needle in the knitting machine affects the durability of the knitted fabric. These are two types:

- WEFT Knitting in which one continuous yarn form courses across the fabrics.
- WARP Knitting in which series of yarns formed wales in the lengthwise direction of the fabric.

- Felting: making clothes from loose fibres is the oldest technique of fibre construction. The primitive nomadic tribes matted the wool fibre into a mass to use it as tent, saddle and other articles because wool is one natural fibre that possesses the properties of being formed directly into a compact mass having strength and durability. the construction of a fabric from loose wool fibres without any interlacing is known as felting.
- Bonding: Bonding fabrics are non-woven fabrics in which techniques other than interlacing are employed for adhering the fibre to each other. The fibre may be held together by some mechanical, chemical, thermal or solvent means or their combination. Felted fabric are also nonwoven fabrics but bonding is different from felting in the sense that in felting, fractional force is used as bonding agent whereas in other techniques of bonding, the fibres are to be applied.
- Knotting: This technique is being used since ages to secure the ends of woven fabric in a decorative way. The ornamental knotting technique is known as MACRAME. It is a craft threads, cords or ropes which are knotted together in different ways to form various designs and patterns.

BASIC WEAVES

PLAIN WEAVE

It is the simplest weave and therefore inexpensive to Produce. Many fabrics that you commonly wear like mulmul dupattas, organdie and chiffon saris are all plain weave. Each and every weft yarn goes alternately under and over the warp yarns across the width of the fabric. If the yarns are close together, the plain weave has a high thread count and the fabric will be firm and will wear well.

Fig. Plain Weave

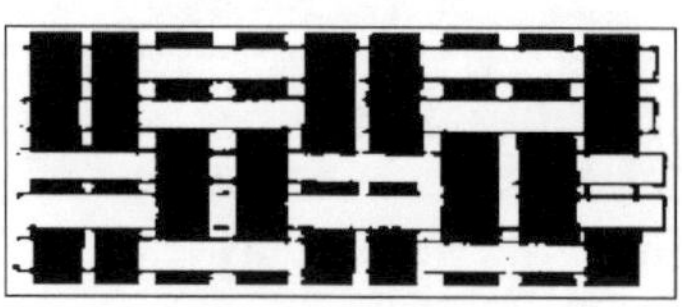

Fig. Basket Weave

Plain weave is of two types:

1. *Rib Weave*: Rib or line effect is created by using thin yarns with thick yarns or single yarns with doubled yarns in any one direction of the fabric.
2. *Basket Weave*: Two or more weft yarns are interlaced as a unit with corresponding number of warp yarns to give a basket like effect. Mattee fabric commonly used for cross stitch embroidery is an example of such a weave.

TWILL WEAVE

This basic weave has a clear diagonal line on the face of the fabric. The denim or jean fabric you wear is twill weave. It is a very strong and durable weave. It is therefore commonly used in men's suit and coat fabrics. Twill weave fabrics show soil less quickly than plain weave.

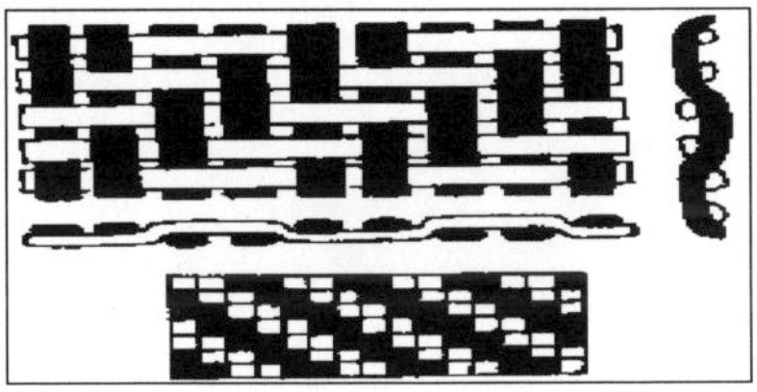

Fig. Twill Weave

SATIN WEAVE

This basic weave has a beautiful shiny surface because of long floats on the surface of the fabric. In the satin weave warp yarns float over several weft yarns before interlacing with a weft yarn and so on. However, the long floats snag easily therefore satin weave is not as strong as plain or twill weave. Corduroy has raised parallel vertical lines. A towel has loops covering its both sides. All these fabrics are made using special looms and weaving techniques. They are obviously expensive fabrics.

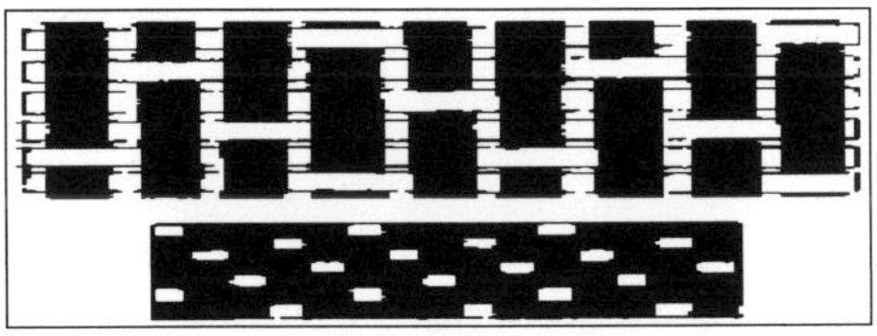

Fig. Satin Weave

NOVELTY WEAVES

Novelty Weaves are also called as decorative, fancy, figure and design weaves. They are formed by predetermined changes in the interlacing of warp and filling yarns. The different weaves include dobby, jacquard, leno, pile and double cloth.

COMMONLY USED NOVELTY WEAVES

Pile Weave

Pile Weave

Pile fabrics are formed by having the basic plain or twill weave as a backing and a third yarn is woven to yield a surface pile. The pile may be warp pile or weft pile. For making ground fabric, plain or twill weave is used, the extra set of filling yarn floats over three or more warp yarns.

The floats are cut and brushed up to form pile. This is called filling pile. Examples are velveteen and corduroy fabrics. If an extra warp yarn floats over the filling yarn, it is called warp pile. Examples are velvet, velour and rug velvet.

Pile weave may be of two types:

1. Cut Pile
2. Uncut Pile

Double Weave

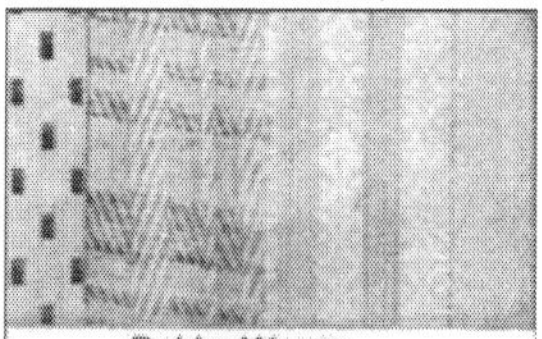
Dobby Weave

Double weave fabrics are obtained by using five or more set of yarns. The most common types of double cloth have two set of warp and two set of weft yarns with an extra yarn interlacing both the cloth. The double cloth has additional bulk, strength and warmth. Examples are coatings, blankets, double brocade and brocatelle.

Leno and Gauze Weave

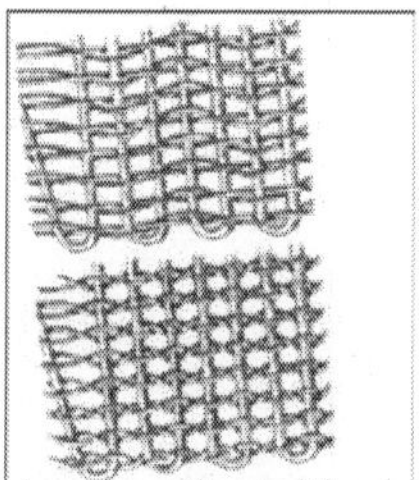
Leno and Gauze Weave

The leno weave produces an open-textured fabric that may be sheer or heavy. It is produced by the doup attachment to the basic loom. The doup attachment controls the warp threads horizontally as well as vertically and the unusual warp interlacing prevents slippage of the filling and reduces shrinkage. Examples include mosquito nets, household bags and curtains.

Dobby Weave

Dobby designs have small figures such as dots, geometric designs and floral patterns woven into the fabric. The design is produced by a combination of two or more basic weaves and the loom may have upto thirty two harnesses. Examples of dobby weave are shirting madras, pique, huck towelling.

Following types dobby weaves are:

- Bird's eye Weave
- Honey comb Weave
- Huchaback Weave

Jacquard Designs

Jacquard Weave

Fabrics with elaborate designs are woven using the Jacquard loom attachment. It was introduced by Joseph Marie Jacquard in 1801. The Jacquard mechanism has the ability to control every warp yarn instead of a series as in regular harness looms. The machine is very big and very expensive. The pattern for the Jacquard loom is transferred to a series of perforated cards, one for each filling pick in the pattern. The card is punched so as to permit the needles on the machine to be raised to pass through the card. The shed is formed and the pick passes through each card stops on the cylinder for its particular pick, moves on, and new card takes its place. This process continues until all the cards are used. Thus when pattern is over the cards start for the next till the entire cloth is woven. Thus extremely complicated and decorative woven designs are produced using jacquard loom attachment. Examples include brocade, damask, tapestry, terry-cloth towels.

Swivel Weave

The Swivel weave differs from lappet in that designs are produced by extra filling yarns. Separate shuttles are placed at each point where the design has to be made. The shed is formed by the pattern, where the shuttle carries the yarn through the shed, the distance of the pattern. The extra filling floats on the back of the fabric, the long floats is cut away after weaving is completed. Example: silk sarees.

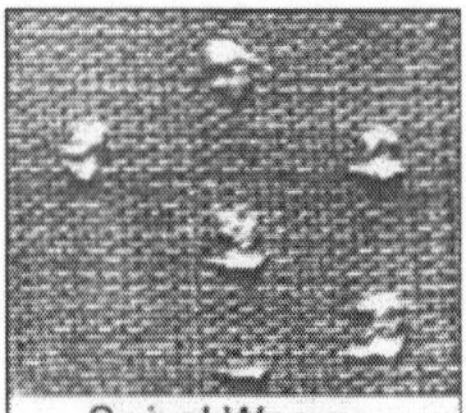

Swivel Weave

Lappet Weave

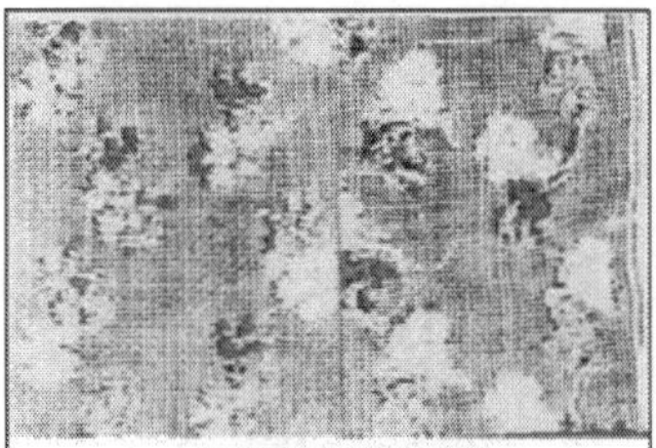

Lappet Weave

In lappet weave an extra warp yarn is introduced in a way so as to create design on the base, fabric. Patterns are woven in the fabric by an attachment to the loom. Long floats formed on the back of the fabric are cut away, if floats are small they are usually left. Example: fancy sarees.

Crepe Weave

It is obtained by making variations in the plain or satin weave with the use of dobby attachment on the loom, to give the surface a crinkled or pebbly appearance. The crape is made by weaving highly twisted yarns alternatively with yarns of ordinary twist, which ordinarily do not have any crepe effect. Both plain weave and satin weave floats in the wrap or filling yarns may be used to form irregular patterns which would produce a rough textured material with crepe weave also known as Granite weave.

DECOLOURISATION OF TEXTILE DYEING EFFLUENTS USING ADVANCED OXIDATION PROCESSES

Textile industry is a leading industry for most countries, such as China, Singapore, UK, Bangladesh, Italy, Turkey etc. But, environmental pollution is one of the main results of this industry. Parralel to usage of huge amounts of water ad chemicals, the textile dyeing and finishing industry is one of the major polluters among industrial sectors, in the scope of volume and the chemical composition of the discharged effluent. Textile industry effluents can be classified as dangerous for receiving waters, which commonly contains high concentrations of recalcitrant organic and inorganic chemicals and are characterised by high chemical oxygen demand (COD) and total organic carbon (TOC), high amounts of surfactants, dissolved solids, fluctuating temperature and pH, possibly heavy metals (*e.g.* Cu, Cr, Ni) and strong colour. The presence of organic contaminants such as dyes, surfactants, pesticides, etc. in the hydrosphere is of particular concern for the freshwater, coastal, and marine environments because of their nonbiodegradability and potential carcinogenic nature of the majority of these compounds. The major concern with colour is its aesthethic character at the point of discharge with respect to the visibility of the receiving waters. The main reason of colour in textile industry effluent is the usage of large amounts of dyestuffs during the dyeing stages of the textile-manufacturing process.

Inefficient dyeing processes often result in significant dye residuals being presented in the final dyehouse effluent in hydrolised or unfixed forms. Apart from the aesthetic problems relating to coloured effluent, dyes also strongly absorb sunlight, thus impeding the photosynthetic activity of aquatic plants and seriously threatening the whole ecosystem. Stricter regulatory requirements along with an increased public demand for colour-free effluent nessesitate the inclusion of a decolourisation step in wastewater treatment plants.

Well known and widely applied treatment method for the treatment of textile industry wastewater is activated sludge process and it's modifications. Combinations of activated sludge process with physical and chemical processes can be found in most applications.

These traditional treatment methods require too many spaces and are affected by wastewater flow and characteristic variations. But, either activated sludge process modifications itself or combinations of this process with physical or chemical processes are inefficient for the treatment of coloured waste streams.

On the other hand, existing physico-chemical advanced treatment technologies such as, membrane processes, ion exchange, activated carbon adsorption etc. can only transfer pollutants from one phase the other phase rather than eliminating the pollutants from effluent body. Recovery and reuse of certain and valuable chemical compounds present in the effluent is currently under investigation of most scientists.

At this point, The AOPs show specific advantages over conventional treatment alternatives because they can eliminate non-biodegradable organic components and avoid the need to dispose of residual sludge. Advanced Oxidation Processes (AOPs) based on the generation of very reactive and oxidizing free radicals, especially hydroxyl radicals, have been used with an increasing interest due to the their high oxidant power. In this chapter, discussion and examples of colour removal from textile effluent will be focused on those of most used AOPs.

ADVANCED OXIDATION PROCESSES: PRINCIPLES AND DEFINITIONS

Advanced Oxidation Processes (AOPs) are defined as the processes which involve generation and use of powerfull but relatively non-selective hydroxyl radicals in sufficient quantities to be able to oxidize majority of the complex chemicals present in the effluent water. Hydroxyl radicals has the highest oxidation potential (Oxidation potential, E_0: 2.8 eV vs normal hydrogen electrode (NHE)) after fluorine radical. Fluorine, the strongest oxidant cannot be used for wastewater treatment because of its high toxicity. From these reasons, generation of hydroxyl radical including AOPs have gained the attention of most scientists and technology developers.

The main and short mechanism of AOPs can be defined in two steps:

1. The generation of hydroxyl radicals,
2. Oxidative reaction of these radicals with molecues. AOPs can convert the dissolved organic pollutants to CO_2 and H_2O. The generation of highly effective hydroxyl radical might possibly be by the use of UV, UV/O_3, UV/H_2O_2, Fe^{+2}/H_2O_2, TiO_2/H_2O_2 and a number of other processes.

AOPs can be classified in two groups:

1. Non-photochemical AOPs,
2. Photochemical AOPs. Non-photochemical AOPs include cavitation, Fenton and Fenton-like processes, ozonation at high pH, ozone/hydrogen peroxide, wet air oxidation etc. Short description of some important AOPs are given below. Photochemical oxidation processes include homegenous (vacuum UV photolysis, UV/hydrogen peroxide, UV/ozone, UV/ozone/hydrogen peroxide, photo-Fenton etc), and heterogeneous (photocatalysis etc) processes.

NON-PHOTOCHEMICAL OXIDATION PROCESSES

Non-photochemical oxidation processes can be classified as:

- Ozonation,
- Ozone/Hydregen Peroxide,
- Fenton Process,
- Electrochemical Oxidation,
- Supercritical water oxidation,
- Cavitataion,
- Elelctrical discharge-based nonthermal plasma,
- Gamma-ray,
- X-ray and
- Electron beam. Ozonation, ozone/hydrogen peroxide and Fenton-process are widely applied and examined processes for the treatment of textile effluent.

Ozonation

Ozone is well known and widely applied strong oxidizing agent for the treatment of both water and wastewater, in literature and on site. Ozone has high efficiency at high pH levels. At these high pH values (>11.0), ozone reacts almost indiscriminately with all organic and inorganic compounds present in the reacting medium. Ozone reacts with wastewater compounds in two different ways namely direct molecular and indirect radical type chain reactions. Both reactions occur simultaneously and hence reaction kinetics strongly depend on the characteristics of the treated wastewater (*e.g.* pH, concentrations of initiators, promoters and scavengers. Simplified reaction mechanisms of ozone at high pH is given in below;

$$3O_3 + H_2O \xrightarrow{OH^-} 2OH\bullet + 4O_2$$

Ozone/hydrogen Peroxide (Peroxone) Process (O_3/H_2O_2)

The combination of ozone and hydrogen peroxide is used essentially for the contaminants which oxidation is difficult and consumes large amounts of oxidant. Because of the high cost of ozone generation, this combination make the process economically feasible. The capability of ozone to oxidise various pollutants by direct attack on the different bonds, aromatic rings is further enhanced in the presence of H_2O_2 due to the generation of highly reactive hydroxyl radicals. The dissociation of H_2O_2 results in the formation of hydroperoxide ion, which attacks the ozone molecule resulting in the formation of hydroxyl radicals. General mechanism of peroxon process is given below:

$$H_2O_2 + 2O_3 \rightarrow 2OH\bullet + 3O_2$$

The pH of solution is also critical for the processs efficiency like other AOPs. Addition of hydrogen peroxide to the aqueous O_3 solution at high pH conditions will result in higher production rates of hydroxyl radicals. Indipendence of peroxone process from any light source or UV radiation gives a specific advantage to this process that it can be used in turbid or dark waters.

Fenton Process

The dark reaction of ferrous iron (Fe(ll)) with H_2O_2 known as Fenton's reaction, which is shown in Eq.-15, has been known for over a century.

$$Fe^{+2} + H_2O_2 \rightarrow Fe^{+3} + OH- + OH\bullet$$

The hydroxyl radical thus formed can react with Fe(II) to produce ferric ion (Fe(III)) as shown in Eq.-16;

$$\cdot OH + Fe^{+2}K \rightarrow Fe^{+3} + OH-$$

Alternatively, hydroxyl radicals can react with and initiate oxidation of organic pollutants in a waste stream,

$$RH + \cdot OH \rightarrow R\cdot + H_2O$$

At value of pH (2.7–2.8), reactions can result into the reduction of Fe^{+3} to Fe^{+2} (Fenton-like).

$$Fe^{+2} + H_2O_2 \Leftrightarrow H^+ + FeOOH^{+2}$$

$$FeOOH^{+2} \rightarrow HO_2 \bullet + Fe^{+2}$$

proceeding at an appreciable rate. In these conditions, iron can be considered as a real catalyst. At pH values <4.0, ferrous ions decompose H_2O_2 catalytically yielding hydroxyl radicals most directly. However, at pH values higher than 4.0, ferrous ions easily form ferric ions, which have a tendency to produce ferric hydroxo complexes.

H_2O_2 is quite unstable and easily decomposes itself at alkaline pH. Fenton process is cost-effective, easy to apply and effective for the degradation of a wide range of organic compounds. One of the advantages of Fenton's reagent is that no energy input is necessary to activate hydrogen peroxide. Therefore, this method offers a costeffective source of hydroxyl radicals, using easy-to-handle reagents (Bautista et al., 2007). The Fenton process consisits of four stages. At first, pH is adjusted to low pH. Then the main oxidation reactions take place at pH values of 3-5. The wastewater is then neutralized at pH of 7-8, and, finally, precipitation occurs (Bigda, 1995, Lee and Shoda, 2008). Furthermore, it commonly requires a relatively short reaction time compared with other AOPs. Thus, Fenton's reagent is frequently used when a high reduction of COD is required (Bigda, 1995, Bautista et al., 2007, Lee and Shoda, 2008, Yonar, 2010).

PHOTOCHEMICAL OXIDATION PROCESSES

Homogeneous Photochemical Oxidation Processes

Vacuum UV (VUV) Photolysis

The Vacuum Ultraviolet range is absorbed by almost all substances (including water and air). Thus it can only be transmitted in a vacuum. The absorption of a VUV photon causes one or more bond breaks. For example, water is dissociated according to;

$$H_2O + hv(<190nm) \rightarrow H \bullet + HO \bullet$$

$$H_2O + hv(<190nm) \rightarrow H + +e^- + HO \bullet$$

Photochemistry in the vacuum-ultraviolet (VUV) spectral domain (approx. 140–200 nm) is of high applicatory interest, e.g:

- In microelectronics, where materials with surface structures of high spatial resolution provide a basis for the fast development of high computational and electronic and optical storage capacities or

- In environmental techniques, in particular for the production of ultra pure water and for the oxidative treatment of waste gas and water (Bolton, 2002, Gonzaleza et al., 2004). VUV-photolysis can be achived by the usage of either a monochromatic (Xe-eximer Xe2*) or polychromatic (Hg) radiation sources. Theses light sources have some limitations such as high price, wave length variations etc. From these reasons application of VUV photolysis are too limited.

Hydrogen Peroxide/UV (H_2O_2/UV) Process

This method is based on the direct photolysis of hydrogen peroxide molecule by a radiation with a wavelength between 200-300 nm region. The main reaction of H_2O_2/UV is given below:

$$H_2O_2 + hv \rightarrow 2HO\bullet$$

The low, medium ad high pressure mercury vapor lamps can be used for this process because it has significant emittance within 220-260 nm, which is the primary absorption band for hydrogen peroxide. Most of UV light can also be absorbed by water. Low pressure mercury vapour lambs usage can lead to usage of high concentrations of H_2O_2 for the generation of sufficient hydroxyl radical. However, high concentrations of H_2O_2 may scavenge the hydoxyle radical, making the H_2O_2/UV process less effective. Some more variables such as temperature, pH, concentration of H_2O_2, and presence of scavengers affect the production of hydroxyl radicals (EPA, 1998, Bolton, 2001, Mandal et al., 2004 Azbar et al., 2005).

Ozone/UV (O_3/UV) Process

Photolysis of ozone in water with UV radiation in the range of 200-280 nm can lead to yield of hydrogen peroxide. Hydroxyl radicals can be generated by these produced hydrogen peroxide under UV radiation and/or ozone as given equations below:

$$O_3 + hv + H_2O \rightarrow H_2O_2 + O_2$$

$$H_2O_2 + hv \rightarrow 2\cdot OH$$

$$2O_3 + H_2O_2 \rightarrow 2\cdot OH + 3O_2$$

Starting from low pressure mercury vapour lamps all kind of UV light sources can be used for this process. Because, O_3/UV process does not have same limitations of H_2O_2/UV process. Low pressure mercury vapor UV lamps are the most common sources of UV irradation used for this process. Many variables such as pH, temperature, scavengers in the influent, tubidity, UV intensity, lamp spectral characteristics and pollutant type(s) affect the effciency of the system (EPA, 1998, Azbar, 2005). Number of laboratory, pilot and full scale applications of Ozone/UV and Hyrdogen peroxide/UV processes can be found in literature. Commercial applications of these processes can also be available.

Ozone/Hydrogen Peroxide/UV (O_3/ H_2O_2/ UV) Process

This method is considered to be the most effective and powerful method which provides a fast and complete mineralisation of pollutants (Azbar, 2005, Mokrini et al., 1997). Similar to other ozone including AOPs, increasing of pH affects the hydroyle radical formation. Additional usage of UV radiation also affects the hydroyle radical formation. Efficiency of ozone/hydrogen peroxide/ UV process is being much more higher with addition of hydrogen peroxide (Horsch, 2000, Contreras et al., 2001). Main short mechanism of O_3/ H_2O_2/ UV process is given below:

$$2O_3 + H_2O_2 \xrightarrow{UV} 2HO\cdot + 3O_2$$

Photo-Fenton Process

The combination of Fenton process with UV light, the so-called photo-Fenton reaction, had been shown to enhance the efficiency of Fenton process. Some reasearchers also attributed this to the decomposition of the photo active $Fe(OH)^{+2}$ which lead to the addition of the HO·radicals (Sun and Pignatello, 1993, He and Lei, 2004). The short mechanism of photo-Fenton reaction is given below:

$$Fe(OH)^{+2} + hv \rightarrow Fe^{+3} + HO^{\cdot}$$

With $Fe(OH)^{2+}$ being the dominant Fe(III) species in solution at pH 2-3. High valence Fe intermediates formed through the absorption of visible light by the complex between Fe(II) and H_2O_2 are believed to enhance the reaction rate of oxidation production (Pignatello, 1992, Bossmann et al., 2001).

Heterogeneous Photochemical Oxidation Processes

Widely investigated and applied Heterogeneous Photochemical Oxidation processes are semiconductor-sentized photochemical oxidation processes. Semiconductors are characterized by two separate energy bands: a low energy valence band (h^+_{VB}) and a high-energy conduction (e^-_{CB}) band. Each band consists of a spectrum of energy levels in which electrons can reside. The separation between energy levels within each energy band is small, and they essentially form a continuous spectrum. The energy separation between the valence and conduction bands is called the band gap and consists of energy levels in which electrons cannot reside. Light, a source of energy, can be used to excite an electron from the valence band into the conduction band. When an electron in the valence band absorbs a photon,' the absorption of the photon increases the energy of the electron and enables the electron to move into one of the unoccupied energy levels of the conduction band (EPA, 1998). Semiconductors that have been used in environmental applications include TiO_2, strontium titanium trioxide, and zinc oxide (ZnO). TiO_2, is generally preferred for use in commercial APO applications because of its high level of photoconductivity, ready availability, low toxicity, and low cost. TiO_2, has three crystalline forms: rutile, anatase, and brookite. Studies indicate that the anatase form provides the highest hydroxyl radical formation rates (Korrmann et al., 1991, EPA, 1998). The photo-catalyst titanium dioxide (TiO_2) is a wide band gap semiconductor (3.2 eV) and is successfully used as a photo-catalyst for the treatment of organic pollutants (Hsiao et al., 1983, Korrmann et al., 1991, Zahhara, 1999). Briefly, for TiO_2, the photon energy required to overcome the band gap energy and excite an electron from the valence band to the conduction band can be provided by light of a wavelength shorter than 387.5 nm. Simplified reaction mechanisms of TiO_2/UV process is given in following equations (eq. 16- eq. 19).

$$TiO_2 + hv \rightarrow e^-{}_{CB} + h^+{}_{VB}$$

$$H_2O + h^+{}_{VB} \rightarrow OH\bullet + H^+$$

$$O_2 + e^-{}_{CB} \rightarrow O_2\bullet^-$$

$$O_2\bullet^- + H_2O \rightarrow OH\bullet + OH^- + O_2 + H_2O^-$$

The overall result of this reversal is generation of photons or heat instead of -OH. The reversal process significantly decreases the photo-catalytic activity of a semiconductor (EPA, 1998). Main advantage of TiO_2/UV process is low energy consumption which sunlight can be used as a light source.

CHARACTERISATION OF TEXTILE INDUSTRY WASTEWATER

Textile industry produces large amounts of liquid by-products. Volume and composition of these waswater can vary from one source to other source. In the scope of volume and the chemical composition of the discharged effluent, the textile dyeing and finishing industry is one of the major polluters among industrial sectors. Textile industry dyes are intentionally designed to remain photolytically, chemically and biochemically stable, and thus are usually not amenable to biodegradation (Pagga and Braun, 1986). Like many other industrial effluents, textile industry wastewater varies significantly in quantity, but additionally in composition (Correira et al., 1994). These wastes include both organic and inorganic chemicals, such as finishing agents, carriers, surfactants, sequestering agents, leveling agents etc. From these reasons, textile effluents are characterized with high COD (H» 400-3.000 mg/L), BOD_5 (H» 200-2.000 mg/L), Total Solids (H» 1.000-10.000 mg/L), Suspended Solids (H» 100-1.000

mg/L), TKN (H» 10-100 mg/L), Total Phosporus (H» 5-70 mg/L), Conductivity (1.000-15.000 mS/cm) and pH (H» 5-10 usually basic) (Grau, 1991, Pagga ad Braun, 1991, Kuo, 1992, Correira et al., 1994, Arslan and Balcioglu, 2000,, Nigam et al., 2000, Azbar et al., 2005, Akal Solmaz et al., 2006, Yonar et al., 2006, Mahmoudi and Arami, 2009, Yonar, 2010,). Another important problem of textile industry wastewater is colour.

Without proper treatment of coloured wate, these dyes may remain in the environment for a long time (Yonar et al, 2005). The problem of coloured effluent has been a major challenge and an integral part of textile effluent treatment as a result of stricter environmental regulations. The presence of dyes in receiving media is easily detectable even when released in small concentrations (Little et al., 1974, Azbar et al., 2004).

This is not only unsightly but dyes in the effluent may have a serious inhibitory effect on aquatic ecosystems.

Definition and determination of colour is another important point for most water and wastewater samples. Some methods can be found in literature for the determination of colour in samples. But, selection of true method for the determination of colour is very important. According to "Standard Methods for the Examination of Water and Wastewater" (APHA- AWWA, 2000), importance of colour is defined with some sentences given below: "Colour in water may result from the presence of natural metallic ions (iron and manganese), humus and peat materials, plankton, weeds, and *industrial wastes*. Colour is removed to make a water suitable for general and industrial applications.

Coloured industrial wastewaters may require colour removal before discharge into watercourses." From these reasons, colour content should be determined carefully.

In Standard Methods, colour content of water and wastewater samples can be determined with four different methods such as:

1. Visual Comparison Method,
2. Spectrometric Method,
3. Tristimulus Filter Method, and
4. ADMI Tristimulus Filter Method. Selection of true and appropriate method for samples is very important. Visual comparison method is suitable for nearly all samples of potable water. This method is also known as Platinum/Cobalt method. Pollution by ceratin industrial wastes may produce unusual colour that can not be easily matched. In this case, usage of instrumental methos are appropriate for most cases. A modification of the spectrometric and tristimulus methods allows calculation of a single colour value representing uniform cromaticity differences even when the sample exhibits colour significantly different from that of platinum cobalt standards (APHA-AWWA, 2000).

COLOUR REMOVAL FROM TEXTILE INDUSTRY WASTEWATER BY AOPS

Most commonly applied treatment flow scheme for textile effluent in Turkey and other countries generally include either a single activated sludge type aerobic biological treatment or combination of chemical coagulation and flocculation + activated sludge process (Yonar et al., 2006). Furthermore, it is well known that aerobic biological treatment option is ineffective removal for colour removal from textile wastewater in most cases and the chemical coagulation and flocculation is also not effective for the removal of soluble reactive dyestuffs. Therefore, dyes and chemicals using in textile industry in effluent may have a serious inhibitory effect on aquatic ecosystems and visual pollution on receiving waters. There are several alternative methods used to decolourize the textile wastewater such as various combinations of physical, chemical and biological treatment and colour removal methods, but they cannot be effectively applied for all dyes and these integrated treatment methods are not cost effective. Advanced Oxidation Processes (AOPs) for the degradation of non-biodegradable organic contaminants in industrial effluents are attractive alternatives to conventional treatment methods and are capable of reducing recalcitrant wastewater loads from textile dyeing and finishing effluents (Galindo et al., 2001, Robinson et al., 2001, Azbar et al., 2004, Neamtu et al., 2004). In this section, applied AOPs for colour removal from textile effluent are given. Technological advantages and limitations of these AOPs is also discussed.

COLOUR REMOVAL WITH NON-PHOTOCHEMICAL AOPS

Ozonation at high pH, ozone/hydrogen peroxide and Fenton processes are widely applied and investigated AOPs for colour removal from textile effluents and tetile dyes. As it can be clearly seen from former sections, ozone can produce hydroxyl radicals at high pH levels. According to this situation, pH is very important parameter for ozonation process. Under conditions aiming hydroxyl free radical (HO•) production (*e.g.*, high pH), the more powerful hydroxyl oxidation starts to dominate (Hoigne and Bader, 1983). Since the oxidation potential of ozone reportedly decreases from 2.07mV (acidic pH) to 1.4mV (basic pH) (Muthukumanar et al., 2001), it is clear that another more powerful oxidant (HO•) is responsible for the increase in the dye degradation, with a consequent colour absorbance decrease. The efficiency of ozonation in the removal of colourand COD from textile wastewater is important to achieve to discharge limits (Somensia et al., 2010). Textile wastewaters is very complex due to the organic chemicals such as many different dyes, carriers, biocides, bleaching agents, complexion agents, ionic and non-ionic surfactants, sizing agents, etc. As a result, it is hard to explain the overall degradation of the organic matter by ozone in textile wastewater individually. Thus, some global textile wastewater parameters such as colour, COD and dissolved organic carbon are used for the degradation kinetic of organic matter by ozonation (Sevimli and Sarikaya, 2002, Selcuk, 2005). Textile wastewaters exhibit low BOD to COD ratios (< 0.1) indicating non-biodegradable nature of dyes and Wilmott et al.(1998) have claimed that aerobic biological degradation is not always effective for textile dye contaminated effluent (Sevimli and Sarikaya, 2002). Somensia et al., (2010), tested pilot scale ozonation for the pre-treatment and colour removal of real textile effluent. Authors have mentioned that the importance of pH on the process efficiency and colour removal efficiencies were determined as 40.6per cent and 67.5per cent at pH 3.0 and 9.1, respectively. COD removal effcieincies ware also determined as 18.7per cent (pH=3) and 25.5per cent (pH=9). On the other hand, toxicity can be reduced significantly compared with raw wastewater. Azbar et al., (2004) carried out a comparative study on colourand COD removal from acetate and fibre dyeing effluent. In this study, various advanced oxidation processes (O_3, O_3/UV, H_2O_2/UV, O_3/H_2O_2/UV, Fe^{+2}/H_2O_2) and chemical treatment methods using $Al2(SO_4)_3.18H_2O$ $FeCl_3$ and $FeSO_4$ for the Chemical Oxygen Demand (COD) and colour removal from a polyester and acetate fibre dyeing effluent is undertaken. Ozonation showed superior performance at pH=9 and 90per cent COD and 92per cent colour can be removed. Akal Solmaz et al., 2006, applied ozonation to real textile wastewaters and found 43per cent COD and 97per cent colour removal efficiencies at pH 9 and C_{O3} 1.4 g/h. In the another study of Akal Solmaz et al., (2009), group has tested different AOPs on two different textile wastewater. 54-70per cent COD removal and 94-96 per cent colour removal efficiencies have been determined at pH = 9. In another study, Selcuk, (2005), have tested coagulation and ozonation for colour, COD and toxicity removal from textile wastewater. Author found that, ozonation was relatively effective in reducing colour absorbances and toxic effects of textile effluents compared with chemical coagulation. Almost complete colourabsorbances (over 98per cent) were removed in 20 min ozone contact time, while COD removal (37per cent) was very low and almost stable in 30 min ozonation period. Yonar et al., (2005), have been studied AOPs for the improvement of effluent quality of a textile industry wastewater treatment plant. Authors were mainly tested homogeneous photochemical oxidation processes (HPOP's) (H_2O_2/UV, O_3/UV and H_2O_2/O_3/UV) for colour and COD removal from an existing textile industry wastewater treatment plant effluent together with their operating costs. At pH=9, 81per cent COD and 97per cent colour removal efficiencies were reported for ozonation process. Ozonation is very effective for the removal of colour from textile wastewater. COD and toxicity can also be removed by ozonation. But, for decision making on these processes advantages and limitations of these processes should be known. Main advantage of ozonation is no need to addition of any chemicals to water or wastewater. Because, ozone is mostly produced by cold corona discharge genertors. And these generators need dry air for the production of ozone. On the other hand, sludge or simiar residues is not produced during this process. At this point, specific advantage can be stated for textile effluents. Mostly, the pH value of textile wastewater are higher than 7 and in some situations higher than 9. Thus, ozonation can be applied to textile effluent without any pH adjustment and chemical addition. But, ozonation process has some disadvantages, such as, inefficient production capacities of cold corona discharge (CCD) generators (2-4per cent), less solubility of gas phase ozone in water, higher energy demads of CCD generators, possible emission problems of ozóne etc. These disadvantages can be overcomed by the production of efficient ozone generators like membrane electrochemical ozone generators. Ozone/Hydrogen peroxide process is onother efficient AOPs for the treatment of recalcitrant organics. Similar to ozonation, ozone including other processes mostly needs alkaline conditions. This argument has been extensively and successfully studied by Hoigne (1998) in the attempt of giving a chemical explanation to the short life time of ozone in alkaline solutions. Hoigné showed that the ozone decomposition in aqueous solution develops through the formation of hydroxyl radicals. In the reaction mechanism OH" ion has the role of initiator:

$$HO^{-} + O_3 \rightarrow OH_{2^-} + O_2$$

$$OH_{2^-} + H^{+} \Leftrightarrow H_2O_2$$

$$OH_{2^-} + O_3 \rightarrow HO_2\bullet + O_2\bullet^{-}$$

$$HO_2\bullet \Leftrightarrow H^{+}O_2\bullet^{-}$$

$$O_2\bullet - + O_3 \rightarrow O_2 + O_3\bullet -$$

$$O_3\bullet - + H^{+} \rightarrow HO_3\bullet$$

$$HO_3\bullet \rightarrow HO\bullet + O_2$$

$$OH\bullet + O_2 \rightarrow HO_2\bullet + O_2$$

It is clear therefore that the addition of hydrogen peroxide to the ozone aqueous solution will enhance the O_3 decomposition with formation of hydroxyl radicals. The influence of pH is also evident, since in the ozone decomposition mechanism the active species is the conjugate base HO_2- whose concentration is strictly dependent upon pH. The increase of pH and the addition of H_2O_2 to the aqueous O_3 solution will thus result into higher rates of hydroxyl radicals production and the attainment of higher steady-state concentrations of hydroxyl radicals in the radical chain decomposition process (Glaze and Kang, 1989). It must be remarked that the adoption of the H_2O_2/O_3 process does not involve significant changes to the apparatus adopted when only O_3 is used, since it is only necessary to add an H_2O_2 dosing system (Andreozzi, 1998). Hydrogen peroxide/ozone (peroxone) process test result for real or synthetic textile wastewater are too limited in literature but ozone and hydrogen peroxide is a very promising technique for potential industrial implementation. Kurbus et al. (2003) were conducted comperative study on different vinylsulphone reactive dyes. For all tested dyes, over 99per cent colour removal can be achieved at pH=12. Kos and Perkovski (2003), were tested different AOPs including peroxone process on real textile wastewater. Textile wastewater initial COD is over 5000 mg/L and authors declared that nearly 100per cent colour removal can be achived with peroxone process. Addition of hydrogen peroxide to ozone is increased colour and COD removal efficiencies nearly 10per cent. Perkovski et al., (2003), were tested peroxone process on anthraquinone dye Acid Blue 62 and they found 60per cent colour removal efficiency. Main advantage and disadvantage of peroxone process is addition of hydrogen peroxide. Addition of hydrogen peroxide is giving higher efficiencies and no need to upgrade the existing ozonation systems. But, addition of hydrogen peroxide means additional costs for the treatment of wastewater. Finally, Fenton process is mostly applied on both textile and other industrial wastewaters. Nevertless, the high electrical energy demand is general disadvantage of most AOPs. The greatest advantages of Fenton process is that no energy input is necessary to activate hydrogen peroxide. Most other AOPs need energy input for this activation such as UV based processes, US based proceeses, wet air oxidation etc. The dark reaction of ferrous ion with hydrogen peroxide was found by Fenton (1894). During the last decades, important scientific studies were carried out on the treatment of most toxic chemicals and waste streamns with this process. Another advantage of Fenton process is the applicability of this process in full scale. Because, this process can be accepted as the modification of traditional physico-chemical treatment. Fenton process can control in different steps of mixing and settling processes. By other words this process does not need specific and complex reactor designs. But, the main important disadventage of this process among all AOPs is sludge production. Ferric salts should be settled and disposed before discharge of the effluent. Treatment efficiencies and results of applied Fenton process results in literature summarized in Table. According to these results, Fenton process is also promising technique for the treatment and decolourisation of textile effluent.

COLOUR REMOVAL WITH PHOTOCHEMICAL AOPS

For the treatment and decolourisation of textile effluent, photochemical oxidation processes are widely investigated in literure. Photochemical oxidation processes are good and emerging alternatives and need UV radiation for the production of hydroxyl radicals. Vakuum UV phooxidation is most powerful member of these processes. Hydroxyl radicals can be produced with VUV with no any chemical addition. Generally Ve-eximer lamp are employed for VUV band radiation. In literature, a number of studies can be found for the treatment of organics with VUV. Despite numerous positive examples, the theory of reactor modelling for sharply nonuniform light distribution is not well developed (Braun et al., 1993). Main reason of this situation is the high price of Xe-eximer lambs. Tarasov et al., (2003) investigated VUV photolysis for dye oxidation. They tested VUV process on 6 different dye solutions (methylene blue (Basic Blue 9), Basic blue Zn-salt; Direct Green 6; fucsine; Acid Yellow 42, Acid Yellow 11). Degradation of all dyes under VUV condition takes place in about a minute. In another study, Al-Momani et al. (2002) studied photo-degradation and biodegradability of three different families of non-biodegradable textile dyes (Intracron reactive dyes, Direct dyes and Nylanthrene acid dyes) and a textile wastewater, using VUV photolysis. Ninety percent of colour removal of dye solutions and wastewater is achieved within 7 min of irradiation. UV/H_2O_2 is one of the popular and commercial advanced oxidation process. Like other AOPs, the reaction pH of the treatment system has been observed to significantly affect the degradation of pollutants (Sedlak and Andren, 1991, Lin and Lo, 1997, Kang and Hwang, 2000, Nesheiwat and Swanson, 2000, Benitez et al., 2001a). The optimum pH has been observed to be 3 in the majority of the cases in which H_2O_2 was used with UV irradiation (Ventakandri and Peters, 1993, Tank and Huang, 1996, Kwon et al., 1999, Benitez et al., 2001b) and hence is recommended as the operating pH. It should be noted here that the intrinsic rates of UV/H_2O_2 process may not be affected much, but at lower operating pH, the effect of the leading to higher overall rates of degradation. Thus, it is better to have lower operating pH (Gogate and Pandit, 2004b). In literature, hydrogen peroxide (H_2O_2) itself acts as an effective hydroxyl radical (OH.) scavenger at high concentrations given in following empirical equation (Arslan, 2000).

$$H_2O_2 + OH^{\cdot} \rightarrow HO_2^{\cdot} + H_2O \quad k = 1.2 - 4.5\ 10^7\,M^{-1}s^{-1}$$

Although HO_2. promoted radical chain reactions and it is an oxidant itself, its oxidation potential is much lower than that of hydroxyl radical (OH.). Thus, the presence of excess hydrogen peroxide (H_2O_2) can lower the treatment efficiency of AOPs and it is very important to optimize the applied hydrogen peroxide (H_2O_2) concentration to maximize the treatment performance of AOPs (Arslan, 2000). The presence in the treated water of carbonate can result in in a significant reduction of the efficiency of abetement of pollutants as explained in some studies (Bhattacharjee and Shah, 1998, Andreozzi et al., 1999). Carbonate acts as radical scavenger;

$$HCO_{3^-} + OH^{\cdot} \rightarrow CO_{3^-}. + H_2O \quad k_{HCO3-,\bullet OH} = 1.5\ 10^7\,M^{-1}s^{-1}$$

$$CO_{3^{-2}} + OH^{\cdot} \rightarrow CO_{3^-.} + OH- \quad k_{CO32-,\bullet OH} = 4.2\ 10^8\,M^{-1}s^{-1}$$

since CO_3-. is much les reactive than hydroxyl radical (OH.) inhibition by carbonate influences the behaviour of most AOPs. At lower operating pH values, the effect of radical scavengers, especially ionic such as carbonate and bicarbonate will be nullified leading to higher overall rates of degradation (Gogate and Pandit, 2004a). Thus, lower operating pH values are recommended for most AOPs in literature. Galindo and Kalt, (1998) documented that the H_2O_2/UV process was more effective in an acid medium (pH H» 3-4) in term of discolouration. On the other hand, the aqueous stream being treated must provide good transmission of UV light, so that turbidity and high suspended solids concentration would not cause interferences. Scavengers and excessive dosages of chemical additives may inhibit the process. Heavy metal ions (higher than 10 mg l^{-1}), insoluble oil and grease, high alkalinity and carbonates may cause fouling of the UV quartz sleeves. Therefore, a good pretreatment of the aqueous stream should be necessary for UV based AOPs (Azbar et al, 2005). Decolourisation and treatment of textile effluent were investigated in most studies (Shu et al., 1994, Galindo and Kalt, 1998,, Arslan and Balcl´o´lu, 1999, Ince, 1999, Neamtu et al., 2002, Cisneros, 2002, Mohey El-Dein et al., 2003, Azbar et al, 2004, Shu and Chang, 2005, Yonar et al. 2005). The use of H_2O_2/UV process seems to show a satisfactory COD (70-95per cent) and colour(80-95per cent) removal performance. Conventional ozonation of organic compounds does not completely oxidize organics to CO_2 and H_2O in many cases. Remaining intermediate products in some solution after oxidation may be as toxic as or even more toxic than initial compound and UV radiation could complete the oxidation reaction by supplement the reaction with it. UV lamp must have a maximum radiation output 254 nm for an efficient ozone photolysis. The O_3/UV process is more effective when the compounds of interest can be degraded through the absorption of the UV irradiation as well as through the reaction with hydroxyl radicals (Rein, 2001; Metcalf and Eddy, 2003). The O_3/UV process makes use of UV photons to activate ozone molecules, thereby facilitating the formation of hydroxyl radicals (Al-Kdasi et al., 2004). Hung-Yee and Ching-Rong (1995) documented O_3/UV as the most effective method for decolourizing of dyes comparing with UV oxidation by UV or ozonation alone. While, Perkowski and Kos (2003) reported no significant difference between ozonation and O_3/UV in terms of colour removal. Even though ozone can be photodecomposed into hydroxyl radicals to improve the degradation of organics, UV light is highly absorbed by dyes and very limited amount of free radical (HO·) can be produced to decompose dyes. Thus same colour removal efficiencies using O_3 and O_3/UV could be expected. In normal cases, ozone itself will absorb UV light, competing with organic compounds for UV energy. However, O_3/UV treatment is recorded to be more effective compared to ozone alone, in terms of COD removal. Bes-Piá et al. (2003) documented that O_3/UV treatment of biologically treated textile wastewater reduced COD from 200-

400 mg/L to 50 mg/L in 30 minutes, while, using ozone alone COD reduced to 286 mg/L in same duration. Azbar et al. (2004) documented that using O_3/UV process high COD removal would be achieved under basic conditions (pH=9). Yonar et al. also repoted that using O_3/UV process showed high COD removal efficiency under similar conditions (pH=9) for physically and biologically treated textile effluent. The addition of H_2O_2 to the O_3/UV process accelerates the decomposition of ozone, which results in an increased rate of OH• generation (Teccommentary, 1996). In literature most AOPs applied for the treatment of textile effluent and, among the all apllied AOPs for dye house wastewater, acetate, polyester fibre dying process effluent and treatment plant outlet of textile industry with the combination of H_2O_2/O_3/UV appeared to be the most efficient in terms of decolouration (Perkowski and Kos, 2003, Azbar et al., 2004, Yonar et al, 2006). The rate of destruction of organic pollutants and the extent of mineralisation can be considerably increased by using an Fe(II,III)/H_2O_2 reagent irradiated with near-UV and/or visible light (Goi and Trapio, 2002, Torrades et al., 2003, Liou et al., 2004, Murugunandham and Swaminathan, 2004), in a reaction that is called the "photo-Fenton reaction". This process involves the hydroxyl radical (HO.) formation in the reaction mixture through photolysis of hydrogen peroxide (H_2O_2/UV) and fenton reaction (H_2O_2/Fe^{+2}.) (Fenton, 1894; Baxendale and Wilson, 1956). Using the photo-fenton process to treat dye manufacturing wastewater, which contains high strength of colour, and the results demonstrated its great capability for colour removal (Kang et al., 2000; Liao et al., 1999). Since the hydroxyl radical is the major oxidant of the photofenton process, the removal behaviour of COD and colouris highly related with the hydroxyl radical formation. However, the relation between the removal of COD and colourwith the hydroxyl radical formation in the decolourisation of textile wastewater by photo-fenton process was rarely found in the literature. The colour removal is markedly related with the amount of hydroxyl radical formed. The optimum pH for both the hydroxyl radical formation and colour removal occurs at pH 3±5. Up to 96per cent of colour can be removed within 30 min under the studied conditions. Due to the photoreduction of ferric ion into ferrous ion, colour resurgence was observed after 30 min. The ferrous dosage and UV power affect the colour removal in a positive way, however, the marginal benefit is less signifcant in the higher range of both (Kang et al., 2000) Liu et al., 2007 investigated the degradation and decolourisation of direct dye (Everdirect supra turquoise blue FBL), acidic dye (Isolan orange S-RL) and vat dye (Indanthrene red FBB) by Fenton and UV/ Fenton processes. A comparative study for Fenton and UV/Fenton reactions by photoreactor has been carried out by scale-up of the optimum conditions, obtained through jar-test experiments. Fenton process is highly efficient for colour removal for three dyes tested and for TOC removal of FBB and FBL. The optimum pH values obtained were all around 3 for FBL, FBB and S-RL.

UV/Fenton process improved slightly for FBB and FBL treatment efficiencies compared to Fenton reaction while S-RL showed much better improvement in TOC removal. The photolysis and photo-catalysis of ferrioxalate in the presence of hydrogen peroxide with UV irradiation (UV/ferrioxalate/H_2O_2 process) for treating the commercial azo dye, reactive Black B (RBB), is examined. An effort is made to decolourize textile effluents at near neutral pH for suitable discharge of waste water. pH value, light source, type of initial catalyst (Fe^{3+} or Fe^{2+}) and concentration of oxalic acid (Ox) strongly affected the RBB removal efficiency. The degradation rate of RBB increased as pH or the wavelength of light declined. The optimal molar ratio of oxalic acid to Fe(III) is three, and complete colour removal is achieved at pH 5 in 2 h of the reaction. Applying oxalate in such a photo process increases both the RBB removal efficiency and the COD removal from 68per cent and 21per cent to 99.8per cent and 71per cent, respectively (Huang et al., 2007) Neamtu et al. (2004) investigated the degradation of the Disperse Red 354 azo dye in water in laboratory-scale experiments, using four advanced oxidation processes (AOPs): ozonation, Fenton, UV/H_2O_2, and photo-Fenton. The photodegradation experiments were carried out in a stirred batch photoreactor equipped with an immersed low-pressure mercury lamp as UV source. Besides the conventional parameters, an acute toxicity test with a LUMIStox 300 instrument was conducted and the results were expressed as the percentage inhibition of the luminescence of the bacteria Vibrio fisheri. The results obtained showed that the decolourisation rate was quite different for each oxidation process. After 30 min reaction time the relative order established was: UV/H_2O_2/Fe(II) > Dark/H_2O_2/Fe(II) > UV/H_2O_2/O_3 > UV/H_2O_2/Lyocol. During the same reaction period the relative order for COD removal rate was slightly different: UV/H_2O_2/Fe(II) > Dark/H_2O_2/Fe(II) > UV/H_2O_2 > UV/H_2O_2/Lyocol > O_3. A colour removal of 85per cent and COD of more than 90per cent were already achieved after 10 min of reaction time for the photo-Fenton process. Therefore, the photo- Fenton process seems to be more appropriate as the pre-treatment method for decolourisation and detoxification of effluents from textile dyeing and finishing processes. Sulphate, nitrate, chloride, formate and oxalate were identified as main oxidation products. Liu et al., (2010), evaluated the photocatalytic degradation of Reactive Brilliant Blue KN-R under UV irradiation in aqueous suspension of titanium dioxide under a variety of conditions. The degradation was studied by monitoring the change in dye concentration using UV spectroscopic technique. The decolourisation of the organic molecule followed a pseudo-first-order kinetics according to the Langmuir–Hinshelwood model. Under the optimum operation conditions, approximately 97.7per cent colour removal was achieved with significant reduction in TOC (57.6per cent) and COD (72.2per cent) within 3 hours. In aother study, Bergamini et al., (2009), investihated photocatalytic (TiO2/UV) degradation of a simulated reactive dye bath (Black 5, Red 239,

Yellow17, and auxiliary chemicals). After 30 min of irradiation, it was achieved 97per cent and 40per cent of colour removal with photocatalysis and photolysis, respectively. No mineralisation occurred within 30 min. According to photocatalytic decolourisation studies, high rate of organic and colour removal can be achieved. The main advantage of these processes is the usage of solar ligh. In another words, there is no energy need for hydroxyl rdical production. But, removal and recycling of semiconductors (TiO_2, ZnO etc.) from aqueous media is very important for both cost minimisation and effluent quality. Finally, for true and good decision making on the treatment process, cost of all the compared processes should be calculated. In next step, cost evaluation of these processes are evaluated.

COST EVALUATION OF AOPS FOR COLOUR REMOVAL

Cost evaluation is an important issue for decision making on a treatment process as much as process efficiency. Actual project costs can not be generalized; rather they are site-specific and thus must be developed for individual circumstances (Qasim et al., 1992). For a fullscale system, these costs strongly depend on the flow rate of the effluent and the configuration of the reactor as well as the nature of the effluent (Azbar et al., 2004). From these reasons, complete cost analysis of an AOP including treatment plant flow chart is too limited in literature. Azbar et al. 2005, Solmaz et al, Ustun et al, Yonar et al and Yonar, 2010 tried to explain the operational costs of examined AOPs. Average costs of applied processes are given in Table.

Table. Average Operational Costs of AOPs

Process Type	Operational Cost (USD/m^3)
Coagulation	0,07-0,20
Ozonation	4,21-5,35
Fenton process	0,23-0,59
Fenton-like process	0,48-0,57
Peroxane	5,02-5,85
UV/H_2O_2	1,26-4,56
UV/O_3	6,38-8,68
$UV/O_3/H_2O_2$	6,54-11,25

In another study of Yonar 2010, treatment plant cost calculations were carried out according to Turkey conditions. The overall costs are represented by the sum of the capital costs, the operating costs and maintenance. For a full-scale system (200 m^3/day, hand-printed textile wastewater), these costs strongly depend on the flow-rate of the effluent and the configuration of the reactor as well as the nature of the effluent (Azbar et al., 2004). Conventional treatment system (physical/ chemical/ biological treatment processes) and Fenton process (physical/Fenton processes) costs are summarized in this section for a meaningful explanation.

Capital Costs

Table. Capital Cost Estimates of Conventional (Physical/Chemical/Biological) and Fenton Process (Physical/Fenton Process) Treatment Plants

Item	Capital Costs(Euro)	
	Conventional Treatment System	**Fenton Proces Treatment System**
Construction Costs		
• *Basin Constructions*[1]	27 500 €	19 000 €
• *Building Constructions*[1,2]	13 000 €	13 000 €
• *Rentals*	6 300 €	4 700 €
• *Electricity*[3]	850 €	700 €
Mechanical Costs		
• *Physical Unit Equipments*[4]	10 800 €	10 800 €
• *Chemical Unit Equipments*[4]	12 800 €	23 500 €
• *Biological Unit Equipments*[4]	12 600 €	–
• *Disinfection Unit Equipments*[4]	700 €	–
• *Sludge Unit Equipments*[4]	15 500 €	15 500 €
• *Piping Costs*[4]	9 500 €	4 000 €
• *Transportation and Rentals*	2 500 €	2 000 €
Electrical Costs		
• *Automation*[5]	7 500 €	7 000 €
• *Wiring*[5]	1 500 €	1 500 €
• *Sensors and Switches*	3 600 €	2 200 €
Sub-Total - A	124 650 €	103 900 €
Other Costs		
• *Engineering Fee (5per cent of sub-total a)*	6 233 €	5 195 €
• *Charges*	2 000 €	2 000 €
• *Profit and overhead (15per cent sub-total a)*	18 698 €	15 585 €
Sub-Total - B	151 581 €	126 680 €
Taxes (Vat: 18per cent of sub-total - b)	27 285 €	22 803 €
Total	178 866 €	149 483 €

Notes:

[1]All construction costs include labour costs

[2]Buildings are designed (pre-fabric 200 m^2 closed area) as same capacity for both treatment plants including a small laboratory, chemical preparation and dosage units, blowers (for biological treatment unit) and sludge conditioning and filter-press units

[3]1 kW = 0.087 Euro

[4]All mechanical costs include labour costs

[5]All electrical costs include labour costs

Capital costs of a treatment plant were calculated in four sub-stages:

1. Constructional,
2. Mechanical,
3. Electrical, and
4. Other costs.

Constructional costs of both treatment flow charts were computed by a Civil Engineering Office according to environmental design results. Constructional costs include excavation, reinforced concrete, buildings, excavator and crane rentals, electricity and labour costs. Land costs were excluded from the computations for the reason of industry own site usage. Mechanical and electrical costs are another chief and important capital costs for a treatment plant. Mechanical costs were determined by summing the costs of mechanical equipment purchase (coarse and fine screens, pumps, dosage pumps, mixers, chemical storage and handling tanks, blowers, diffusers, tank skimmers, filter-press etc.), pipes and fittings, material transportation and mechanical labour. Electrical costs contain automation, wiring, sensors (flow, pH, oxygen, ORP, level switches etc.) and electrical labour. Finally, other costs incorporate engineering design fee, charges and taxes, and profit and overhead. All equipment and materail prices and labour costs were collected from different treatment plant equipment suppliers and engineering offices in Turkey.

Table presents capital cost estimates for the conventional and Fenton process treatment plants designed on the basis of 200 m^3/day. The total capital cost estimates for conventional treatment plant and Fenton process treatment plant are 178 866 and 149 483 Euro, respectively. All equipment costs were provided including 2 years nonprorated warranty by all suppliers. But, sensors, switches and other spare parts were excluded from warranty. It can clearly be observed from the cost analysis that the specific costs for Fenton process treatment plant are about 16per cent lower than that of the conventional treatment plant alternative. On the other hand, constructional costs of the conventional treatment system are higher than Fenton process treatment alternative. But, mechanical and electrical capital cost trends can be regarded identical for both treatment alternatives. These cost differences originate from biological treatment unit, because activated sludge tank entails great construction area and more mechanical work effort.

Operation and maintenance costs

Operation and maintenance costs (OandM) include power requirement, chemicals, spare parts, wastewater discharge fees, plant maintenance and labour. Textile industry wastewater treatment plant sludges are accepted as a toxic and hazardous waste in Turkish Hazardous Wastes Control Regulations Anonimous, 2005). Therefore, toxic and hazardous waste disposal costs and charges strongly depend on disposal technology and locations of the treatment plant and hazardous waste disposal plants. For these reasons, only sludge disposal costs were excluded from OandM cost estimations.

Table. OandM Costs of the Studied Treatment Methods (Cost of Sludge Disposal was Excluded).

Item	Operating and Maintanence Costs (Euro/m^3)	
	Conventional TreatmentSystem	**Fenton Process Treatment System**
• Electrical power for processes and other facilities	0.24	0.17
• Spare part costs	0.04	0.03
• Chemicals	0.70	0.81
• Labour	0.34	0.34
• Sub-Total	1.32	1.35
• Equipment repair, replacement and overhead (10per cent of sub-total)	0.13	0.135
Total	1.452	1.485
Reagent Prices	***Unit***	***Price (Euro)***
• *Hydrogen peroxide*	Kg	0.55
• *Sulphuric Acid*	Kg	0.85
• *Sodium Hydroxide*	Kg	0.75
• *Ferric Chloride*	Kg	1.95
• *Ferrous Sulphate*	Kg	1.05
• *Sodium Hypochlorite*	Kg	0.20
• *Polymer*	Kg	2.35
• *Electricity*	kWh	0.087

On the other hand, labour costs are very important part of OandM costs. Labour costs are facility-specific, and depend on the size, location and plant design. Therefore, labour costs may vary substantially (Pianta et al., 2000). These treatment plants can be considered as small treatment plants because of 200 m^3/day flow capacity. Nine working hours per day and a salary of 18 Euro/day (equal to minimum wage) for 2 workers and 32 Euro/day for operator and/or engineer were presumed, and the labour costs were calculated using a fixed rate of 0,34 Euro/m^3. Similar to labour costs, electricity and chemical prices are also country-specific. The total costs of both conventional and Fenton process treatment plants were estimated as 1.452 Euro/m^3 and 1.485 Euro/m^3. According to these results, Fenton process treatment system OandM costs are slightly (3per cent) higher than conventional treatment system owing to relatively higher chemical usage of Fenton process treatment system. However, capital cost difference of both systems may afford operating cost difference for 15 years. The labour costs constitute about 23per cent of the overall OandM costs. On the other hand, electricity appears to be another important cost value for conventional system. Consequently, Fenton process has shown superior treatment and colour removal performances, and can be accepted as more economical choice for handprinted textile wastewater treatment.

AZO DYES AND THEIR METABOLITES: DOES THE DISCHARGE OF THE AZO DYE INTO WATER BODIES REPRESENT HUMAN AND ECOLOGICAL RISKS?

Colourants (dyes and pigments) are important industrial chemicals. According to the technological nomenclature, pigments are colourants which are insoluble in the medium to which they are added, whereas dyes are soluble in the medium. The world's first commercially successful synthetic dye, named mauveine, was discovered by accident in 1856 by William Henry Perkin. These synthetic compounds can be defined as coloured matters that colour fibres permanently, such that they will not lose this colour when exposed to sweat, light, water and many chemical substances including oxidizing agents and also to microbial attack (Rai et al., 2005; Saratele et al., 2011). By the end of the 19th century, over ten thousand synthetic dyes had been developed and used for manufacturing purposes (Robinson et al., 2001a; Saratele et al., 2011), and an estimate was made in 1977 that approximately 800,000 tons of all recognized dyestuffs had been produced throughout the world (Anliker, 1977; Combes and Haveland-Smith, 1982). The expansion of worldwide textile industry has led to an equivalent expansion in the use of such synthetic dyestuffs, resulting in a rise in environmental pollution due to the contamination of wastewater with these dyestuffs (Pandey et al., 2007; Saratele et al., 2011). The Ecological and Toxicological Association of the Dyestuffs Manufacturing Industry (ETAD) was inaugurated in 1974 with the goals of minimizing environmental damage, protecting users and consumers and cooperating with government and public concerns in relation to the toxicological impact of their products (Anliker, 1979; Robinson et al., 2001a). A survey carried out by ETAD showed that of a total of approximately 4,000 dyes that had been tested, more than 90per cent showed LD_{50} values above 2×10^3 mg/kg, the most toxic being in the group of basic and direct diazo dyes (Shore, 1996; Robinson et al., 2001a). Thus it appears that exposure to azo dyes does not cause acute toxicity, but with respect to systemic bioavailability, inhalation and contact with the skin by azo dyes is of concern, due to the possible generation of carcinogenic aromatic amines (Myslak and Bolt, 1988 and Bolt and Golka, 1993 as cited in Golka et al., 2004). Of the approximately 10^9 kg of dyestuffs estimated to be manufactured annually throughout the World, the two most widely used in the textile industry are the azo and anthraquinone groups (Kri•anec and Marechal, 2006; Forss, 2011). Thus, this chapter is a comprehensive review on the azo dyes and their effects on human and environmental health.

AZO DYES

Azo dyes are diazotized amines coupled to an amine or phenol, with one or more azo bonds (–N=N–). They are synthetic compounds and account for more than 50per cent of all the dyes produced annually, showing the largest spectrum of colours (Carliell et al., 1995; Bae and Freeman, 2007; Kusic et al., 2011). Nearly all the dyestuffs used by the textile industry are azo dyes, and they are also widely used in the printing, food, papermaking and cosmetic industries (Chung and Stevens, 1993; Chang et al., 2001a). An estimate was made in the 80's, that 280,000 t of textile dyes were annually discharged into industrial effluents worldwide (Jin et al., 2007; Saratale et al., 2011). Since the azo dyes represent about 70per cent by weight of the dyestuffs used (Zollinger, 1987), it follows that they are the most common group of synthetic colourants released into the environment (Chang et al., 2001b; Zhao and Hardin, 2007; Saratale et al., 2011). One only needs very small amounts of dyes in the water (less than 1 ppm for some dyes) to cause a highly visible change in colour (Banat et al., 1996), and coloured wastewater not only affects the aesthetic and transparency aspects of the water being received, but also involves possible environmental concerns about the toxic, carcinogenic and mutagenic effects of some azo dyes (Spadaro et al., 1992; Modi et al., 2010; Lu et al., 2010). It can also affect the aquatic ecosystem, decreasing the passage of light penetration and gas dissolution in lakes, rivers and other bodies of water (Saranaik and Kanekar, 1995; Banat et al., 1996; Modi et al., 2010). The more industrialized the society, the greater the use of azo dyes, and hence the greater the risk of their toxic effects affecting the society. It has already been noted that, as from the 70's, intestinal cancer has been more common in highly industrialized societies, and therefore there may be a connection between the increase in the number of cases of this disease and the use of azo dyes (Wolff and Oehme, 1974; Chung et al., 1978). Bae and Freeman (2007) already demonstrated the biological toxicity of the direct azo dyes used in the textile industry. The results indicated that C.I. Direct Blue 218 was very toxic to daphnids, with a 48-h LC_{50} between 1.0 and 10.0 mg/L. It must be remembered that toxicity to daphnids is sufficient to suggest potential damage to every receptor ecosystem, and emphasizes the need for the synthetic dye manufacturing industry to carry out toxicological studies (Bae and Freeman, 2007).

AZO DYES AND THEIR MUTAGENIC EFFECTS

The azo dyes show good fibre-fixation properties as compared other synthetic dyes, showing up to 85per cent fixation, but nevertheless this explains why so much dye is released into the environment, representing the other 10 to 15per cent of the amount used. Most of the synthetic dyestuffs found in this class are not degraded by the conventional treatments given to industrial effluents or to the raw water (Nam and Reganathan, 2000; Oliveira et al., 2007). Shaul et al. (1991) studied 18 azo dyes, and found that 11 passed practically unchanged through the activated sludge system, 4 were adsorbed by the activated sludge and only 3 were biodegraded, resulting in the release of these substances into bodies of water. Oliveira et al. (2007) showed that even after treatment, effluent from dyeing industries was mutagenic and contained various types of dye. Such data are of concern, especially when one considers that the effluent from the same industry was studied by Lima et al. (2007), who found an increase in the incidence of aberrant crypt in the colon of rats exposed to this sample, this being an early biomarker of carcinogenesis (Lima et al., 2007). Azo dyes can also be absorbed after skin exposure, and such dermal exposure to azo dyes can occur as an occupational hazard or from the use of cosmetic products. It was postulated in the 80s that the percutaneous absorption of azo dyes from facial makeup could even be a risk factor in reproductive failures and chromosomal aberrations in a population of television announcers (Ku´erová et al., 1987; Collier et al., 1993). Various azo dyes have been shown to produce positive toxic results for different parameters. Tsuboy et al. (2007) analyzed the mutagenic, cytotoxic and genotoxic effects of the azo dye CI Disperse Blue 291, and the results clearly showed that this azo dye caused dose-dependent effects, inducing the formation of micronuclei (MNs), DNA fragmentation and increasing the apoptotic index in human hepatoma cells ($HepG_2$). A variety of azo dyes have shown mutagenic responses in Salmonella and mammalian assay systems, and it is apparent that their potencies depend on the nature and position of the aromatic rings and the amino nitrogen atom. For instance, 2-methoxy-4-aminoazobenzene is an extremely weak mutagen, whereas under similar conditions, 3-methoxy-4- aminoazobenzene is a potent hepatocarcinogen in rats and a strong mutagen in *Escherichia coli* and *Salmonella typhimurium* (Hashimoto et al., 1977; Esancy et al., 1990; Garg et al., 2002, Umbuzeiro et al., 2005a). The azo dyes Disperse Red 1 and Disperse Orange 1 increase the frequency of MNs in human lymphocytes and in $HepG_2$ cells in a dosedependent manner. The azo dyes Disperse Red 1 and Disperse Red 13 showed mutagenic activity in the Salmonella/microsome assay with all the strains tested and in the absence of metabolic activation, except for Disperse Red 13, which was negative with respect to strain TA100. After adding the S9 mix, the mutagenicity of the two azo dyes decreased (or was eliminated), indicating that the P450-dependent metabolism probably generated more stable products, less likely to interact with DNA. It was also shown that the presence of a chlorine substituent in Disperse Red 13 decreased its mutagenicity by a factor of about 14 when compared with Disperse Red 1, which shows the same structure as Disperse Red 13, but without the chlorine substituent. The presence of this substituent did not cause cytotoxicity in $HepG_2$ cells, but toxicity to the water flea *Daphnia similis* increased in the presence of the chlorine substituent (Ferraz et al., 2010). Chung and Cerneglia (1992) published a review of several azo dyes that had already been evaluated by the Salmonella/microsome assay. All the azo dyes evaluated that contained the nitro group showed mutagenic activity. The dyes Acid Alizarin Yellow R and Acid Alizarin GG showed this effect in the absence of metabolic activation (Brown et al., 1978). The dyes C.I. Basic Red 18 and Orasol Navy 2RB, which also contained nitro groups, were shown to be mutagenic both in the presence and absence of metabolic activation (Venturini and Tamaro, 1979; Nestmann et al., 1981). This review also showed the results obtained in the Salmonella/microsomal test of azo dyes containing benzeneamines, and found that Chrysodin was mutagenic in the presence of a rat-liver preparation (Sole and Chipman, 1986; Chung and Cerneglia, 1992). Another study applied the micronucleus assay in mouse bone marrow to the azo dye Direct Red 2 (DR2) and the results identified DR2 as a potent clastogen and concluded that excessive exposure to this chemical or to its metabolites could be a risk to human health (Rajaguru et al., 1999). Al-Sabti (2000) studied the genotoxic effects of exposing the Prussian carp (*Carassius auratus gibelio*) to the textile dye Chlorotriazine Reactive Azo Red 120, and showed its mutagenic activity in inducing MNs in the erythrocytes. They also showed that the dye had clastogenic activity, a potent risk factor for the development of genetic, teratogenic or carcinogenic diseases in fish populations, which could have disastrous effects on the aquatic ecosystem since the fate of compounds found in effluents is to be discharged into water resources (Al- Sabti, 2000). In addition to the effects caused by exposure to contaminated water and food, workers who deal with these dyes can be exposed to them in their place of work, and suffer dermal absorption. Similarly, if dye-containing effluents enter the water supply, possibly by contamination of the ground water, the general population may be exposed to the dyes via the oral route. This latter point could be of great importance in places where the existent waste treatment systems are inefficient or where there is poor statutory regulation concerning industrial waste disposal (Rajaguru et al., 1999).

EFFECTS OF THE AZO DYES METABOLITES

Sisley and Porscher carried out the earliest studies on the metabolism of azo compounds in mammals in 1911, and found sulphanilic acid in the urine of dogs fed with Orange I, demonstrating for the first time that azo compounds could be metabolized by reductive cleavage of the azo group (Sisley and Porscher, 1911 as cited in Walker, 1970). The mutagenic, carcinogenic and toxic effects of the azo dyes can be a result of direct action by the compound itself, or the formation of free radicals and aryl amine derivatives generated during the reductive biotransformation of the azo bond (Chung et al., 1992; Collier et al., 1993; Rajaguru et al., 1999) or even caused by products obtained after oxidation via cytochrome P450 (Fujita and Peisach, 1978; Arlt et al., 2002; Umbuzeiro et al., 2005a). One of the criteria used to classify a dye as harmful to humans is its ability to cleave reductively, and consequently generate aromatic amines when in contact with sweat, saliva or gastric juices (Pielesz et al., 1999, 2002). Some such aromatic amines are carcinogenic and can accumulate in food chains, for example the biphenylamines such as benzidine and 4- biphenylamine, which are present in the environment and constitute a threat to human health and to the ecosystems in general (Choudhary, 1996; Chung et al., 2000). After an azo dye is orally ingested, it can be reduced to free aromatic amines by anaerobic intestinal microflora and possibly by mammalian azo reductase in the intestinal wall or the liver (Walker, 1970; Prival and Mitchel, 1982; Umbuzeiro et al., 2005a). Such biotransformations can occur in a wide variety of mammalian species, including both *Rhesus* monkeys and humans (Rinde and Troll, 1975; Watabe et al., 1980; Prival and Mitchel, 1982;).

Aromatic Amines

As early as the late nineteenth century, a doctor related the occurrence of urinary bladder cancer to the occupation of his patients, thus demonstrating concern about the exposure of humans to carcinogenic aromatic amines produced in the dye manufacturing industry, since his patients were employed in such an industry and were chronically exposed to large amounts of intermediate arylamines. Laboratory investigations subsequently showed that rats and mice exposed to specific azo dye arylamines or their derivatives developed cancer, mainly in the liver (Weisburger, 1997, 2002). Briefly, in 1895, Rehn showed concern about the urinary bladder cancers observed in three workers from an 'aniline dye' factory in Germany. This led to the subsequent testing in animals of various chemicals to which these workers were exposed, and, as a result, the carcinogenic activity of the azo dye, 2,3-dimethyl-4-aminoazobenzene for the livers of rats and mice was discovered (Yoshida, 1933 as cited in Dipple et al., 1985). An isomeric compound, N,N-dimethyl-4-aminoazobenzene was also found to be a liver carcinogen (Kinosita, 1936 as cited in Dipple et al., 1985). Only in 1954 was the cause of the bladder tumors observed in the workers in the dye industry established to be 2-naphthylamine. This aromatic amine induced bladder cancer in dogs, but not in rats (Hueper et al., 1938 as cited in Dipple et al., 1985). In addition, workers in textile dyeing, paper printing and leather finishing industries, exposed to benzidine based dyes such as Direct Black 38, showed a higher incidence of urinary bladder cancer (Meal et al., 1981; Cerniglia et al., 1986). Cerniglia et al. (1986) demonstrated that the initial reduction of benzidine-based azo dyes was the result of azoreductase activity by the intestinal flora, and the metabolites of Direct Black 38 were identified as benzidine, 4-aminobiphenyl, monoacetylbenzidine, and acetylaminobiphenyl (Manning et al., 1985; Cerniglia et al., 1986). Furthermore, these metabolites tested positive in the Salmonella/microsome mutagenicity assay in the presence of S9 (Cerniglia et al., 1986). In the opinion of Ekici et al. (2001), although general considerations concerning the kinetics of azo dye metabolism indicate that an accumulation of intermediate amines is not very likely, this possibility cannot be excluded under all conditions.

According to legislation passed in the European Community on 17th July 1994, the application of azo dyes in textiles is restricted to those colourants which cannot, under any circumstances, be converted to any of the following products: 4-Aminodiphenyl; 4-Amino-2',3-dimethylazobenzene (*o*aminoazo-toluene); 4-Aminophenylether (4,4'-oxydianiline); 4-Aminophenylthioether (4,4' thiodianiline); Benzidine; Bis-(4-aminophenyl)-methane (4,4' diaminodiphenylme- thane); 4- chloroaniline (*p*-chloroaniline); 4-Chloro-2-methylaniline (4-chloro-*o*-toluidine); 2,4-Diaminotoluene (2,4-toluylenediamine); 3,3'-Dichlorobenzidine dihydrochloride; 3,3'-Dimethoxybenzidine (*o*-dianisidine); 3,3'-Dimethylbenzidine (*o*-toluidine); 3,3'-Dimethyl- 4,4'-diamino-diphenyl methane; 2-Methoxy-5-methylaniline (*p*-kresidine); 4-Methoxy-1,3- phenylenediamine sulfate hydrate (2,4-diaminoanisole); 4,4'-Methylene-bis (2- chloroaniline); 2-Methyl-5-nitroaniline (2-amino-4-nitrotoluene); 2-Naphthylamine; *o*- Toluidine; 2,4,5-Trimethylaniline (Bundesgesetzblatt, 1994 and Directory of Environmental Standards, 1998 as cited in Ekici et al., 2001).

More recently, the scientific community has come to consider the possibility of manufactured azo dyes breaking down generating amines to be a health hazard.

The International Agency for Research on Cancer only includes benzidine-based dyes in Group 2A and eight other dyes in Group 2B. Nevertheless, the possibility of azo bond reduction leading to the production of aromatic amines has been demonstrated under a variety of conditions, including those encountered in the digestive tract of mammals (Chung and Cerneglia, 1992; Pinheiro et al., 2004). Therefore, the majority of the attention concerning possible hazards arising from the use of azo dyes is now being directed at their reduction products (Pinheiro et al., 2004).

Nitroanilines are aromatic amines that are commonly generated during the biodegradation of azo dyes under anaerobic conditions, formed by reductive cleavage of the azo bonds (–N=N–) by the action of microorganisms present in the wastewaters (Pinheiro et al., 2004; Van der Zee and Villaverde, 2005; Khalid et al., 2009).

Depending on the individual compounds, many aromatic amine metabolites are considered to be non-biodegradable or only very slowly degradable (Saupe, 1999), showing a wide range of toxic effects on aquatic life and higher organisms (Weisburger, 2002; Pinheiro et al., 2004; Khalid et al., 2009).

METABOLIC PATHWAYS INVOLVED IN THE REDUCTION AND OXIDATION OF AZO DYES

Following oral or skin exposure to azo dyes, humans can subsequently be exposed to biotransformation products obtained by the action of intestinal microorganisms or that of others present on the skin, or due to reactions in the liver (Esancy et al., 1990; Chadwick et al., 1992; Chung et al., 1992; Stahlmann et al., 2006). Therefore it is extremely important to study the metabolic pathways of azo dyes that can contaminate the environment, in order to understand the overall spectrum of the toxic effects.

The metabolic pathways the azo dyes actually follow depend on several factors, such as:

- The mode of administration;
- The degree of absorption from the gastro-intestinal tract after oral ingestion;
- The extent of biliary excretion, particularly after exposure to different routes other than the oral one;
- Genetic differences in the occurrence and activity of hepatic reducing-enzyme systems;
- Differences in the intestinal flora; and
- The relative activity and specificity of the hepatic and intestinal systems, particularly those responsible for reducing the azo link, and all these factors are interrelated (Walker, 1970).

Azo dyes behave as xenobiotics, and hence after absorption, they are distributed throughout the body, where they either exert some kind of action themselves or are subjected to metabolism. Biotransformation may produce less harmful compounds, but it may also form bioactive xenobiotics, ie, compounds showing greater toxicity (Kleinow et al., 1987; Livingstone, 1998). The main routes involved in the biotransformation of dyes are oxidation, reduction, hydrolysis and conjugation, which are catalyzed by enzymes (Zollinger, 1991; Hunger, 1994), but in humans, biological reductions and oxidations of azo dyes are responsible for the possible presence of toxic amines in the organism (Pielesz et al., 2002). Orange II can be reductively metabolized producing 1-amino-2-naphthol, a bladder carcinogen for rats (Bonser et al., 1963; Chung et al., 1992). This suggests that any toxicity induced by unchanged azo dye molecules should not be accepted as the only effect of these compounds, since the reductive cleavage products from these dyes can be mutagenic/ carcinogenic (Field et al., 1977; Chung et al., 1992).

Oxidative Metabolism

Highly lipid-soluble dyes such as azo dyes, with chemical structures containing amino groups, either alkylamino or acetylamino, but without sulfonated groups, are preferentially biotransformed by oxidative reactions

(Hunger, 1994). Oxidation processes are mainly catalyzed by a microsomal monooxygenase system represented by cytochrome P450 (Hunger, 1994), which belongs to a superfamily of heme proteins, present in all living organisms and involved in the metabolism of a wide variety of chemical compounds (Denisov et al. 2005; Mansuy, 2007). The general mechanism of metabolic oxidation involves an electron transport chain, which first transfers an electron to the P-450-Fe^{3+} complex, which, on reduction, receives an oxygen atom and in the final steps, leads to the formation of an oxidation product in the organism (Furhmann, 1994 as cited in Hunger, 1994). There are three different oxidation pathways of importance for azo dyes: I) CHydroxylation, ring hydroxylation in the case of azo dyes, probably via an epoxidation mechanism and subsequent rearrangement to a phenol. II) N-Hydroxylation at primary or secondary amino groups, or with acetyl amino groups in the liver. This reaction is followed by esterification with glucuronate or sulfate. The activated esters, which are water-soluble, can be excreted, or the ester group can split off with the formation of a nitrenium compound -NH^+, which can covalently bind to a nucleophilic group of the DNA. III) Demethylation, which is the stepwise oxidation of the methyl groups of dialkylamino compounds, and the N-hydroxy derivative so formed can be further demethylated or react to form a nitrenium compound (Hunger, 1994). Studies on the metabolism and carcinogenicity of N,N-dimethylaminoazobenzene (Butter Yellow), a classical hepatocarcinogen in rats, have shown that N-methylaminoazobenzenes are mainly metabolized by N-demethylation. In this way, Butter Yellow was first reversibly demethylated to the mono-N-methyl compound, which, in turn, was irreversibly demethylated to form p-aminoazobenzene. These changes were shown to precede reduction of the azo link, by isolating N-methyl-p-amino azobenzene and p-aminoazobenzene from the animal tissues (Miller et al., 1945; Walker, 1970). Radiotracer studies have shown (Miller et al., 1952) that demethylation occurs via the formation of a hydroxymethyl compound, followed by elimination of the methyl group in the form of formaldehyde (Mueller and Miller, 1953; Walker, 1970). Hydroxylation of the aromatic ring can occur before reductive fission of the azo group, and also on the amines produced by such a reduction, and this pathway appears to be very important in compounds which contain an unsulphonated phenyl moiety (Walker, 1970).

Reductive Metabolism

Trypan Blue has been shown to have carcinogenic and teratogenic properties (Field et al., 1977). Although original Trypan Blue is not mutagenic, it was reduced by the cell-free extract of an intestinal anaerobe, *Fusobacterium* sp.2, to a mutagenic product, O-toluidine (3,3'-dimethylbenzidine) (Hartman et al., 1978; Chung et al., 1992). In addition to Trypan Blue, Benzopurpurine 4B and Chlorazol Violet N were also shown to be Ames-positive frame-shift mutagens, but only in the presence of metabolizing systems capable of effecting azo reduction. The activity of these dyes may therefore be attributed to the benzidine metabolite, O-toluidine, which is generated because these amines are themselves indirect frame-shift agents (Hartman, et al., 1978; Matsushima et al., 1978). The benzidine produced after the reduction of some dyes can induce bladder cancer in humans and tumors in some experimental animals (Combes and Haveland-Smith, 1982; Chung, 1983). Some azo dyes, such as Brown FK, have been shown to be directly mutagenic in bacterial tests (Haveland-Smith and Combes, 1980a, b; Rafii et al., 1997). However many other azo dyes, such as Congo Red and Direct Black 38, only give a positive result for mutagenicity after chemical reduction or incubation with the contents of the human intestinal tract (Haveland- Smith and Combes, 1980 a,b; Reid et al., 1983; Cerniglia et al., 1986;;Chung and Cerniglia, 1992; Rafii et al., 1997). Reductive cleavage of the azo linkages is probably the most toxicologically important metabolic reaction of azo compounds. This reaction can be catalyzed by mammalian enzymes, especially in the liver (Walker, 1970; Kennelly et al., 1982) or by intestinal (Chung et al., 1978; Hartman et al., 1978) or skin bacteria such as *Staphylococcus aureus* (De France, 1986; Platzek et al., 1999; Golka et al., 2004). Azo compounds can reach the intestine directly after oral ingestion or via the bile after parenteral administration. They are reduced by azo reductases produced by intestinal bacteria, and to a lesser extent by enzymes from the cytosolic and microsomal fractions of the liver. The first catabolic step in the reduction of azo dyes is the cleavage of the azo bond, producing aromatic amines (Cerniglia et al., 1986), accompanied by a loss of colour of the dye, and bacterial azoreductases show much greater activity than hepatic azoreductases (Watabe et al., 1980; Collier et al., 1993; Raffi et al., 1997). This reduction process may produce compounds that are more or less toxic than the original molecule (Collier et al., 1993; Rafii et al., 1997), depending on the chemical structure of the metabolite generated. Although its occurrence in the liver has been regarded as the result of a detoxification reaction, azo reduction may be the first step in azo dye carcinogenesis (Chung et al., 1992). In addition, Nam and Reganathan (2000) demonstrated that both nicotinamide adenine dinucleotide (NADH) and nicotinamide adenine dinucleotide phosphate (NADPH) are capable of reducing azo dyes in the absence of any enzyme, under mildly acidic conditions. The reduced forms of NADH and NADPH are ubiquitous sources of electrons in biological systems, and these function as cofactors in many reductive enzyme reactions. This suggests that the introduction of methyl and methoxy substituents at the 2-,2,3-,2,6-, or 2,3,6-positions of the aromatic ring, accelerates the reduction of phenolic azo dyes by NADH, as compared to that of unsubstituted dyes (Nam and Reganathan, 2000). It is possible that both the mutagenicity and carcinogenicity of azo dyes are in fact frequently due to the generation of aromatic amines, with subsequent N- and ring hydroxylation and

N-acetylation of the aromatic amine (Chung and Cerniglia, 1992). If the azo dyes contain nitro groups, they can also be metabolized by the nitroreductases produced by microorganisms (Chadwick et al., 1992; Umbuzeiro et al., 2005a). Mammalian enzymes in the liver and in other organs can also catalyze the reductive cleavage of the azo bond and the nitroreduction of the nitro group. However, it has been shown that the intestinal microbial azoreductase and nitroreductase play a more important role in this type of metabolism. In both cases, the formation of N-hydroxylamines can cause DNA damage, and if the dyes are completely reduced to aromatic amines, they can then be oxidized to Nhydroxyderivates by P450 enzymes. In addition, N-hydroxy radicals can be acetylated by enzymes such as O-acetyltransferase, generating nitrenium electrophilic ions which are able to react with DNA forming adducts (Chung et al., 1992; Arlt et al. 2002; Umbuzeiro et al., 2005a). Research carried out by Zbaida (1989) showed that the hydroxylation of non-reactive azo dyes such as azobenzene, increased their binding to microsomal cytochrome P-450 and consequently their rate of reduction (Zbaida, 1989). Studies with mono-azo dyes have indicated that low electron densities close to the azo bond favour reduction (Walker and Ryan, 1971; Combes and Haveland-Smith, 1982), and this may occur due to hydrogen bonding of an azo N atom together with a proximal naphthol group, producing a keto-hydrazone configuration. It is possible that such a structure, which is present in many food colours, may generate dyes which are more resistant to hepatic and microbial reduction (Parke, 1968 cited as in Combes and Haveland-Smith, 1982). Kennelly et al. (1984) showed that Direct Black 38, Direct Brown 95, Direct Blue 6, Congo Red, Trypan Blue and Chicago Sky Blue were easily reduced by the intestinal microflora when orally administered to rats by gavage, however when administered via the hepatic portal vein, only Direct Black 38, Direct Brown 95 and Direct Blue 6 were reduced, all of which are potent liver carcinogens (Robens et al., 1980; Chung et al., 1992). Sweeney et al. (1994) tested azo dyes for genotoxicity following bacterial reduction of the dye. They found that both reduced amaranth and reduced sunset yellow induced cytotoxicity when incubated with a repair deficient E. coli strain in the absence of hepatic enzymes, indicating DNA damage. On the other hand they failed to mutate the *S. typhimurium* strains TA98 and TA1OO, but in contrast, strain TA102, which detects oxidative mutagens (De Flora et al., 1984), was mutated by reduced amaranth and reduced sunset yellow (Sweeney et al., 1994). Reduction can also modify the type of activity observed. The direct mutagenicities of Alizarin Yellow GG and Acid Alizarin Yellow R were eliminated by reduction, but in the presence of the exogenous metabolic system (S9), the resulting products were mutagenic and exhibited frame-shift activity (Brown et al., 1978; Combes and Haveland-Smith, 1982). The Sudan dyes I, II, III and IV are oil-soluble azo dyes (1-amino-2-naphthol-based azo dyes), widely used in colouring plastics, leather, fabrics, printing inks, waxes and floor polishes (An et al., 2007; Xu et al., 2010). Sudan I is a liver and urinary bladder carcinogen in mammals and is also considered as a possible human mutagen, since it can produce the benzenediazonium ion during metabolism catalyzed by cytochrome P450, which could be the mechanism by which Sudan I is activated leading to a carcinogenic final product (Stiborová et al., 2002, 2005; Xu et al., 2010).

An et al. (2007) found a dose-dependent increase in DNA migration in the comet assay, and in the frequency of micronuclei with all the concentrations of Sudan I tested (25–100 mM). These data suggest that Sudan I caused breaks in DNA strands and chromosomes. Sudan II causes mutations in Salmonella Typhimurium TA 1538 in the presence of a rat liver preparation (Garner and Nutman, 1977; Xu et al., 2010).

Concern about the safety of Sudan III, which is used in cosmetics, has arisen from its potential metabolic cleavage by skin bacteria producing 4-aminoazobenzene and aniline (Pielesz et al., 2002), and Sudan IV has been shown to require reduction and microsomal activation in order to be mutagenic (Brown et al., 1978).

These are important mechanisms, since, with the exception of Sudan II, Xu et al. (2010) showed that the bacteria found in the human colon are frequently able to reduce Sudan dyes (Xu et al., 2010). Almost all the azo dyes are reduced *in vivo*, but the reduction of the ingested dose is frequently incomplete, and thus a certain amount of the dye can be excreted in the unchanged or conjugated form.

For instance, in the case of orally dosing rats with Sudan III, none of the expected reduction products was excreted, although p-aminophenol could be detected in the urine after i.p. injection (IARC, 1975; Combes and Haveland-Smith, 1982). This may have been due to the formation of hydrazones at one of the azo bridges of this diazo dye, which might make it resistant to intestinal reduction (Combes and Haveland-Smith, 1982). Stahlmann et al. (2006) reported investigations made to evaluate the sensitizing and allergenic potentials of two metabolites expected to be formed by the metabolic activity of skin bacteria and/or by metabolism in the skin. Two metabolites (4-aminoacetanilide and 2-amino-*p*-cresol) of Disperse Yellow 3, an azo dye widely used in the textile industry, were tested using modified local lymph node assay protocols in NMRI mice.

The metabolite 2- amino-*p*-cresol gave a clearly positive response in the sensitisation protocol, showing marked increases in lymph node weight and cell proliferation, accompanied by a relative decrease in T-cells and relative increases in B-cells and 1A+ cells. Hence, 2-amino-*p*-cresol can be considered to be a stronger allergen in this model. In contrast, 4-aminoacetanilide only led to an increase in lymph node weight and cellularity at the higher concentration of 30per cent, with no consistent changes in the phenotypic analysis, indicating that this metabolite alone was a weak sensitizer (Stahlmann et al., 2006).

DYING PROCESSING PLANTS EFFLUENTS AND THEIR TREATMENTS

The textile industry accounts for two-thirds of the total dyestuff market. Part of dye used in the textile dyeing process does not attach to the fibres, remaining in the dye baths and eventually being discharged in the wastewater (Fang et al., 2004). The resulting wastewater is usually treated with activated sludge, and the liquid effluent is released to adjacent surface waters (Umbuzeiro et al., 2005 b). Many dyes do not degrade easily due to their complex structure and textile dye effluent does not decolourize even if the effluent is treated by the municipal wastewater treatment systems (Shaul et al., 1991; Robinson et al., 2002; Forgacs et al., 2004). A study carried out in 1989 showed that the commercial aminoazobenzene dye, C.I. Disperse Blue 79, was not degraded by a conventionally operated activated sludge process and that 85per cent of the dye remained in the system. Of this 85per cent, 3per cent was retained by the primary sludge, 62per cent by the activated sludge and 20per cent was found in the final liquid effluent released into the environment (US EPA, 1989; Umbuzeiro et al., 2005 b). The use of an anaerobic system before the activated sludge treatment can result in cleavage of the azo bonds and the release of the corresponding aromatic amines. However, the colourless aromatic amines produced by these anaerobic microorganisms can be highly toxic and carcinogenic (Hu, 1994; Banat et al., 1996; Robinson et al., 2002). Ekici et al. (2001) tested the stability of selected azo dye metabolites in both activated sludge and water and concluded that they were relatively stable in the aquatic environment and could not be efficiently degraded in wastewater plant systems. With respect to their mutagenicity, Fracasso et al. (1992) showed that dye factory effluents from primary and secondary biological treatments increased their levels of mutagenic activity as compared to the raw (untreated) effluent. The use of activated carbon filtration was beneficial but did not completely remove the mutagenic activity of the final effluent (Fracasso et al., 1992; Umbuzeiro et al., 2005 b). Azo dyes are usually designed to resist biodegradation under aerobic conditions, the recalcitrance of these compounds being attributed to the presence of sulfonate groups and azo bonds. On the other hand, the vulnerability of reducing the azo bonds by different mechanisms (*e.g.* biotreatment in anaerobic conditions) could result in the generation of aromatic amines, which are somewhat toxic and carcinogenic (Öztürk and Abdullah, 2006; Bae and Freeman, 2007; Kusic et al., 2011). It should also be mentioned that azo dyes are associated with various health risks to humans, and therefore coloured wastewaters should be efficiently treated prior to discharge into the natural water bodies (Kusic et al., 2011). Several methods are used to decolourize textile effluents including physicochemical methods such as filtration and coagulation, activated carbon and chemical flocculation (Gogate and Pandit, 2004). These methods involve the formation of a concentrated sludge, which, in turn creates a secondary disposal problem (Maier et al., 2004; Elisangela et al., 2009), since these methods merely transfer the pollution from one phase to another, which still requires secondary treatment (Gogate and Pandit, 2004; Kusic et al., 2011). Recently, new biological processes have been developed for dye degradation and wastewater reuse, including the use of aerobic and anaerobic bacteria and fungi (Elisangela et al., 2009). The decolourizing of azo dyes using a fungal peroxidase system is another promising method (Hu, 1994). The ligninolytics are the most widely researched fungi for dye degradation (Elisangela et al., 2009) and of these, the white-rot fungi have been shown to be the most efficient organisms for the degrading of various types of dye such as azo, heterocyclic, reactive and polymeric dyes (Novotný et al., 2004). These fungi produce lignin peroxidase, manganese-peroxidase and laccase, which degrade many aromatic compounds due to their nonspecific systems (Forgacs et al., 2004; Revankar and Lele, 2007; Madhavi et al., 2007). However, all these processes for the mineralization of azo dyes need to be carried out in a separate process, since the dye compounds cannot be incorporated into the medium, and this would be impractical due to the great volume of wastewater requiring treatment (Hu, 1994). In addition, the long growth cycle and complexity of the textile effluents, which are extremely variable in their compositions, limit the performance of these fungi. Although the stable operation of continuous fungal bioreactors for the treatment of synthetic dye solutions has been achieved, the application of white-rot fungi for the removal of dyes from textile wastewaters still confronts many problems due to the large volumes produced, the nature of the synthetic dyes and the biomass control (Nigam et al., 2000; Mielgo et al., 2001; Robinson et al., 2001b; Elisangela et al., 2009). Of the chemical methods under development, advanced oxidation processes (AOPs) seem to be a promising option for the treatment of toxic and non-biodegradable organic compounds in various types of wastewater, including the coloured ones (Forgacs et al., 2004; Gogate and Pandit, 2004; Kusic et al., 2011). AOPs have received considerable attention due to their potential to completely oxidize the majority of the organic compounds present in the water. AOPs could serve as oxidative pretreatment method to convert non or low-biodegradable organic pollutants into readily biodegradable contaminants (Mantzavinos and Psillakis, 2004; Kusic et al., 2011). The electron beam (EB) treatment is also included in the class of AOPs, and laboratory investigations, pilot-plant experiments and industrially established technology have shown the efficiency of the EB treatment in destroying textile dyes in aqueous solutions (Han et al.,2002; Pálfi et al., 2011). Chlorine has been extensively used as a complementary treatment to remove or reduce the colour of industrial effluents containing dyes, and also to disinfect the water in drinking water treatment plants (Sarasa et al., 1998; Oliveira et al., 2010). The discolouration process using sodium hypochlorite (NaOCl) or chlorine gas, is based on the electrophilic attack of the amino group, and subsequent cleavage of the chromophore group (responsible for the dye colour) (Slokar and Marechal, 1998). However, the treatment of textile effluents using the conventional activated sludge method followed by a chlorination step, is not usually an effective method to remove azo dyes, and can generate products which are more mutagenic than the original untreated dyes, such as PBTAs (chlorinated 2-phenylbenzotriazoles). It has been reported that conventional chlorination should be used with caution in the treatment of aqueous samples contaminated with azo dyes (Umbuzeiro et al., 2005b; Oliveira et al., 2010). Another alternative could be the use of photoelectrocatalysis on titanium supported nanocrystalline titanium dioxide thin film electrodes, where active chlorine is produced promoting the rapid degradation of reactive dyes (Carneiro et al., 2004; Osugi et al., 2009). Osugi et al. (2009) investigated the decolourizing of the mutagenic azo dyes Disperse Red 1, Disperse Red 13 and Disperse Orange 1 by chemical chlorination and photoelectrochemical oxidation on Ti/TiO_2 thin-film electrodes using NaCl and Na_2SO_4 media. After 1 h of treatment, 100per cent decolourizing was achieved with all the methods tested. After 1 h of photoelectrocatalytic oxidation, all the dye solutions showed complete reduction of the mutagenic activity using the strains TA98 of *Salmonella* in the absence or presence of the S9 mix, suggesting that this process could be a good option for the removal of disperse azo dyes from aqueous media. The results involving conventional chlorination showed that this method did not remove the mutagenic response from the dyes, and in fact promoted an increase in mutagenic activity in the presence of metabolic activity for Disperse Red 13 (Osugi et al., 2009).

CONCLUSIONS

The discharge of azo dyes into water bodies presents human and ecological risks, since both the original dyes and their biotransformation products can show toxic effects, mainly causing DNA damage. Azo dyes are widely used by different industries, and part of the dyes used for colouring purposes is discharged into the environment. The azo dyes constitute an important class of environmental mutagens, and hence the development of nongenotoxic dyes and investment in research to find effective treatments for effluents and drinking water is required, in order to avoid environmental and human exposure to these compounds and prevent the deleterious effects they can have on humans and aquatic organisms.

5

Department of Textile Technologies

IRONING

Ironing is the work of using a heated tool, or tools, (an iron) to remove wrinkles from fabric. The heating is commonly done to a temperature of 100°Celsius. Ironing works by loosening the bonds between the long-chain polymer molecules in the fibres of the material. While the molecules are hot, the fibres are straightened by the weight of the iron, and they hold their new shape as they cool. Some fabrics, such as cotton, require the addition of water to loosen the intermolecular bonds. Many modern fabrics (developed in or after the mid-twentieth century) are advertised as needing little or no ironing. Permanent press clothing was developed to reduce the ironing necessary by combining wrinkle-resistant polyester with cotton.

Ironing may also be used as a germ/parasite killing hygienic operation, such as in areas where the Tumbu fly is prevalent. While nobody knows precisely when the practice began, people have been smoothing clothes with a variety of items, including glass, wood, and stones, since before recorded history. The first known use of metal to "iron" clothes, however, is known to have occurred in China. The electric iron was invented in 1882, by Howard Seeley. Seeley patented his "electric flatiron" on June 6, 1882.

PURPOSE

Some people consider ironed clothes to be more aesthetically appealing than wrinkled clothes. After stitching or sewing new clothes, especially pants and suits, tailors iron them to give the clothes the appropriate shape. Fabrics such as linen are considered to be more comfortable following ironing.

Ironing will also kill vermin such as body lice, scabies mites and Tumbu fly, although the same thing could be accomplished more efficiently by machine drying.

RECOMMENDED IRONING TEMPERATURES

Lower temperature than above may be advised in some cases where the colour might be sensitive.

Textile	Temperature	Dot mark
Toile	240 °C	
Triacetate	200 °C	
Cotton	204 °C	* * *
Linen		* * *
Viscose/Rayon	190 °C	* *
Wool	148 °C	* *
Polyester	148 °C	*
Silk	148 °C	*
Sympatex		*
Acetate	143 °C	*
Acrylic	135 °C	
Lycra/Spandex	135 °C	
Nylon	135 °C	
	< 110 °C	*
	< 150 °C	* *
	< 200 °C	* * *

IRONING BOARDS

Ironing boards don't have to be hidden away in a closet or cloistered in the laundry room.

As mundane as it may seem, there's a lot to learn about selecting a quality ironing board for great sewing and pressing results.

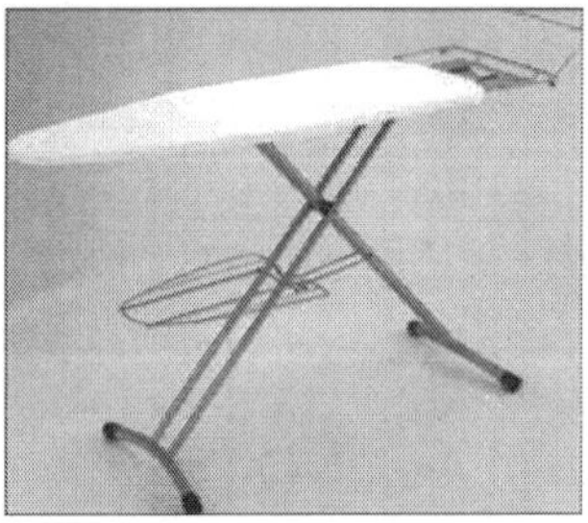

BOARD BASICS

Ironing boards can be freestanding or mounted, depending on the available space and frequency of use. Look for over the door models that fold down for use, wall-mounted fold-ups, hide-away boards protected in their own slim cabinets, tabletop versions, portable take-to-class types and the traditional two-legged versions.

TAKING SHAPE

Traditional ironing boards are long and skinny, about 15" wide and 4 1/2' long. They offer a narrower end for pressing hard-to reach areas. Sizes vary depending on the brand and some companies offer much larger and wider boards for pressing quilt tops and home décor projects. Others offer board toppers to retro-fit a traditional size ironing board to increase the surface area. Ironing board heights should be adjustable to avoid back strain. Most offer multiple heights than snap into position for stability.

TERRIFIC TOPS

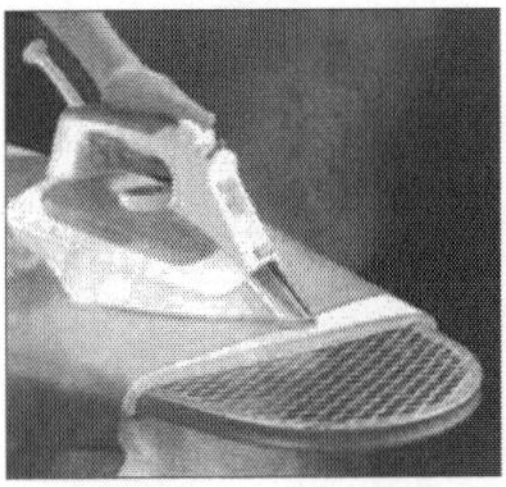

A good ironing board should have a smooth, flat metal surface with vent or grid holes to allow steam to escape and not be trapped in the project being pressed. Padding should be provided to cover the metal and it should be made from a natural fibre fabric like cotton or wool for best pressing. Foam pads are available for most ironing boards, but they do not allow for moisture absorption. A cotton or wool cover is advisable for sewers, though many boards come with a Teflon-coated drawstring cover which reflects the heat. Because of this, many sewers prefer to tailor their own covers to fit their sewing room ironing surfaces. Ready-made covers come in prints, solids and some are printed with grids for sewers and quilters to use in squaring pieces while pressing.

ADD-ONS

Good quality ironing boards may sport a rack underneath the board surface for storing laundry, press cloths or pressing equipment, etc. Others offer an attached iron rest at the end of the board, so the iron can be taken off the actual pressing surface and stored when not in use.

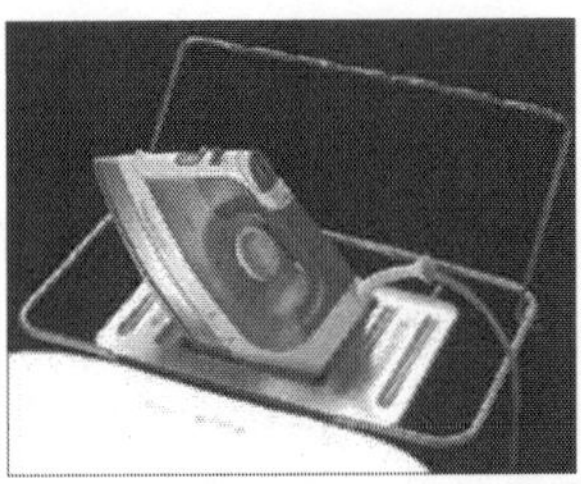

A garment rack may be incorporated at the end of the ironing surface for hanging freshly pressed items before they make it to the closet or to a sewing client. Iron cords can get in the way and some ironing boards have a cord holder which elevates the cable to keep it off the pressing surface. These usually screw on to the underside of the board, so can be removed when pressing extra-large projects. Non-skid feet are a handy feature if the ironing board lives on a hard-surface floor.

VACUUMTABLES

When is an ironing board also a vacuum? It's not a silly joke, but a reality that sewers love. Vacuum ironing tables use a fan to pull steam down through the surface (and the project being pressed) for quick cooling and skilled shaping on tailored projects. The suction also helps to hold slippery fabrics in place.

Most vacuum tables also have a reversible motor so that air can be pumped upward through the project. This is a helpful feature for pressing delicate areas like puffed sleeves, fabric that can be easily damaged like velvet or embroidery, and also to raise nap on fabrics like velvet and corduroy so that the iron doesn't leave lasting impressions on the pile surface.

Coupled with steam generator ironing systems, these boards are the ultimate in pressing perfection and maximum steaming for tailoring and other sewing tasks requiring moisture and quick cooling. Most vacuum boards have wheels for easy mobility and they fold up for storage.

PORTABLE PADS

If pressing on the go is something you do often, check out the myriad portable pressing pads that fold up for travel. Whether you're taking them to class or on a family vacation, these handy surfaces substitute for a full-blown ironing board. Some come with gridded surfaces for being sure your bias is straight or other tasks where measuring is key, and still others flip to a rotary cutting mat on the reverse side.

4 REASONS TO LOVE IRONING

1. WELL PRESSED =WELL DRESSED

A well-ironed shirt means you care about the details. Details make a difference. The details get you noticed. With the best tools available, you take pride in doing things yourself – like preparing a memorable meal or creating the perfect garden. Ironing your own clothes shows that same pride. Remember, *how* you wear your clothes is just as important as what you wear.

2. CLOTHES HAVE FEELINGS TOO

Well, maybe not feelings, but they'll definitely let you know when they've been mistreated. Sending your cherished clothes to most dry-cleaners often gives you average results: shine, shrunken collars and broken buttons. Doing it yourself at home with the new style, pressurized steam irons and specialty ironing tables will give you professional quality results. Whether it's your favourite gingham shirt or that custom silk blouse from a trip abroad, your clothes will look better and last longer, without the added cost.

3. THREE SAVES

Laundering and ironing clothes at home won't save the planet but it will reduce the use of energy and exposure to harmful chemicals like Perclorethylene – the traditional agent used in dry cleaning, now banned in many areas. Most of your clothes can be laundered at home with excellent results using an energy-efficient front-load washing machine, then finished with a pressurized steam iron and vacuum/blowing ironing table. For garments that must be 'dry' cleaned, we recommend one of the new 'wet' cleaning processes that use 100per cent water instead of chemicals like "Perc." It costs a bit more, but aren't your family and the environment worth it? Save money, Save energy, And help save the planet.

4. ZEN AND THE ART OF IRONING

The kids are asleep. Your home phone, cell phone and hand-held devices are turned off: Peace and quiet. You may not want to talk to anyone else today, but you do want to get something done in this perfect state of calm. Enter ironing. Many people swear by its meditative qualities: the repetitive movement of the iron; the hiss of steam; the disappearance of wrinkles – one by one. If that's too quiet, listen to some great music while you treat your clothes to a steam – generated by a premium Reliable™ iron – or turn on the TV for company. Before you know it, a pile of wrinkled laundry turns into a crisp, clean wardrobe for the week ahead. And you... are as refreshed as your clothes.

STEAM IRON

A steam iron uses superheated water to eliminate wrinkles in clothes and fabrics which may not be suitable for traditional dry ironing. Distilled water is usually poured into a holding tank and special heating elements convert it to steam. This hot mist comes out through a number of holes in the soleplate (bottom heating element) of the steam iron. As the steam loosens the individual fibres of the clothes, the steam iron's pressing action smooths out wrinkles.

Many traditional dry irons contain a steam iron option. Water is poured into a small reservoir and superheated until it becomes usable steam. A mechanical switch on the handle of the iron allows the user to select the steam setting. The steam itself should come out of small holes located near the tip of the pressing plate. For most conventional ironing needs, this combination of dry and steam iron settings should suffice. The steam generated by a traditional dry iron may not be overwhelming, but it will loosen wrinkled fibres.

Some ironing jobs such as curtains or quilts require a much more generous supply of steam. This is where a true steam iron can be useful. Distilled water is placed in a larger reservoir in a steam iron. Users can push a button to receive a burst of steam when needed. More holes in the soleplate means a more generous supply of steam while ironing heavier materials like suits or curtains.

Advanced steam iron systems may also feature handheld wands which can direct clouds of steam in and around vertically-hung garments or curtains. The steam itself should eliminate wrinkles without the need for a pressing element. Consumer rating boards suggest that the ability to generate 'vertical steam' is a major plus for a steam iron system. Some dry iron/steam iron combinations can only generate steam when the iron is held horizontally over an ironing board. Good vertical steaming means the steam iron can be used to iron curtains and quilts in place.

Another important feature to look for in a steam iron is a means to control the level of steam. Some models only allow for an on/off situation—either the user has steam or not. Better steam iron systems have a dialing mechanism which can adjust the intensity of the steam produced. Consistency of steam is also a consideration. Once the water has become superheated, the steam should be steady and have some power behind it. Effective steaming means penetrating individual fibres, not merely wetting the surface.

IRONING-MACHINE

The ironing-machine, or, as it is often called, the mangle, is another device for making laundry work easier. It may successfully take the place of the hand iron for a larger part of the family ironing. There are two types of ironing-machines on the market:

1. Cold-roll ironing-machines, in which the rollers between which the garment passes are made of wood and are unheated, depending on their weight and pressure to remove wrinkles;
2. Hot-roll ironing-machines, in which one roll is cold and is covered with a blanket and cloth, just as for an ironing-board, and the other roll or concave plate is made of smooth iron and is heated.

The cold roll revolves against the heated metal plate. This is the more economical and satisfactory ironing-machine, although its original cost is greater. The hot-roll machine may be heated by gas, or gasoline by a slight change involving a small expense. If power is available, the machine may be run by power. The use of one of these machines reduces greatly the time required to iron in the usual way. Garments with gathers and sleeves cannot be thus ironed to look perfectly smooth and well shaped, but all bed and table linen, towels, handkerchiefs, stockings, such underwear as may not require perfect smoothness, kitchen aprons and the like, may be done successfully and satisfactorily.

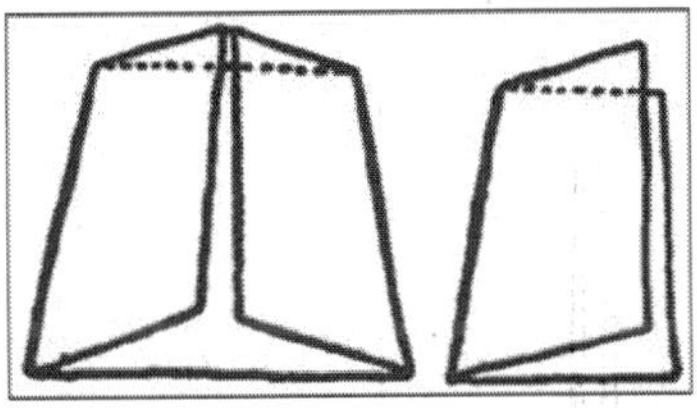

Fig. Another Method of Folding Shee.

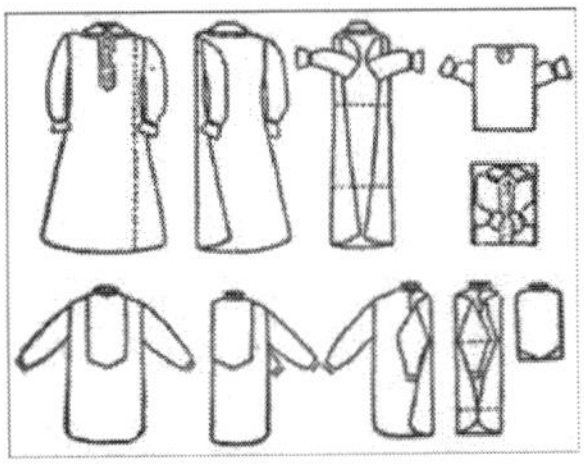

Fig. A Method of Folding Nightdresses and Shirts.

ALTERNATIVE CLOTHES CLEANING

WET CLEANING

Wet cleaning is both an old method and a new method for cleaning garments. Water has been used to clean clothes for centuries. Inventors, scientists and engineers have also been working for centuries to discover new methods to clean clothes quickly and efficiently. By the 1930's, the dry cleaning method became very popular as an affordable and efficient method for cleaning a wide variety of garments.

One concern with dry cleaning, however, is that it uses chemicals that can be hazardous to garment care workers and to the environment. Because of these hazards, scientists and engineers began looking for a cleaning method that does not use hazardous chemicals. They came up with a number of new methods which are addressed in Module 5. One new method is machine wet cleaning, which uses water and detergents to clean clothes.

It is not the use of water that makes wet cleaning a new method for cleaning clothes. New computer-controlled wet cleaning machines, new dryers, new detergents and new spot removers are what make wet cleaning a new method. The new washing machines and dryers have controls that allow them to safely and efficiently clean a wide variety of garments in water. New detergents and spot removers are made of ingredients that are safer for workers and the environment, yet are as safe and effective at removing soils, stains and odours as dry cleaning solvents. Equipment, detergents and skill all contribute to successful wet cleaning In this course, participants will learn about cleaning garments using the the wet cleaning process. It is hoped that after taking this course, participants will be motivated to add wet cleaning methods to their daily operations.

WHY DO WE NEED A NEW WAY TO CLEAN?

The professional garment care industry is currently experiencing one of the greatest periods of change since it began. Two forces are driving this change: technology and regulations.

The garment care industry is working on new equipment, new detergents and new cleaning methods. Some of this new technology is available now. Some of it is still being developed and tested. Environmental regulations are forcing many industries to change their processes. These regulations have also affected dry cleaners.

Dry cleaning using organic solvents such as camphene, benzene, gasoline and Stoddard solvent began in the mid-1800's. Since then, the garment cleaning industry has continually changed to keep up with consumer demand, new developments, and new regulations.

The 1930's and 1940's were profitable years for the industry. It wasn't until the 1950's that the industry hit its first growth challenge: the automatic home washer and dryer. These new machines allowed consumers to clean more of their clothes at home.

The 1960's dealt the biggest blow to the industry with the introduction of wash and wear clothing. In the 1970's the popularity of easily cleaned polyester fabrics meant that even fewer clothes were being taken to the cleaners. The coin-operated and discount dry cleaner arrived to challenge neighbourhood businesses in the 1980's.

In the 1990's, casual office wear has become more acceptable. This has caused the market for dry cleaning services to shrink again. Although casual office wear reduces the market for traditionally dry cleaned garments, it also provides a new marketing opportunity for wet cleaners. The dry cleaning industry is also being affected by environmental regulations that have been introduced since the 1970's. These regulations are driving the industry to develop cleaning processes that do not use hazardous chemicals.

Several new cleaning processes are currently being researched. The goal of this research is to create new processes that do not harm the environment, but are also efficient, effective and profitable. These new processes include cleaning with liquid carbon dioxide, ultrasonics, ozone, and wet cleaning.

At the time this manual was written, wet cleaning was the only practical, environmentally sound and potentially profitable new method. Therefore, the focus of this manual is on wet cleaning. In this country, 80per cent of professional cleaners use *perchloroethylene* (or "perc") as their dry cleaning solvent and the remaining 20per cent use petroleum. Because most cleaners use perchloroethylene, this manual compares wet cleaning with perchloroethylene dry cleaning.

A BRIEF HISTORY OF WET CLEANING

In May 1992, an International Roundtable on Pollution Prevention and Control in the Dry Cleaning Industry was assembled by the United States Environmental Protection Agency (EPA) and its Design for the Environment Programme (DfE). The roundtable was created to consider the health and environmental concerns associated with perchloroethylene dry cleaning.

The roundtable participants included professionals from the dry cleaning industry and allied trades, researchers, environmentalists and government officials. These participants formed a partnership that evaluated many methods for reducing the risks to garment care workers and the environment that are associated with the use of perchloroethylene. In November and December of 1992, the EPA's Design for Environment (DfE) staff conducted a test study of *multiprocess wet cleaning* at the New York School of Dry Cleaning in New York City. In multiprocess wet cleaning, garments are inspected and pre-treated if necessary. The garments are then spot cleaned, steam cleaned or hand washed. In this test, the cost and effectiveness of wet cleaning was compared to perchloroethylene dry cleaning.

The results of the test are described in the EPA's report "Multiprocess Wet Cleaning— Cost and Performance Comparison of Conventional Dry Cleaning and an Alternative Process," EPA 744-R-93-004, September 1993.

The results stated that under certain situations, multiprocess wet cleaning can be technically feasible and economically competitive with perchloroethylene dry cleaning.

One possible problem this study identified was that multiprocess wet cleaning may require more manual labour than dry cleaning. Additionally, the long term effects on garments were not evaluated. For this reason, dry cleaners may not want to use the multiprocess wet cleaning method. Since the test was done, however, new wet cleaning machines have become readily available.

The wet cleaning machines reduce the amount of cleaning that needs to be done by hand. This makes *machine wet cleaning* more efficient than multiprocess wet cleaning, and more attractive to dry cleaners. A second problem related to wet cleaning has been Federal labeling requirements. Current United States Federal Trade Commission (FTC) care labeling rules discourage dry cleaners from wet cleaning garments labeled "dry clean only."

The FTC published two requests for public comment on their labeling rules, one in June, 1994, and another in December, 1995.

The FTC wanted to know if it should change these rules in order to (among other things) help the EPA's goal of pollution prevention in the dry cleaning industry. These rules are currently under review.

The DfE roundtable was also given the job of developing pollution prevention awareness and practices in the dry cleaning industry. One of the products of this group is the Cleaner Technology Substitutes Assessment (CTSA). This study is designed to evaluate garment cleaning methods that could be used instead of perchloroethylene dry cleaning. The study looks at these alternative cleaning methods from environmental, health, and economic points of view. Another role of the study is to communicate information about alternative cleaning methods.

In order to gather information for the study, a demonstration project has been established with partial funding from the EPA. The goals of the demonstration are to examine some of the issues and concerns surrounding garment wet cleaning.

This project has resulted in a partnership between a private entrepreneur and the Centre for Neighbourhood Technology (CNT) in Chicago. CNT is a Chicago- based non-profit organization dedicated to environmentally sound, local economic development.

The demonstration involves a garment care shop, "The Greener Cleaner," which opened to the public in May 1995.

"The Greener Cleaner" uses machine wet cleaning. A garment wet cleaning business depends on customer acceptance in order to be successful. CNT evaluated "The Greener Cleaner" by using customer survey cards and by having volunteers evaluate how well garments were cleaned. Complete results of the CNT study can be obtained by contacting CNT directly.

Wet cleaning technologies have proven to be viable in Europe. In the past few years, these wet cleaning technologies have begun to spread in the United States. Many commercial cleaners use wet cleaning as a critical supplement to their cleaning process, and there are a growing number of wet clean-only facilities across the U.S. The USEPA's Design for Environment programme has a brochure on wet cleaning which list wet cleaners nationwide.

WHAT DOES THAT MEAN

Multiprocess wet cleaning: A wet cleaning process in which garments are inspected and pre-treated if necessary. The garments are then spot cleaned, steam cleaned or hand washed.

Machine wet cleaning: A process for cleaning textiles in water by professionals using special equipment and detergents. Machines carefully control mechanical action, temperature and humidity. It is normally followed by restorative finishing procedures.

Organic solvent: A solvent that contains carbon in addition to other substances.

Perchloroethylene: An organic solvent frequently used in dry cleaning and as a spot cleaner. Other names for perchloroethylene are: tetrachloroethylene, PCE, and perc.

Pollution prevention: Solving pollution problems by preventing the pollution from occurring in the first place. This is frequently done by eliminating or reducing the amount of materials, hazardous or otherwise, that are used in a process, so that less pollution is created. Elimination is the preferred method of pollution prevention.

Solvent: A liquid that is able to dissolve other materials. Water is the most common solvent. Most solvents are only able to dissolve certain types of materials.

HEALTH, SAFETY AND ENVIRONMENTAL REGULATIONS

HAZARDS

Most of the regulations that are currently in effect are designed to limit the amount of perchloroethylene that comes in contact with workers and the environment.

Perchloroethylene, also known as *perc*, can be a hazard to workers and to the environment if it is not used and disposed of properly.

Acute and Chronic Exposure

There are two types of exposure to perc: acute and chronic. Acute exposure to perc occurs when people are exposed to perc for a short amount of time (under 24 hrs). Acute exposure to perc may cause drowsiness, nausea, and irritation of the eyes, skin and respiratory tract.

Chronic exposure may occur when people are exposed to perc repeatedly over a long period of time. Garment care workers have the biggest risk of chronic exposure, because they can be repeatedly exposed to perc at work.

The public can also be at risk, however, when perc from a dry cleaning plant gets into the local air and water. Chronic exposure to perc can cause central nervous system damage, and to a lesser degree, lung, liver and kidney damage. The central nervous system effects include irritability, headaches, fatigue and *vertigo*.

There is ongoing debate about whether or not perc is a *carcinogen*, or cancer-causing substance. Different agencies have taken different positions on this question. The U.S. Environmental Protection Agency (EPA) considers perc a suspected carcinogen.

The National Institute for Occupational Health and Safety (NIOSH) recommends that perc be handled as a human carcinogen. The International Agency for Research on Cancer (IARC) classified perc as a probable carcinogen in 1995.

Exposure to Workers and others

Workers are at an especially high risk of occupational exposure in shops where "transfer machines" are used. Transfer machines are dry cleaning machines where the washing and drying take place in two physically separate machines.

After the wash cycle is finished, workers have to manually transfer clothes saturated with perc from washers to dryers. During this transfer, workers may be exposed on the surface of the skin, such as arms and hands, when their skin comes in contact with the wet clothes.

This is called *dermal exposure*. Even more important, workers may be exposed to perc when they breathe in the fumes while transferring the clothes. This is called *inhalation exposure*. Currently, there are about 10,000 transfer machines in use. Carbon adsorbers and refrigerated condensers are used to control *emissions* and are currently installed in about half of the transfer machines in operation.

Newer machines do not require workers to physically transfer clothes from the washer to the dryer. These newer "closed loop dry-to-dry" machines also have many safety features that help prevent perc from coming in contact with workers or the environment.

Dry-to-dry machines allow a dry cleaner to process a load of garments without coming in contact with perc-saturated clothes. Transfer machines can no longer be sold in the U.S., so all the dry cleaning equipment currently available includes these new controls.

People who work in dry cleaning shops have the highest risk of exposure to perc. People living near or above these shops who spend a great deal of time at home, such as the elderly or pregnant women, also can be exposed to perc for extended periods of time.

Environmental Effects

The use of perc, regardless of the type of equipment used, may result in some emissions to the air, water or soil. Perc evaporates quickly when released to soil or surface water (rivers, lakes, oceans, etc). When exposed to sunlight, it breaks down rapidly. When perc gets into water, it does not significantly affect organisms that live in the water. In general, it is not absorbed by aquatic organisms and does not tend to build up in their bodies.

A small percentage of the perc in water, however, will be absorbed by microorganisms and enter the food chain. Currently, there are over 30,000 dry cleaners in the United States, 80per cent of whom use perc. The total amount of perchloroethylene used for dry cleaning in this country in 1994 was estimated at over 11 million gallons (approximately 147 million pounds).

After it has been used, perc may end up as:

- Emissions into air
- Emissions into water and soil
- Hazardous waste

Perc is one of the ten most common contaminants found in groundwater. A Federal survey reported that perc, from all commercial and industrial sources, is present in about 26per cent of U.S. groundwater supplies. The California Air Resources Board estimates that 30per cent of the total perc used—3.3 million gallons—ends up as hazardous waste.

It is estimated that recyclers can recover up to 80per cent of the waste generated—2.64 million gallons. Although these recovery figures are impressive, the fact remains that hundreds of thousands of gallons of waste are generated each year which could potentially contaminate the environment.

Environment, Health and Safety in Wet Cleaning

Because perc is not used in the wet cleaning process, wet cleaners do not need to worry about perc related environmental health and safety problems. Wet cleaning uses water as the solvent. The detergents, stain removal agents and finishes used in wet cleaning are generally non-toxic. Companies that produce wet cleaning detergents, stain removers and finishes are sensitive to environmental and t*oxicological* concerns and are producing products that are phosphate free and *biodegradable.*

Data collected by the Centre for Neighbourhood Technology at the Greener Cleaner show that wet cleaning does not produce hazardous waste or air emissions. Workers, customers and people who live and work near a dry cleaner might benefit from the introduction of wet cleaning and the resulting decrease in the use of perchloroethylene.

One environmental health and safety concern that does apply to wet cleaning is the use of toxic solvents during stain removal. A new shop with a 100per cent wet cleaning operation would probably not use toxic stain removal agents, but a shop that has used perc and converted totally or partially to wet cleaning may still use these chemicals at the stain removal board. Non-toxic stain removal agents are currently available from several different sources and should be used on garments that will be wet cleaned. If the plant operator or owner continues to use toxic stain removal agents, the waste generated by the process should be treated as hazardous waste.

WHO MAKES THE RULES?

The federal Occupational Safety and Health Administration (OSHA) and state labour departments are in charge of ensuring the safety and health of workers inside the plant.

The federal Environmental Protection Agency and the state departments of environmental protection are in charge of regulating pollutants emitted or transferred from the plant. This is because the legislature and regulatory agencies have always considered worker health and safety issues to be separate from environmental protection issues.

Some state agencies are starting to combine environmental protection with worker health and safety. This has resulted in state programmes to reduce the use of toxic chemicals in processes. However, the regulations that are currently in effect are still mainly divided into environmental regulations and health and safety regulations.

The minimum compliance requirements of a dry cleaner based on the Federal regulations. Individual states can make regulations of their own that are even stricter than the Federal regulations. When states make their own regulations, the garment cleaning industry only needs to report to and follow the rules of that state.

For example, New York and California are two states that have decided to make their own regulations. Anyone who intends to own or operate a professional garment care facility should research state and local regulations to determine which regulations apply in his or her state.

Since 1970, the U.S. Congress has passed several laws designed to protect and improve environmental and public health. The laws set standards for limits of chemical releases to air, water and soil.

These laws originally targeted industries that use large quantities of toxic substances. As these industries have come into compliance, the focus has turned towards industries that use small quantities of toxic substances.

WHAT RULES ARE CURRENTLY IN EFFECT?

The federal regulations outlined here apply to dry cleaning operations. Perc is listed as a hazardous substance under the Occupational Safety and Health

Act (OSHA), the Resource Conservation and Recovery Act (RCRA) the Clean Water Act (CWA), the Clean Air Act (CAA), and the Comprehensive Emergency Response, Compensation and Liability Act (CERCLA). A dry cleaner that uses perc must comply with these regulations. Complying with these regulations can affect a dry cleaner's profit margins.

The costs associated with compliance include:

- Liability insurance
- Permitting fees
- Hazardous waste disposal fees
- Equipment upgrades
- Contaminated site clean up
- Time spent learning and complying with regulations

Shifting to a mix of wet cleaning and dry cleaning can provide partial relief from some of the regulatory requirements imposed on dry cleaners. For example, some of the Clean Air Act regulations are based on how much of a toxic chemical is used by a facility.

Small garment cleaning businesses with a dry-to-dry machine installed before 1991 are required to upgrade their equipment under the CAA if they purchase more than 140 gallons of perc each year. By switching some cleaning to wet cleaning, a business may be able to reduce its purchases below the 140 gallon limit. Once below 140 gallons, the business would be exempt from the upgrade requirement.

Occupational Safety and Health Act

The Occupational Safety and Health Act of 1970 was passed to protect the health of the worker in the workplace environment. The Act is administered by the Occupational Safety and Health Administration (OSHA). The Act sets limits on the amount of toxic materials that workers can be exposed to in their workplaces.

Permissable Exposure Limits (PELs)

Both long-term and short-term exposure limits have been established. These limits are called the Permissible Exposure Limits (PELs). The long-term limits refer to a worker's allowable average exposure during a normal 8-hour workshift of a 40-hour work week.

The average exposure limit takes into account how long a worker is exposed to different levels of the chemical. This average is call the Time Weighted Average (TWA). The overall average exposure for any 8-hour workday of a 40-hour work week cannot be greater than the PEL.

The current long-term PEL for perc is 100 parts per million (ppm). In 1989, because of concerns about whether the 100 ppm limit is low enough to protect workers, OSHA reduced the limit to 25 ppm. Dry cleaners had until December 31, 1991 to meet the new standard.

In fact, many dry cleaners have installed control equipment to control emissions so that they meet the 25 ppm PEL. However, soon after the 25 ppm standard was issued, a suit was filed by industry groups to challenge the new limits and a court decision threw out the 25 ppm limit. As a result, currently OSHA can only enforce a PEL of 100 ppm but still recommends that dry cleaners limit indoor air emissions of perc to 25 ppm.

Several state occupational agencies continue to enforce the 25 ppm limit. In addition, the American Conference of Governmental Industrial Hygienists (ACGIH) and the perc manufacturers also recommend 25 ppm as an 8-hour TWA. Currently, negotiations are under way to reduce the PEL again, with some public interest groups recommending levels as low as 5 ppm. In addition to the long-term PEL, two short-term exposure limits have been established by OSHA. These are designed to protect workers from high exposures that might occur occasionally throughout the work day. In dry cleaning, this might occur when opening the machine to remove the clothes, when cleaning the still or filter housings, when cleaning out the button trap, or at any other time when perc vapours are in close contact with the worker.

The OSHA short-term PELs for perc are 200 ppm averaged over 15 minutes, and 300 ppm averaged over 5 minutes. The 5 minute maximum (peak) exposure can only occur once in any 3 hour period. ACGIH recommends a short-term exposure limit of 100 ppm, averaged over 15 minutes.

Resource Conservation and Recovery Act

Subtitle C of the Resource Conservation and Recovery Act (RCRA) of 1976 regulates what is done with hazardous waste that is generated at facilities. The regulations only apply to facilities that generate more than 100 kilograms (220 pounds) per month of hazardous wastes.

Facilities that generate less than this amount are considered small quantity generators and are exempt from this regulation. Waste that contains perc is one of the hazardous wastes regulated under RCRA. In the case of a dry cleaner, this waste could consist of still bottoms, cartridge filters and filter muck. This waste must be disposed of at a licensed hazardous waste facility.

Some water used in the dry cleaning process becomes contaminated with perc. This occurs when water is used by emission control devices and perc filtration devices, and when water is used to maintain these devices. Waste water generated by dry cleaners, even though only slightly contaminated, is considered hazardous waste under RCRA.

The hazardous waste code for this type of waste is F002. Separator water is also considered hazardous waste because it may contain as much as 150 ppm of perchloroethylene. Solid wastes and waste water from wet cleaning processes are not considered hazardous under RCRA. Some stain removal agents may also generate hazardous waste.

Clean Air Act (1977) and Clean Air Act Amendments (1990)

The Clean Air Act of 1977 (CAA) and the Clean Air Act Amendments of 1990 (CAAA) also affect dry cleaners. The CAAA added perc to the list of regulated hazardous air pollutants.

In September 1993 the CAAA put out the National Emission Standards for Hazardous Air Pollutants (NESHAP), which regulates emissions of perc by dry cleaners. These standards limit the amount of emissions of regulated hazardous air pollutants including perc.

The CAAA requires that dry cleaners keep track of their use of perc and its disposal. It requires dry cleaners to change their operating procedures and modify their existing equipment to meet new air emissions standards. It also requires that all newly purchased dry cleaning machinery be equipped with control technology. These standards also prohibit the sale of transfer machines. Federal law requires that the industry be in compliance with the perc NESHAP by September 23, 1996.

New Source Performance Standards (NSPS) were put into effect in 1984 for petroleum based dry cleaners. The NSPS's are enforced only in Clean Air Act non-attainment areas—those areas that do not meet national standards for ambient air quality. The NSPS's set limits on solvent loss in drying, set standards on the use of filters, and require leaks to be repaired in a timely fashion. Wet cleaning processes do not emit hazardous air pollutants that are regulated under the Clean Air Act.

The contamination may have resulted from a variety of situations, such as:

- Contaminated waste water leaking through sewer pipes
- Perc leaks during operation of a dry cleaning machine
- Dumping of wastes

The materials used in wet cleaning are not a source of liability under the CERCLA.

Sewer or Septic Discharge

Water discharge from a dry cleaning or wet cleaning facility is subject to restrictions depending on the size of the facility and whether the discharge is to a publicly owned treatment facility or to a septic system.

Specifically, the following restrictions apply:

- Contact water (separator and vacuum water) must be disposed of properly as hazardous waste or may be evaporated under an EPA policy decision
- Boiler blowdown water is regulated on a state by state basis
- Vacuum water from the finishing equipment must be disposed of as hazardous waste

- Waste from the stain removal board must be treated as hazardous waste

The waste water generated by a wet cleaning operation is discharged either to a municipal sewer—Publicly-Owned Treatment Works or POTW—or to a private septic system. Cleaners should always check with their local POTW, state agency or municipality before discharging into either a sewer or septic system since they may require that waste water must be non-toxic in order to avoid polluting the environment.

This will ensure that it is legal to discharge their waste water and that no permits are required. For average size wet cleaning facilities, a discharge permit, a pretreatment system, or monitoring of discharge may not be required by the Publicly- Owned Treatment Works. Discharge to septic systems is regulated by the local municipalities and the state. Waste from a professional cleaner is considered industrial waste, and, in many states, discharge into a septic system is prohibited. The local regulating authorities, appropriate state agencies and the POTW should be notified during the planning stages of any wet cleaning or dry cleaning facility to determine any applicable regulations or permitting requirements.

WHAT ARE THE POTENTIAL LIABILITIES?

When new regulations are passed that restrict dry cleaning facilities, the public becomes more aware of the public health and environmental risks of perc. The professional fabricare industry, primarily composed of very small "mom and pop" operations, has been hard hit by these new regulations. Landlords are wary of renting to dry cleaners because of the potential liability associated with perc. The cost of compliance and the potential for retroactive liability has affected the profit margin of many operations, forcing some owners to go out of business. Some families that in the past might have passed the family dry cleaning business down to the next generation must now be mindful that current retroactive liability laws may transfer past liability to the new owner.

These regulations have had far-reaching effects on the dry cleaning industry. Not only do dry cleaners need to comply with new Federal regulations by implementing expensive pollution control technology, they need to consider that their home state could pass even stricter regulations in the future. A cleaner who makes a partial or total switch to machine wet cleaning (or water based cleaning) is subject to limited regulations, and in some cases, no regulations at all.

What does that Mean?

Acute exposure: contact with a substance that occurs over a short period of time

- *Biodegradable:* Able to be broken down into non—harmful substances by natural processes in the environment, such as sunlight, rain and wind

- *CAA:* Clean Air Act of 1977
- *CAAA:* Clean Air Act Amendments of 1990
- *Carcinogen:* A substance that causes cancer
- *CERCLA:* Comprehensive Emergency Response, Compensation, and Liability Act
- *Chronic exposure:* Contact with a substance that occurs over a long period of time
- *CWA:* Clean Water Act
- *Dermal exposure:* Exposure to a substance when the substance comes in contact with the skin
- *Emission:* The release of a substance into the air, water or ground
- *Exposure:* Coming in contact with a substance. The contact can be on the skin, through breathing into the lungs, or by eating or drinking.
- *Inhalation exposure:* Exposure to a substance that occurs when something in the air is breathed into the lungs
- *NESHAP:* National Emission Standards for Hazardous Air Pollutants
- *NSPS:* New Source Performance Standards
- *OSHA:* Occupational Safety and Health Admini-stration; the part of the Federal government that administers the Occupational Safety and Health Act of 1970
- *PEL:* Permissible Exposure Limit; the maximum amount of a substance that a worker is allowed to come in contact with based on a daily or weekly average
- *Perc:* A short name for perchloroethylene
- *RCRA:* Resource Conservation and Recovery Act of 1976
- *Toxicology:* The science that studies the effects, antidotes and detection of poisons or adverse effects of chemicals on living organisms.
- *SARA:* Superfund Amendments and Reauthorization Act
- *Superfund:* Another name for CERCLA
- *TWA:* Time Weighted Average; the method used to calculate a worker's average exposure to a substance
- *Vertigo:* Dizziness

FIBRES, YARNS, AND FABRICS

A customer walks into a cleaning shop and puts a garment on the counter. To the customer, the garment is only a shirt, or a pair of pants, or a dress. When a cleaner looks at that same garment, he or she sees much more than that. The cleaner sees that the garment is made from a particular type of fabric, and the fabric is made from a particular type of fibre.

To successfully wet clean the garment, the cleaner must understand how the fabric and fibre will react to the cleaning process. Fabrics and fibres react differently in water than they do in perchloroethylene or other dry cleaning solvents. Even an experienced fabricare professional must pay careful attention to fabric and fibre when wet cleaning. This module describes the characteristics of fibres and fabrics. This module also explains the effect of the cleaning process on fibres and fabrics.

WHAT IS A FIBRE?

A fibre is a fine hair-like structure that is either natural or man-made. Natural fibres come primarily from plants and animals. Plant fibres come from the seeds of plants like cotton, or from stems like linen, ramie, hemp, or jute. Animal fibres come from the hair of certain animals. For example, the hair of sheep is used to make wool and the hair of goats is used to make cashmere and mohair.

Silk fibres are spun by silkworms. Manufactured fibres, also called man-made fibres, were developed to provide properties that natural fibres do not have. Manufactured fibres are divided into two main categories: synthetic fibres and regenerated fibres.

Synthetic fibres are created from polymers that come from chemicals like petroleum. Nylon, polyester and acrylic are all synthetic fibres. Regenerated fibres are created from polymers that come from natural materials. Rayon and acetate are both regenerated from *cellulose*, a material that occurs naturally in plants.

Manufactured fibres are made using a process that is very similar to the process a spider uses to make strands for its web. Long chains of molecules called *polymers* are made into a liquid. This liquid is forced through tiny holes. As the liquid comes out of the holes, it hardens into fibre strands called *filaments*, or strands of "endless" length.

Both natural and manufactured fibres can be made directly into fabric, such as felts and nonwoven textiles. However, fibres are usually twisted or grouped together into long strands called yarns. The yarns are then used to make a wide variety of knit and woven fabrics, as well as lace.

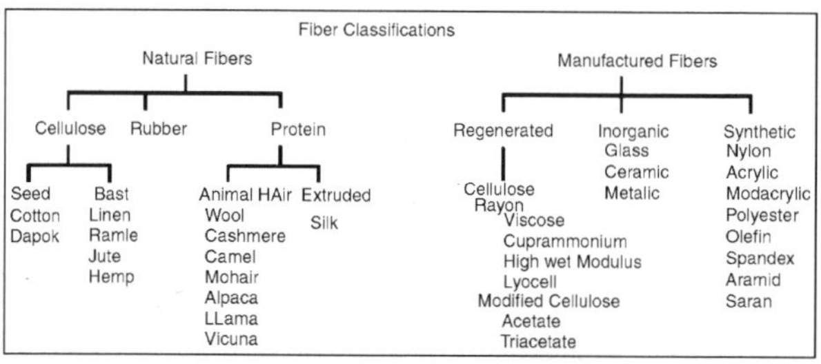

Fig. Fibre Classifications.

The length of the fibre helps to determine the properties of the yarn and of the fabric that is made from that yarn. Almost all natural fibres range in length from less than one inch up to 36 inches.

These are called *staple fibres*. The exception to this rule is silk. The silk worm spins filaments of silk that can be up to 1,600 yards long. Like silk, all manufactured fibres are also spun as filaments. Unlike silk, there is no limit to the length of manufactured fibres. However, these fibres can be cut into short or "staple" lengths if necessary.

YARNS

Yarns are used to make a wide variety of knit fabrics, woven fabrics and lace. The type of fibre used to make a yarn and the way the yarn is made determine the properties of the yarn and of the final fabric. These properties include strength, *abrasion resistance, elasticity, luster, hand* and overall surface characteristics. In many cases, fibres are twisted to make yarn. Increasing the twists or turns in a yarn will increase strength, crease resistance and abrasion resistance. Yarns that are more loosely twisted make softer fabrics. Some yarns are made from a single type of fibre, while other yarns are made from more than one type of fibre.

Multiple ply yarns are stronger than single yarns. Fancy yarns, like boucle or chenille, can be made to create decorative effects, and may take special care in cleaning. Figure shows several different types of fibres and yarns. Filament fibres are typically sleek and smooth. The surface of fabric made from filament fibres is therefore also sleek and smooth.

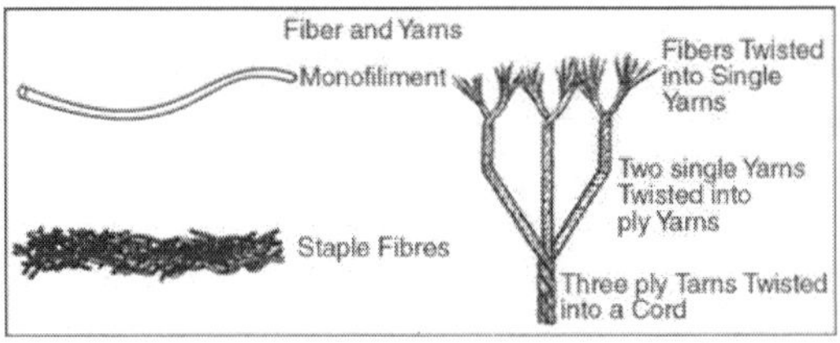

Fig. Fibre and Yarns.

Natural staple fibres tend to be rougher. The surface of fabric made from staple fibres will also be rougher, due to the exposed fibre ends. For example, nylon and silk fibres have round, smooth surfaces, and therefore nylon and silk fabric may feel smooth and slippery. Wool fibres have a round scaly surface, making wool fabrics feel bulky and slightly rough.

FABRIC TYPES

The basic fabric types can be classified as woven, knitted, lace and nonwoven. Woven fabrics consist of two sets of parallel yarns that are interlaced to form a fabric. The lengthwise yarns are called "warp." Warp yarns are held under tension throughout the weaving process.

The crosswise yarns are called "weft" or "filling." Some examples of woven fabrics are plain, twill and satin weaves, as well as certain pile fabrics like velvet, velveteen and corduroy. Figure shows some examples of woven fabrics.

The weaving process tends to create fabrics that are strong and do not lose their shape. The heavier the fabric and the tighter the weave, the more durable the fabric will be. This assumes, of course, that the yarn and fibre are also strong and stable. Fabrics that are tightly woven with delicate yarns, for example, can still be very fragile.

Knitted fabrics consist of one or more continuous yarns that are looped through themselves to form interconnected chains. Weft knits are the most common type of knits. Weft knits use one continuous yarn to form the fabric through a series of loops, and are easy to snag and run. Warp knits use a series of continuous yarns to form the fabric, giving the fabric more stability and less tendency to run. Figure shows examples of warp and weft knits.

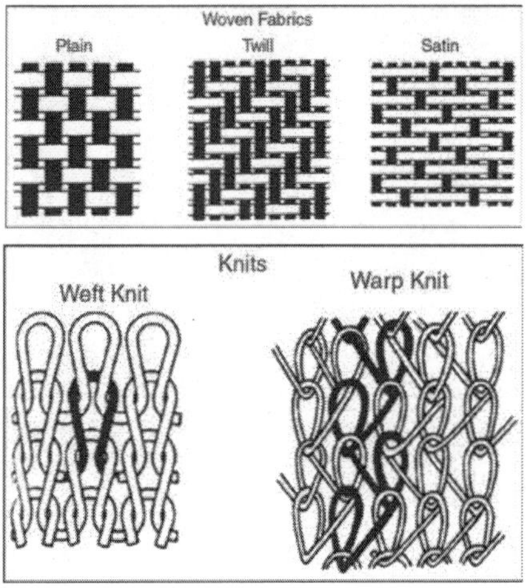

Knit fabrics tend to be more comfortable than woven fabrics because they stretch. This is an advantage for the person who wears the fabric. It can cause problems for the cleaner, however, because this tendency to stretch can cause the fabric to lose its shape when it is handled. Knitted fabrics also tend to shrink. Fortunately, knitted fabrics can be 'blocked' or stretched to recover their original shape and size. Because some fabrics are constructed of loosely twisted yarns, pilling is often a problem. When a knitted fabric is rubbed repeatedly, small balls of fibre called "pills" build up on the surface of the fabric. This is called "pilling." Cleaners need to be careful not to cause pilling when they rub the surface of a knit fabric to remove soil or stains.

Lace is made of yarns that are knotted, twisted, or looped to provide a fragile, sheer pattern. When cleaning garments with lace, great care must be taken not to break the delicate yarns. Nonwovens are fabrics made by bonding together a web of fibres through physical or chemical means. Felt is an example of a nonwoven fabric. Felt is made primarily from wool fibres. The felting process uses moisture, heat, mechanical action and chemical action to take advantage of the natural tendency of wool fibres to stick together. The result is a dense, bulky fabric. Imitation suede is another example of a nonwoven fabric.

Nonwoven fabrics tend to be weaker and more easily damaged by rubbing than woven or knit fabrics of the same weight. This is because nonwoven fabrics are made directly from fibres, and therefore do not benefit from the strength of yarns.

Nonwoven fabrics are often used for interfacing. Nonwovens can be bonded to a fabric to make it more stable. This can cause problems for the cleaner. One problem is that the cleaning process can dissolve the adhesive and cause the two fabrics to separate, or "delaminate." Another problem is that one fabric may shrink more than the other, which can cause puckering.

THE EFFECT OF CLEANING ON FIBRES

Each type of fibre reacts differently to the wet cleaning process. How a fibre reacts depends on a number of factors, including strength, elasticity, dimensional stability, absorbency and the effect of chemicals and heat. These factors have a clear influence on the way a garment should be handled in the wet cleaning process.

Fibre Strength

	Dry (Grams/Denier)	Wet (Grams/Denier)
Acetate	1.2 - 1.5	0.8 - 1.2
Acrylic	2.0 - 3.5	1.8 -3.3
Cotton	3.0 -5.0	3.3 - 6.4
Linen	Stronger than Cotton	
Modacrylic	2.0 3.5	2.0 - 3.5
Nylon	3.0 -6.0*	2.6 - 5.4
Olefin	4.8 - 7.0	4.8 - 7.0
Polyester	4.0 - 5.0*	4.0 - 5.0
Rayon	0.73 - 2.6*	0.70 - 1.8
Silk	2.4 - 5.1	1.8 - 4.2
Wool	1.0 - 1.7	0.8 - 1.6

* High Tenaclty Nylon High as 9.5 g/den
High Tenacity Polyester as High as 8.0 g/den
High wet modulus Rayon- 4.5 g/den

Fig. Fibre Strength.

Fibre strength is a measure of the amount of force needed to cause a fibre to break. It is also called *tenacity*, and it is measured in grams/ denier. Fibre strength can change when the fibre is exposed to water. Cotton and linen are the only fibres that get stronger when exposed to water.

Rayons and acetates are examples of fibres that get significantly weaker in water. Polyesters and olefins, which are among the strongest of the common fibres, do not get weaker or stronger. The strength of these fibres is not affected by water at all. Figure shows the wet and dry strength of different fibres.

The effect of water on fibre strength is very important for the wet cleaner in the stain removal and cleaning stages of garment care. Because the strength of fibres can change when the fibres are wet, care must be taken when applying any force to wet fibres. For example, rayons and acetates, since they get significantly weaker in water, could be damaged by scrubbing or harsh agitation. Figure shows the resistance of different fibres to abrasion.

Elasticity and Dimensional Stability

Elasticity is the ability of a fibre to go back to its original length after being stretched. Fibres that have good elasticity are those that return to their original length when stretched. Cotton, linen and rayon have poor elasticity, while wool, nylon, polyester and olefins are very elastic fibres.

The *dimensional stability* of a fibre is its ability to maintain its original shape. Fibres with good dimensional stability do not shrink or stretch during cleaning. Fibres with poor dimensional stability may shrink or stretch during cleaning.

Abrasion Resistance

Acetate	Fair to low
Acrylic	Good to Sufficient
Cotton	Good*
Linen	Good, but Damaged by Repeated Flexing
Nylon	Good to Excellent
Polyester	Good to Excellent
Rayon	Poortofair
Silk	Poor
Wool	Good*

* Wool and cotton are rated or lower than each other in differents situations

Wool and rayon fibres have poor dimensional stability. They tend to shrink if not treated with care when exposed to water. When a rayon garment shrinks, it can often be stretched back to its original shape in the finishing process.

However, when a wool garment shrinks, it can be difficult to return it to its original shape. Proper care when cleaning garments can prevent shrinkage.

For example, wool fabric should never be cleaned using moisture with heat, or moisture with vigorous agitation, because these combinations can cause felting and thus permanent shrinkage. Even some fibres with good dimensional stability can lose their shape during the cleaning process.

Synthetic fibres tend to have good dimensional stability, but exposure to high temperatures can permanently change their shape. Figure shows the elasticity and dimensional stability of common fibres.

Absorbency and the Effect of Water on Fibres

A fibre's ability to take up moisture is referred to as absorbency. In technical terms, absorbency is called *moisture regain*. Moisture regain is a measure of the weight of the moisture a fibre absorbs compared to the weight of the fibre itself. It is measured at a standard temperature and humidity, usually 70°F and

65per cent humidity. If 100 grams of a fibre absorbs 5 grams of moisture, then the fibre has a moisture regain of 5per cent. Cotton, for example, has a high moisture regain of 7per cent. One hundred grams of cotton will absorb 7 grams of moisture. Polyester, on the other hand, has a low moisture regain of only 0.2per cent. One hundred grams of polyester will only absorb 0.2 grams of water. Figure shows the standard moisture regain for common fibres.

Fibres with a high moisture regain are also called *hydrophilic* or "water-loving" fibres. These fibres possess the ability to absorb water with ease. All natural fibres, as well as rayon and acetates, are examples of hydrophilic fibres. These fibres tend to swell when exposed to water. The fact that these fibres tend to absorb water is both an advantage and a disadvantage to the cleaner. The primary advantage is that stain removal is easier, because water and detergent are easily absorbed into and flushed out of the fibre. The primary disadvantage is that swollen, saturated fibres can also cause increased shrinkage, wrinkling and loss of shape of the garment.

Elasticity and Dimensional Stability

	Elasticity	Dimensional Stability
Acetate	Poor	Fair
Acrylic	Good	Poor
Cotton	Poor	Good
Linen	Poor	Good
Modacrylic	Fair	Good if not Exposed to High Temperatures
Nylon	Excellent	Good if not Exposed to High Temperatures
Olefin	Excellent	Good if not Exposed to High Temperatures
Polyester	Excellent	Good if not Exposed to High Temperatures
Rayon	Poor	Poor
Silk	Good	Good
Wool	Excellent	Poor

Fibres with a low moisture regain do not absorb water or absorb only small amounts of water. These fibres are referred to as *hydrophobic* or "water-fearing."

Synthetic and inorganic fibres such as polyester and nylon are hydrophobic. These fibres tend to wrinkle less when processed in water.

In fact, synthetic fibres keep their original shape when processed in water, just as natural fibres tend to keep their shape when processed in dry cleaning solvents.

The majority of garments today consist of blends of natural and manufactured fibres. Because synthetic fibres do not tend to be affected by water, these garments can be easier to clean than garments made entirely of natural fibres.

Some natural fibres are treated with special processes to make them easier and safer to clean. Two of these processes are mercerizing and Sanforizing™. Mercerizing is a process in which cotton fibres, yarns or fabrics are held under tension in a cold alkali bath.

This strengthens the cotton fibres, makes them shinier, and allows them to absorb dye better. The Sanforizing™ process is a pre-shrinking process. This means that the fabric manufacturer shrinks the fabric on purpose. Once the fabric is shrunk, it is sent to a clothing manufacturer to be made into garments. The result is that the garments will not shrink more than two per cent when washed and dried in home machines.

Absorbency

	Moisture Regain (%)*
Acetate	6.0 - 6.5
Acrylic	1.0 - 2.5
Cotton	7.0 - 11.0
Linen	8.0 - 12.0
Modacrylic	0.4 - 4.0
Nylon	4.0 - 4.5
Olefin	0.01 - 0.1
Polyester	0.2 - 0.8
Rayon	11.0 - 15.0
Silk	11.0
Wool	13.0 - 18.0

* As a Persentage of the dry Weight at 70.F and 65% Relative Humidity

Effect of Chemicals

A number of different chemicals are useful for getting out soils and stains, and for identifying fabric types. However, these same chemicals can also damage certain fibres. It is important for the garment care professional to understand the effect of certain chemicals on fibres. These chemicals include bleaches, alkalies, acids, and organic solvents.

Bleaches: Bleaches are used to remove colour or stains from fabrics. There are two types of bleaches commonly used in fabric care: *oxidizing bleaches* and *reducing bleaches*. Oxidizing bleaches include hydrogen peroxide (H_2O_2), sodium perborate, and sodium hypochlorite (NaClO), also known as chlorine bleach. Reducing bleaches include sodium bisulfite, sodium hydrosulfite and titanium stripper. Of all the bleaches used in garment cleaning, chlorine is the most harmful to fibres. Most other bleaches could be used safely.

Acrylics and polyesters are resistant to damage from all types of bleaches. Cotton, linen, and rayon fibres are not harmed by reducing bleaches, but can only withstand weak chlorine bleaches for short periods of time. Wool and silk— and animal fibres in general— will deteriorate and dissolve in chlorine bleaches, but are more resistant to other bleaches. Some synthetic fibres, such as nylon and modacrylic, may also be harmed by chlorine bleaches. Figure shows the resistance of different fibres to bleaches.

Acids and Alkalies: *Acids* and *alkalies* are used in laboratories for fibre identification and in cleaning to remove difficult stains like rust or blood. As might be expected, acids and alkalies each have different effects on fibres. Synthetic fibres are generally resistant to alkalies and acids, with some exceptions. Acrylic and polyester are sensitive to strong alkalies at high temperatures. Acetate, nylon 66 and polyester will decompose in strong acids. The natural fibres are affected differently by acids than alkalies. Plant fibres that contain cellulose are sensitive to acids, but are typically not damaged by alkalies. In fact, exposing cotton to alkalies strengthens the cotton fibres. This is the basis of a process called mercerization. The protein, or animal fibres, are generally unaffected by acids, but are damaged by even weak alkalies. Figures show the effect of acids and alkalies on different fibres.

Organic Solvents: Organic solvents, like perchloroethylene, petroleum and acetone, are used to remove a variety of oil-based stains from fabrics. Natural fibres, both animal fibres and cellulose fibres, either resist organic solvents or are unaffected by them. This fact has allowed these fibres to be cleaned in dry cleaning solvents like perc and petroleum with little or no damage to the fibres.

Synthetic fibres, however, may be damaged by organic solvents. For example, acetate and modacrylic dissolve in acetone, which is used to remove nail polish and paint stains.

The olefin fibres (polypropylene, polyethylene) will dissolve in solvents like perc in temperatures above 160°F. Since temperatures in a perc dry cleaning machine do not get this high, it is possible to dry clean these fibres. Figure shows the effects of organic solvents on different types of fibres.

Effect of Bleaches	
Acetate	Resistant to Oxidizing Bleaches
Acrylic	Highly Resistant
Cotton	Resistent, but chlorine beaches will Destroy if Uncontrolled
Linen	Similar to Cotton
Modacrylic	Some Fibers may be harmed by Chlorine Bleach
Nylon	May be Harmed by Chlorine Bleach
Polyester	Highly Resistant
Rayon	Similar to Cotton
Silk	Deteriorates in Chlorine Bleach
Wool	Will Yellow and Dissolve in Chlorine Bleach

There is one major difference between the effects of organic solvents and the effects of water on fibres. The application of some organic solvents to some synthetic fibres, may cause the fibres to melt or dissolve. If this happens, there is no way to reverse this effect.

Effect of Acids	
Acetate	Soluble in Acetic acid (33% or Stronger) Decomposed by strong Acids
Acrylic	Resistant to Most Acids
Cotton	Similar to Rayon
Linen	Similar to Rayon
Modacrylic	Resistsnt to most Acids
Nylon 66	Decomposed by Strong Mineral Acid Resistant to Weak Acids
Olefin	Very Resistant
Polyester	Resistant to most Mineral Acids; Disitegreted by 96% Sulfuric
Rayon	Disintegrates in Cold more Sensitive
Silk	Similar to Wool, but more Sensitive
Wool	Generally Unaffected by Acid

Effect of Alkalies	
Acetate	Little Effect from Weak Alkalies
Acrylic	Resists weak Alkalies
Cotton	Swells when Treated with Caustic soda, but is not Damaged
Linen	Very Resistant
Modacrylic	Resistant to alkalies
Nylon 66	litter or no Effect
Olefin	Very resistant
Polyester	Resistant to cold Alkalies
Rayon	No effect by week Alkalies; Swells and loses Strength in concentrated Alkalies
Silk	Similar to Wool, but Damage is Slower
Wool	Attacked by weak Alkalies; destroyed by Strong Alkanes

Effecct of Organic Solvents	
Acetate	Soluble in Acetone
Acrylic	May swell in some Solvents
Cotton	Resistant
Linen	Unaffected
Modacrylic	Soluble in Warm Acetone, Otherwise Unaffected
Nylon 66	Generally Unaffected
Olefin	Soluble in Chlorinated Hydrocarbons Above 160°F
Polyester	Generally Unaffected
Rayon	Unaffected
Silk	Unaffected
Wool	Unaffected

When water is applied to natural fibres, the fibres may shrink or lose their shape. Some of these effects can be reversed, and both of these effects can be avoided.

Effect of Heat

The effect of heat on fibres is an important factor in both the cleaning and finishing of garments.

A cleaner should ask himself or herself a number of questions before applying heat to a fibre:

- Does this fibre melt or decompose, and if so, at what temperature?
- Does this fibre shrink when it is heated?
- Does this fibre stretch when heated under tension?

Can this fibre be heat-set, or become fixed in a certain state after heating? If one heat this fibre for a long time, will it damage the fibre and make the fabric useless? Natural fibres do not melt or soften when they are heated. This allows them to be exposed to high temperatures.

In fact, fabrics made from natural fibres may actually need these high temperatures to remove wrinkles in the finishing process. Dry heat has very little effect on natural fibres. However, these fabrics may shrink or stretch if exposed to hot water or to moist heat, and must be closely monitored to achieve the desired result. Wools and rayons require particular care in moist heat.

Synthetic fibres have softening or sticking points ranging from 260°F (olefins) to 490°F (acrylic). The lower this softening point, the more care must be taken when using heat in the care of a garment. Too much heat can cause a fabric to stick to pressing equipment or to become permanently heat-set in an undesirable state.

THE EFFECT OF CLEANING ON FABRICS

Wet cleaning can negatively affect fabrics in three main ways:

- Fabrics can shrink or stretch
- Colours can bleed
- Textile finishes may be removed

Understanding these effects can make a cleaner more successful.

Shrinkage

The most common form of shrinkage is called "relaxation shrinkage." During the manufacturing process, fibres and fabrics are constantly stretched in the warp, or length, direction. After manufacturing, the fabric wants to "relax" and return to its natural length. During cleaning, water or solvent can cause the fabric to "relax." This can result in shrinkage.

Relaxation shrinkage usually occurs the first time the garment is cleaned. Woven fabrics generally shrink more lengthwise. This is the direction of the warp yarns when the fabric was woven. Knitted fabrics tend to shrink in both directions.

There are three other main types of shrinkage: natural shrinkage, progressive shrinkage, and differential shrinkage. "Natural shrinkage" depends on the properties of fibres. For example, wool and rayon tend to shrink, while polyesters do not. Shrinkage in wool is often called "felting." Felting can occur when heat, moisture, and mechanical action are applied to wool fabric.

This causes the wool fibres to interlock and results in shrinkage. A "felted" fabric feels thicker and more compact than the original fabric. "Progressive shrinkage" occurs when a fabric shrinks more each time it is cleaned. This is an important point to note, because even if a garment has been cleaned before, it may continue to shrink.

Differential shrinkage may occur when a garment is constructed of two or more fabric types. Each fabric may shrink differently, causing puckering of the seams or baggy linings.

Colourfastness

Colourfastness is the tendency of a fabric to hold or lose its colour. A fabric that is "colourfast" will not lose its colour and "bleed" on other garments during cleaning. Colour can be added to a polymer, a fibre, a yarn, a fabric, or a garment. Colour is added using dyes or pigments. Different dyes are used for different fibres, providing varying levels of colourfastness. The more completely the dye penetrates the fibre, the better the colourfastness of the fabric will be. Manufactured fibres tend to be the most colourfast.

Because the fibre is dyed as a polymer, the colour is deeply imbedded in the fibre structure.

When dyeing is done at a later stage in the manufacturing process, colourfastness may suffer. Some fabrics are printed using pigments, not dyes. These pigments sit on the surface of the fibre and may be sensitive to rubbing or abrasion.

Fabric can lose its colour for a number of reasons, including:

- Exposure to water or chemicals such as acids, alkalides, bleaches or perc.
- Exposure to light, especially sunlight
- Exposure to atmospheric gases
- Exposure to heat
- Perspiration
- Abrasion or rubbing

Garments must be tested for colourfastness before they are cleaned. This is especially important for dark fabrics and for garments that are made of both dark and light fabrics. Experienced garment care professionals, no matter what the cleaning method, will learn which types and colours of fabrics are the problems or "bleeders."

Textile Finishes

Finishes are applied to fibres, yarns and fabrics for a variety of reasons. They can be used to modify the appearance of the fabric to provide a shine or soft feel to the fabric surface.

Finishes can also be applied to fabric to reduce its shrinkage, help prevent it from wrinkling, or improve its resistance to water or soil. Finishes can be applied to fabric by mechanical or chemical means.

The finish can be:

- Permanent, which means it lasts as long as the fabric lasts
- Durable, which means it lasts for a relatively long time
- Temporary, which means it lasts for a relatively short time

One of the finishes that is used on fabric is called "sizing." Sizing is a chemical treatment that adds body and luster to fabrics. In this process the microscopic spaces between the individual fibres are filled with starch or resin. This reduces the amount of water the fabric will absorbed, which means the fabric will not swell as much when it is cleaned with water.

The starch and gelatin sizings used on cottons and rayons are usually temporary and may be removed by water.When the sizing is removed, fabrics can shrink, lose their shape, or lose their surface finish. Some more durable resin sizings last somewhat longer on the fabric surface, and are not affected as much by water.

GARMENT CONSTRUCTION

Garments are made of many different parts. All these parts must be considered when cleaning the garment. The first part to consider is the fibre and fabric that the garment is made of. A garment can be made from a single-fibre fabric, a multi-fibre fabric (or blend), or more than one type of fabric:

- A garment that is made from only one type of fibre should be cleaned just as to the properties of that fibre. For example, a wool sweater should not be cleaned in hot water with vigorous agitation. A rayon blouse should be treated very delicately and may need some reshaping or stretching during the finishing process.
- A garment made from a fabric blend—such as wool/polyester or polyester/rayon—should be cleaned with the properties of the more delicate or sensitive fibre in mind. For example, a wool/polyester blend should be treated as if it were wool, and a polyester/rayon blend should be treated as if it were rayon.
- Garments that are made of several types of fabrics can present two major problems. If the fabrics are very different colours— such a navy blue and white—the darker colour must be tested for colourfastness before the garment is cleaned. If the fabrics are made of different fibres, shrinkage may occur at different rates in the different fabrics. The garment should be cleaned with the properties of the more delicate or sensitive fibre in mind, to best prevent either fabric from shrinking. In addition to the fibre and fabric, other parts of the garment also need to be considered:
- Care must be taken on certain garments that seams, collars, and lapels do not pucker. These parts of the garment are often made of different materials that shrink at different rates. For example, when the fabric shrinks at a different rate than the thread or fused interfacing, puckering can occur. Often, this puckering can be fixed during finishing, but it is better to avoid the problem than to fix it.
- Delicate garments may be prone to seam ravelling and edge fray, especially if the seams are not secured with proper tailoring. Seams should be inspected before the garment is cleaned in order to determine if this will be a problem.
- Plastic buttons and non-fabric trims are typically a problem for dry cleaners because plastic and non-fabric trims can melt or stick when they come in contact with perc. This is not a problem with the wet cleaning process because water will not damage these types of materials.

HOW TO IDENTIFY FIBRES

Garment care professionals should know how to identify the fibre content of a garment. Knowing the types and percentages of fibres in a garment will help the cleaner determine the best care and treatment. Federal law requires that certain information be provided with each garment, including fibre content and care labelling.

These labels are usually attached permanently to the garment. This is not required by law in all circumstances, however, and consumers will often remove labels for reasons of comfort. If a garment has no identifying label, the cleaner should take the time to identify the garment's fibre content. The burn test is a simple method of fibre identification. This test can be done on almost any garment where there is an unexposed area from which a small sample can be clipped. Never try to do the burn test on a piece of fabric that is still attached to the garment.

Different fibres have distinct odours and appearances when they are burned. Exact identification can be difficult, but the cleaner will almost always be able to narrow the identification down to one of three categories: cellulose, protein, or synthetic. Based on this information, the cleaner can determine the best method for cleaning the garment. Figure shows the burning characteristics, odour of residue, and appearance for several cellulose, protein and synthetic fibres.

SOILS, ODOURS, STAINS AND SEPARATION

SOILS,ODOURS AND STAINS

Soils, odours and stains are some of the most challenging problems that cleaners must solve. It is important to get as much information as possible from the customer about the source of the soil, odour or stain. If the source is known, it is much easier to choose the correct process to remove the problem.

Soils

Garments can become soiled from many different sources.

These include:

- Inorganic dust, dirt and particulates from the air
- *Inorganic* dirt from rubbing the garment on dirty surfaces
- *Organic* oils from soot and pollen
- Food and drink
- Blood and urine
- Greases, tars, adhesives or other thick, sticky substances
- Aged skin cells that have flaked off the body

With the exception of the organic oils and greases, most of these soils do not penetrate deeply into the fibres of the fabric. However, many of these soils may be tightly attached to the garment.

Soils can be hard to remove when:

- They have been ground into the fabric structure by a strong force
- They are attached to the fabric by *electrostatic bonding* with the fibres
- The ragged surfaces of the soil particles cling to the surface of the fabric

Soil can be removed by:

- Dry mechanical agitation such as beating or scrubbing
- Dry mechanical spinning
- Flushing with a water, *surfactant* or solvent solution

The fabricare professional can also combine these methods if a single method does not remove the soil.

Odours

There are several sources of objectionable odours. These include:

- Decomposing perspiration and body oils ground into the garment
- Organic contaminants from foods, solvents, greases or oils
- Bacteria or mites attracted to organic contaminants on the garment
- Particles of smoke and soot from cigarettes or fires
- Synthetic *volatile organic compounds*—such as fuels, paints, and solvents—that have been transferred to the garment

Odours typically suggest that some organic process is taking place on the garment. Either a volatile compound is evaporating, or a food particle or body cell is decomposing, or bacteria are *metabolizing*. The odour itself is not on the garment, but the source of the odour is.

The techniques that can remove odour are:

- Removing the source of the odour by agitation, steaming or washing with a water, surfactant or solvent solution
- "Airing out" the garment over time in an oxygen rich atmosphere to allow the source of the odour to completely decompose
- Disinfecting the garment with a bleach or peroxide
- Heating the garment to promote rapid oxidation
- Masking the odour with a more dominant odour or perfume (this method does not remove the odour; it only covers up the odour temporarily)

The cleaner can also combine these methods if a single method does not remove the source of the odour.

Stains

Stains tend to colour a fibre in much the same way that a commercial dye does. Sources of stains include:

- Ink
- Dye
- Solvents
- Food and drink
- Perspiration
- Blood

Some stains are drawn up through the fibres by capillary action, the same way juice or coffee is soaked up by a paper towel. These stains stick to the surface of the garment and yarns, but do not penetrate the fibres. Other stains soak into the fibres themselves. Depending on the type of fibre, stains that penetrate into the chemical structure of a fibre may "set" over time.

These stains can be very difficult to remove. The more a stain contrasts with the colour of the garment, the more of a problem it is. A red ink stain on a white shirt must be completely removed, but a red ink stain on a red, floral garment may only need to be hidden. The stain can be hidden by reducing its colour so that it blends into the background.

Stains can be removed or hidden by:

- Flushing with a water, surfactant or solvent solution
- Bleaching or dyeing

The cleaner can also combine these methods if a single method does not treat the stain successfully.

SEPARATING SOILS, ODOURS AND STAINS FROM FABRICS

Separating soil, odour sources and stains from fabric is a two-step process: "get it off" and "keep it off."

From the point of view of wet cleaning, this process consist of:

- Wetting a fabric and the soil, odour source or stain in order to separate the problem from the fabric
- Holding the soil, odour source or stain away from the fabric to prevent it from soiling the fabric again during the cleaning process

Different separation techniques are effective at separating different soils, odour sources and stains. Some of these techniques are mechanical processes; others involve the use of chemical agents.

Processes

Mechanical agitation removes dirt from a garment by either direct force or by breaking and pulverizing dirt particles. Breaking and pulverizing results in smaller particles, which are able to leave the fabric structure more easily. These smaller particles are able to leave the fabric more easily than larger particles. Mechanical agitation does not work for all soils, however. For example, mechanical agitation alone may remove sand from a garment but may not be as effective at removing ground-in dirt. Other techniques are also required for removing stains.

Water is a powerful solvent that can help separate certain soils and stains from fabrics. Some soils dissolve in water that is used to clean a garment. The mechanical force of water flushing can also remove some soils, even though the soils themselves do not dissolve in water. Adding certain chemicals to water can help water clean more effectively. These chemical additives decrease the surface tension of the water. This allows the water to penetrate the fabric more rapidly and thoroughly. These chemical additives are called "surfactants" because of their effects on surface tension.

Steaming and heating a garment also assist in separating soils and stains from a fabric. Heat can melt or soften particles. Steam can cause particles to expand or lose their electrostatic charges. Finally, drying can assist in soil separation as particles curl, break or change shape during the evaporation process. A dehydrated particle or cell is less likely to decompose rapidly or to attract bacteria and, therefore, less likely to generate odours.

Chemical Agents

Detergents. Detergents are a particular kind of surfactant. Detergents penetrate into tight spaces such as those between fibres and soils. Surfactants ease the process of separation by surrounding the soil particle and easing it away from the fabric and into the water.

Once a surfactant has surrounded a soil particle and released it from a fabric, the problem is to keep the soil particle from attaching itself to the fabric again. Electrostatic charges help solve this problem. Most surfactants and most fabrics carry negative electrostatic charges in water. These charges repel each other. Every time a soil particle surrounded by surfactant comes close to the fabric, the negative charges repel each other. This keeps the soil particles from reattaching themselves to the fabric. The soil particles stay suspended in water until they are flushed away in a rinse. Figure illustrates this process.

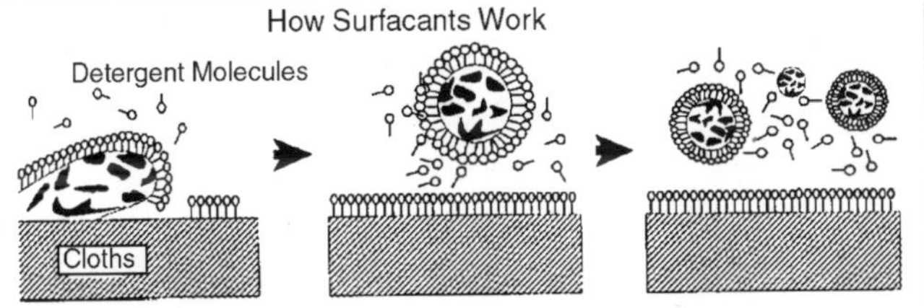

Detergent Molecules Surround the Dirt, and Weaken its Attachment to the Clothes | This Chemical Action Separated the Dirt from the Clothes. | The Separated Dirt does not Attach to the Clothes again Because it is Surrounded by Detergent Molecules

Surfactants (detergents) are made up of two different parts. One portion has an attraction to water which allows the detergent to be soluble in water. The other repels water but is attracted to fatty and oily compounds. When surfactants are dissolved in water that contains fabric with fat or oil spots, the surfactants are attracted to the fat or oil. Because surfactants are also attracted to water, they tend to pull the fat or oil away from the fabric and into the water. This helps remove these spots from fabrics.

Soaps: Soaps are a surfactant that is made when fatty acids are combined with alkali. Many soaps include extra alkali. This increases the effectiveness of the soap. Because most dirt is slightly acidic, the alkali in soaps tends to neutralize the dirt. This releases the chemical bonds that exist between the dirt and the fabric.

Synthetic Solvents: There are many synthetic surfactants and solvents on the market. Many of these solvents are made from petroleum. The most common synthetic solvent used for garment cleaning is perchloroethylene. Stoddard solvent, which is distilled from petroleum, is also commonly used in dry cleaning.

Bleaches and Oxidizers: Bleaching does not clean garments; it merely disinfects them. Whitening, odour removal and some organic stain removal is often accomplished by disinfection.

Disinfection kills bacteria and mites that may accumulate on organic soils. Disinfection is accomplished by oxidation. Chlorinated bleaches oxidize and dissolve organisms that can then be flushed away in rinses. The speed of oxidation can be increased by exposing fabrics to certain forms of oxygen such as ozone or peroxide.

Digestors: The separation of some organic materials such as food or blood from a fabric can be accomplished by biological means. Enzymes are biologically active organisms that eat and digest organic soils such as food and blood. This loosens the soils and breaks them down into other materials that can be rinsed away.

Finishing Agents: Finishing agents are used for a number of purposes. For example, they restore body, resilience and smoothness to garments after processing. They also make fabric more soil resistant. This is accomplished by forming a film around the surface of the fibre. The film makes it more difficult for soil to become attached to the fibre. If soil does become attached to the fibre, the film also makes it easier to remove that soil.

GARMENT CLEANING METHODS

CURRENT AND ALTERNATIVE GARMENT CLEANING METHODS

Most of the commercial garment cleaning in the United States is currently done using the dry cleaning process. In the future, however, new cleaning processes may also be widely used. Some of these new processes are:

- Carbon dioxide
- Ozone
- Ultrasonic cleaning
- Wet cleaning

All of these processes have been or are currently being studied as methods for commercial garment cleaning. Only wet cleaning is currently ready to be used as an alternative to dry cleaning on a large commercial scale. The other processes require further testing and development.

DRY CLEANING

Conventional dry cleaning treats garments in either perchloroethylene or petroleum. Perchloroethylene is the solvent used in 80per cent of dry cleaning plants. Petroleum is used in the remaining 20per cent. The machinery used with these solvents has been modified over time resulting in four "generations" of machines. All four generations are still in use at this time. The first generation has separate washers and dryers, and operators transfer garments from the washer to the dryer. For this reason, these are called "transfer machines."

The second generation combined the washer and dryer into one unit. The third generation added controls to help reduce perchloroethylene vapour emissions. The fourth generation included additional controls that recycle the air in the machine in order to further reduce vapour emissions. As a result of these improvements, workers are exposed to less perchloroethylene vapour and the machines themselves use less perchloroethylene.

Dry cleaning has proven itself to be an efficient, affordable, costeffective process for cleaning a wide variety of fibres and fabrics without causing damage to garments. The primary drawback to dry cleaning is that perchloroethylene and petroleum solvents are hazardous substances. Because of the hazards, the use of these solvents is becoming increasingly regulated.

In order to comply with present and future regulations, the garment care industry has two options. First, the manufacturers can develop another generation of dry cleaning equipment with more effective controls. These machines would use less perchloroethylene and petroleum solvents, and release less of these solvents into the workplace and the environment.

Second, the industry can invest in alternative cleaning processes that do not depend on hazardous solvents. In both cases, dry cleaning facilities will need to replace older equipment with new equipment.

Carbon Dioxide: Supercritical and Liquid

Research is being conducted using carbon dioxide (CO_2) in both the *supercritical* and liquid state as a method of cleaning clothes.

Carbon dioxide possesses several qualities that make it a desirable alternative for clothes cleaning:

- It is not *toxic* to humans
- It is easily recycled
- It is an inexpensive solvent
- It does not cause damage to *ozone* in the atmosphere
- It does not cause metal to corrode
- It is non-flammable
- It does not pollute ground water or soil
- It is readily available

Liquid carbon dioxide has become the primary focus of much of the research because it is safer to use and handle than supercritical carbon dioxide.

Carbon dioxide is a molecule formed naturally when two oxygen atoms join one carbon atom ($C+O+O=CO_2$). Under normal circumstances, carbon dioxide is an *inert gas*. Carbon dioxide is familiar to everyone. When we breath in, our bodies absorb oxygen from the air. When we breath out, much of what we exhale is carbon dioxide gas. Carbon dioxide gas is typically found in very low concentrations in the atmosphere.

Under conditions of high temperature and pressure, CO_2 reaches a state called the supercritical phase. Grease and oils dissolve when they come in contact with CO_2 that is in the supercritical phase. This makes supercritical CO_2 an effective cleaner. Supercritical CO_2 has other advantages as well. When the pressure is reduced, grease and oil that are dissolved in the supercritical CO_2 separate from the CO_2. The grease and oil can then be disposed of, and the CO_2 can be reused.

Supercritical CO_2 has been used by the manufacturing industry to clean metal parts for some time. It has been tested in limited situations to determine if it can be used to clean garments.

Further testing is required to determine if supercritical CO_2 is effective, affordable, and safe to use as a garment cleaner.

Liquid CO_2 exists at lower temperature and pressure than supercritical CO_2. Liquid carbon dioxide was originally developed by the aerospace industry for cleaning optical and electrical parts. It has been modified to dry clean garments. Liquid CO_2 has no surface tension, which helps it separate dirt and grease from fabric.

It can also be combined with surfactants to make it more effective. Like supercritical CO_2, liquid CO_2 can be recycled during the cleaning process. It has been reported that liquid CO_2 can clean leather, furs and fine garments without causing damage or degradation.

Ozone Cleaning System

The Ozonated Water Continuous Cleaning System is a new process that is still being studied but is available commercially on a limited basis. The theory behind the process is that ozone (O_3), working as an oxidizing agent, has the ability to disinfect, deodorize and bleach fabrics. Ozone can be added to water, in much the same way that carbon dioxide is used in carbonated soft drinks. Ozonated water can be combined with detergents to produce a garment that is cleaned, disinfected and deodorized.

The garment remains on a hanger throughout the wash, rinse, extraction and drying cycles. The ability to let the garment remain on the hanger during the cleaning process reduces the amount of labour used for finishing. The tendency of the garment to crease and wrinkle is decreased because each garment is processed individually.

The company conducting the research suggests that with this process there is a reduced amount of labour required.

The researchers report no shrinkage, change of shape, or wrinkling and claim that ornamental buttons do not need to be removed. They also state that it is cost effective to process clothes using this process.

Ultrasonic Cleaning

Recent research cooperatively funded by the U.S. Department of Energy and Kansas City Plant and Garment Care, Inc. determined that agitation for a continuous-flow, water-based clothes washing process could be provided by ultrasound.

When clothes cleaned with ultrasound were examined, the clothes were clean enough to encourage researchers to continue developing this process. Ultrasonic cleaning works by the pressure of ultrasonic waves directed at the fabric through a solution of water and soaps. Ultrasonic waves are sound waves at a frequency that is higher than the human ear can hear. Ultrasound works because molecules of soil are more dense than molecules of fibre.

The energy from the ultrasound ex- cites the molecules of soil and causes them to separate from the fabric. Ultrasonic cleaning is currently being investigated as a possible alternative to conventional dry cleaning.

The method has been shown to be effective, but more testing is required to determine the effect of ultrasound on various fabrics.

Wet Cleaning

Cleaning garments with water is an age-old technology which is currently attracting attention from consumers and cleaners interested in using an environmentally friendly method.

Unlike traditional laundering methods, wet cleaning combines the gentle mechanical action of hand washing with the convenience of machine laundry. The success of wet cleaning is dependent upon more than just the machine itself. Wet cleaning machine operators need to be knowledgeable in the area of fibre science because wet cleaning requires a certain amount of observation and skill. Success is also dependant upon the combination of quality detergents, proper training and commitment to the process.

Wet cleaning uses a wet cleaning machine, detergents, additives, stain removal agents and a specialized dryer. Conventional additives to the wet cleaning process are not considered environmentally hazardous.

The cleaning process consists of essentially the same steps as conventional dry cleaning. Basically the process can be broken down into sorting, stain removal, wet cleaning, drying and finishing.

A COMPARISON OF DRY CLEANING AND WET CLEANING

Dry cleaning and wet cleaning are currently the two most efficient, affordable and cost-effective processes for cleaning garments.

Figure illustrates the dry cleaning process. Figure illustrates the wet cleaning process. In general, these two cleaning processes are similar. The basic steps in each process are preparation, washing, drying, and finishing and assembly.

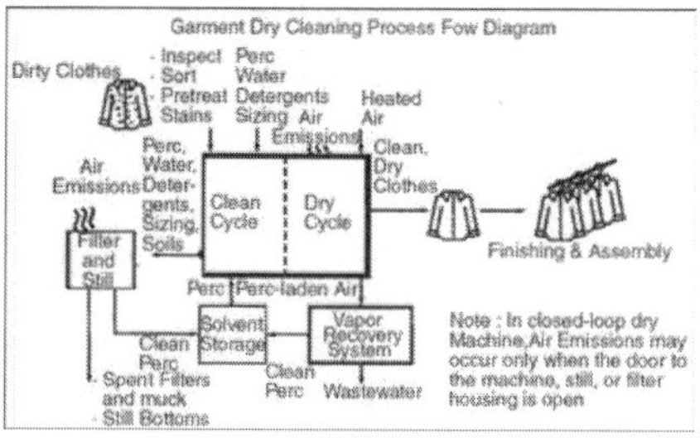

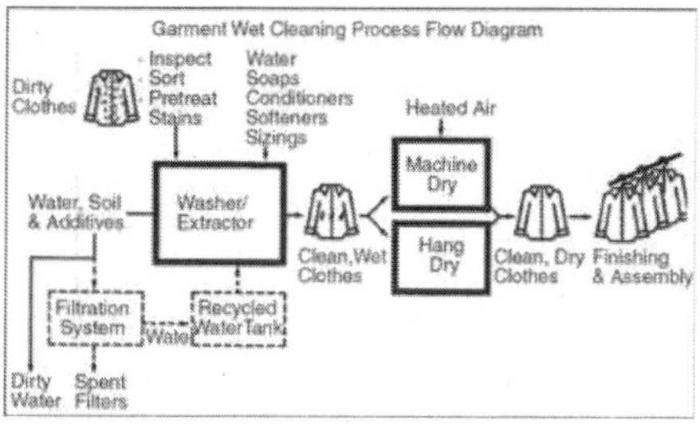

Preparation

In both processes, the cleaner inspects, sorts and removes stains from the garments before washing. Inspection is basically the same for dry and wet cleaning. In both cases, the cleaner inspects the garments to determine what problems may occur in the cleaning process and to determine the best way to clean the garments.

If two dry cleaners were given the same garments and asked to sort them, they probably would not sort them the same way.

If two wet cleaners were given the same test, they probably would not sort the garments the same way, either. Even so, the dry cleaners would be using one set of general rules for sorting, and the wet cleaners would be using another set of rules. The general rules for sorting are different because the cleaning processes are different.

The differences between cleaning with solvents and cleaning with water also result in differences in the stain removal process:

- Because the wet cleaning machine removes some water-based stains, these stains do not have to be removed before washing.
- Pretreated garments need to be dry before they can be put in a dry cleaning machine. Pretreated garments do not need to be dry before going into a wet cleaning machine. In fact, it is better in some cases to make sure that pretreated garments do not dry out, because this can cause rings to form on the fabric. This means that the wet cleaner can remove stains and immediately put garments in the wet cleaning machine.
- Several manufacturers have developed stain removers specifically for wet cleaning. Many of these products are less toxic than stain removing agents that are used for dry cleaning.

Washing and Drying

The washing and drying processes are the heart of the differences between dry cleaning and wet cleaning.

These differences include:

- Dry cleaning machines and wet cleaning machines have different controls.

- With the exception of older transfer machines, most dry cleaning machines both wash and dry. Wet cleaning machines only wash garments. Separate dryers are used for drying garments. Many of these dryers have been developed specifically for wet cleaning.
- Some wet cleaners choose to partially or completely air dry certain garments. This can require more space in which to hang the garments. It can also increase turnaround time.
- Somewhat more care is initially required to monitor the wet cleaning process in order to ensure that garments are cleaned successfully. The cleaner must have a good knowledge of fibres, fabrics, and the effects of the wet cleaning process.
- Dry cleaning machines use organic solvents like perchloroethylene or petroleum as the primary cleaning agent. Water, detergent and sizing are added to increase the efficiency of the dry cleaning process. Wet cleaning machines use water and detergent as the primary cleaning agent. Soap, sizing, conditioner and softener are added to the wet cleaning process.
- The waste products produced by dry cleaning machines include air and water that are hazardous. Special filters and control devices are required to minimize the amount of toxic waste that comes in contact with workers and the environment. These hazardous wastes also expose dry cleaners to liability if the waste is not disposed of properly. Wet cleaning machines do not produce hazardous waste, though some stain removal agents may be hazardous.

Finishing and Assembly

Garments that have been wet cleaned will require more work in the finishing step. This happens most often when the cleaner is learning the wet cleaning process. Less time will be required to finish these garments as the cleaner and the finisher gain more experience with the wet cleaning process. The purchase of state-of-the-art finishing equipment can help shorten finishing time.

HANDS-ON WET CLEANING

SORTING

Here it is covers general information on the sorting process.

The following additional steps will be covered at the sorting station:

- Sort loads just as to machine manufacturers' and detergent manufacturers' instructions
- Suit pieces should be sorted into same load
- Sort complete full loads and half loads
- Ties should be pinned inside pant leg or put in a mesh bag
- Put all knit sweaters in mesh bag

STAIN REMOVAL

Here we are covers general information on testing for colourfastness, and covers stain removal. The following information and steps will be covered at the stain removal station:

- Start in the morning with dark loads, because dark loads tend to have fewer stains. This means that dark loads can be put in the washer more quickly. While the dark loads are being washed, the cleaner can remove stains from light-coloured garments.
- Stain removal agents should be biodegradable and non-toxic. If dry cleaning solvents are used, waste should be collected and disposed of as hazardous waste.
- Excellent lighting will enhance results
- Test for colourfastness

WET CLEANING

The following information and steps will be covered at the wet cleaning station: n Match machine programme to load size and type n If using for both wet cleaning and laundry, switch machine to wet clean cycle n Use only EPA-approved soaps and detergents

- If the water is very soft, decrease detergent and increase sizing to protect fabric
- Watch for cracks in the door tubing that can cause leaking
- Take clothes out upon completion of cycle
- If using for both wet cleaning and laundry, rinse between cycles
- Clean drum after waterproofing or fire retardant cycle
- Clean suit pieces in the same load
- Process jackets in the afternoon so the jackets have time to hang overnight
- When cleaning down garments use a quick fill and extract cycle to completely wet the down

DRYING

The following information and steps will be covered at the drying station:

- Use a large enough dryer for adequate tumbling
- Use the shortest cycle possible to prevent relaxation damage
- Take clothes out immediately after processing
- Do not dry clothes to less than 10per cent humidity
- Air dry garments when necessary
- Very fine and loose weave sweaters should be dried flat
- Molded black plastic form hangers work best

FINISHING

The following information and steps will be covered at the finishing station:

- Finish garments only when the amount of moisture in the garment is the same as the amount of moisture in the air of the cleaning facility

- Less finishing time is required if the finisher uses stretching equipment
- Conditioners, including starches, can reduce finishing time on some garments
- To bring back nap, brush suedes with a copper wire brush towards the nap
- Heavy wool coats should be brushed with a carding brush to pick up the pile
- Wipe velvet with a velvet brush to get all the pile going in the same direction

WET CLEANING EQUIPMENT

WET CLEANING MACHINES

Wet cleaning machines have been designed to process a wide range of garments, including garments that are usually dry cleaned. Manufacturers have focused on mechanical agitation and programmable controls to give the wet cleaner flexible options for safe, efficient cleaning. Some of the new wet cleaning machines use a gentle, rolling action to minimize the risk of damage to fabric. These machines provide wash speeds as low as 5 to 10 rpm and programmable wash action that allows for minimal agitation, such as 1 second on and 60 seconds off.

Other wet cleaning machines use air bubbles to provide gentle agitation. Gentle wet cleaning machines also have programmable spin speeds for extracting water from garments. Extraction cycles can be set to match the garment and fabric weight, so that the maximum amount of water is removed for each type of garment.

Additional controls are available for water use, water level, temperature, heating, and automatic chemical injection. These features allow wet cleaners to precisely control the wet cleaning process. These controls also allow the wet cleaner to automatically process garments that would ordinarily be hand washed.

The drum capacity or load size of wet cleaning machines ranges from 10 to 135 pounds. An important fact to keep in mind is that the wet cleaning capacity of the machine is 50-60per cent of the laundry capacity. This means that if the cleaner plans to routinely process 25 pound loads on a wet cleaning cycle, the machine purchased should have a minimum capacity of 50 pounds.

The amount of water used per wet cleaning cycle should be considered when choosing a wet cleaning machine, particularly in areas where water and sewer costs are high.

Studies done in Chicago and Canada show that switching from dry to wet cleaning may increase water use. However, the studies also show that wet cleaning was still economically favourable to dry cleaning.

Machines with automatic detergent feeds should also have either an alarm on the detergent and/or additive tanks, or an automatic shut off. These features ensure that detergents and finishes are added when garments are processed. Processing garments without detergents and finishes can result in wasted labour, detergents and utility costs. The machine should also be equipped with enough detergent dispensers to allow the cleaner to process the full range of garments that he or she wishes. If skins are to be processed, the machine needs to be equipped with 2 dispensers in addition to the dispensers needed for normal garments on the wet clean cycle.

All currently available wet cleaning machines can be used as either a wet clean system or a standard laundry machine. This allows the cleaner to process shirts and "dry clean only" garments in one machine. Another advantage to purchasing a wet clean system is the ability to process skins. Leathers and furs are generally a high profit item for the professional cleaner and the ability to process them inhouse can increase profits.

DRYERS

A number of dryers have been developed specifically for wet cleaning. Most machines have programmable temperature and humidity controls that help the cleaner precisely set the amount of moisture that remains in the garment at the end of the drying process.

Even though the dryers have sophisticated temperature and humidity controls, the cleaner must pay careful attention to the load size and the mix of garments and fabric weights. Because different fabrics dry at different rates, sorting garments by fabric type and weight is even more important for drying than it is for wet cleaning. Cleaners can also programme drying cycles specifically for loads that contain a mix of fabrics, but care is needed to ensure that all garments dry properly.

Both electric and steam dryers use considerable energy. An electric dryer may need up to a 100 amp service to run properly. Cleaners who are considering installing a new dryer need to find out if they need to install 100 amp service to run it. One dryer manufacturer will be modifying their machine from a three phase to a single phase in the future, but for now the 100 amp service is necessary. If the cleaner is considering a steam dryer, he or she must determine if the current boiler has the capacity to support the dryer. A gas dryer should not be used in a shop that uses perc. Though garments should be removed from the dryer when they are still slightly moist to prevent shrinkage, it is often necessary to let the garment dry completely before finishing.

It is possible to let the garments air dry, or to purchase equipment to speed up this final drying process. Wind whips are available to blow dry garments, especially those with heavy areas like shoulder pads and thick waist bands. Drying cabinets can be used to dry garments that may be damaged by agitation of any sort.

WET CLEANING EQUIPMENT FEATURES AND SPECIFICATIONS

Wet cleaning technology is developing rapidly and machines available today may have different features. For a copy of the full report contact Environment Canada.

DYEING

Dyeing of textiles has been practiced for thousands of years with the first written record of the use of dyestuff dated at 2600 BC in China. All dyes were natural substances obtained from plant, animal or mineral sources. In 1856, William Henry Perkin, while searching for a cure for malaria, discovered the first synthetic dye, Mauve.

The mauve dye was a brilliant fuchsia colour, but faded easily. Since that time, a great number of synthetic dyes have been manufactured and their resistance to running and fading has been almost eliminated. Almost all garments purchased today are dyes with synthetic dyes.

Dyes used for fabric such as cotton, wool, and silk are complex organic molecules that contain what is known as a chromophore group, that is, they contain some type of conjugated, alternating double and single bonds in part of the molecule. These molecules can absorb certain wavelengths of visible light and reflect the remaining light and, thus, give a fabric its colour.

Not only do the dyes have polar or ionic groups, but fabrics such as cotton and wool also contain polar groups such as –OH (hydroxyl) and –NH (amide) which help the dye attach to the fabric. Sometimes chemical bonds are formed between the dye and the fabric molecules which hold the two together. Another process involves the use of *a mordant,* which serves as sort of an intermediary that bonds the dye and the fabric. If the dye molecules attach firmly to the fabric, the colour will be "fast", that is, it does not run when wet or washed, after the initial rinsing of excess dye. This experiment will deal with four types of dyeing processes, direct dyes, mordant dyes, developed dyes, and vat dyes, along with an example of how dyes can be used to identify different fabrics. Direct Dyes are molecules that adhere to the fabric molecules without help from other chemicals. In this experiment as suggested, use a dye known as malachite green to illustrate this process.

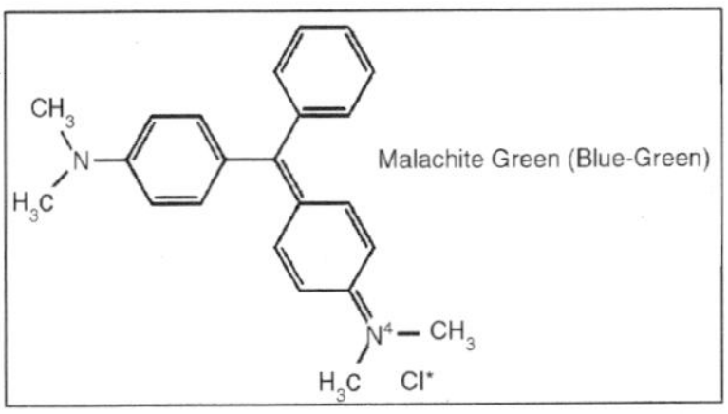

Mordant Dyes are dyes that do not adhere to fabrics directly. These need a chemical intermediate, known as a mordant, to attach themselves to the fabric.

In this process, the mordant is applied to the fabric and then the dye is applied, which then bonds to the mordant. In this experiment, as suggested, use tannic acid as a mordant to dye a sample of cotton with malachite green dye.

Developed Dyes are those where the dye-forming chemical reaction is carried out on the fabric. This process is commonly done with diazo dyes which are characterized by containing a –N=N– structure in the molecule. In this experiment, as suggested, use a dye known as para red which will be formed on the cotton fabric from the reaction between *p*-nitroaniline and 2-naphthol.

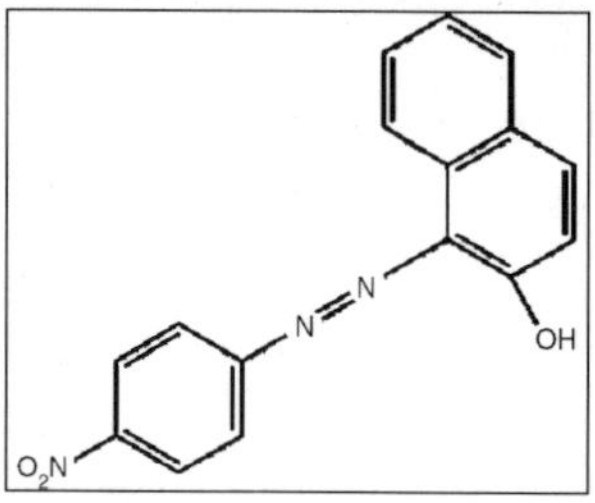

Vat Dyes are insoluble in their coloured form. They are reduced by another chemical and converted to a soluble form. The reduced dye is applied to the fabric, and then exposed to the air which oxidizes he dye back to its coloured form. In this experiment indigo dye will be used to dye a sample of cotton fabric. Indigo is the dye used to make blue jeans blue.

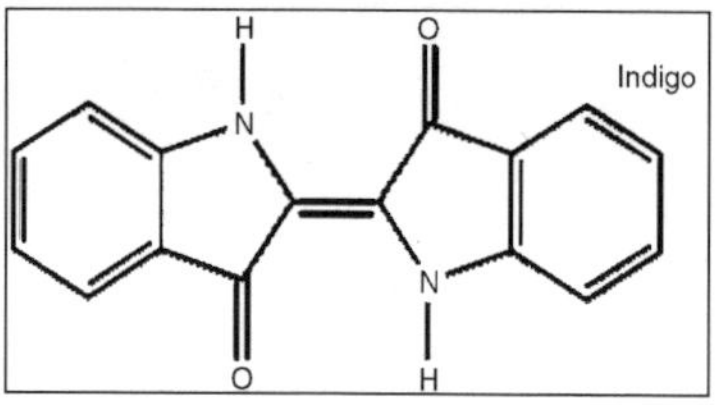

MATERIALS NEEDED

- 6 beakers, 400 mL
- Beaker, 20 mL or 50 mL
- Erlenmeyer flask, 125 mL, with stopper
- 3 stirring rods
- Graduated cylinders, 10 mL and 100 mL
- Tongs or forceps
- Beaker tongs
- Metal spatula
- Malachite green
- P-nitroaniline
- Tannic acid
- Sodium nitrite, NaNO2
- 2-naphthol
- Indigo powder

- TIS Identification stain No. 1
- TIS Identification stain No. 3A
- Iron(II) sulfate, $FeSO4$
- Sodium hydrosulfite, $Na2S2O4$
- $.2H2O$
- Hydrochloric acid, HCl, concentrated
- Sodium hydroxide, NaOH, pellets
- 5per cent sodium hydroxide solution, NaOH
- 10per cent acetic acid solution, $HC2H3O2$
- Ammonia, $NH3$, concentrated
- Thermometer, -10 to 120°C
- Ice
- Cotton, strips, approximately 2 cm x 5 cm
- Wool, strips, approximately 2 cm x 5 cm
- silk, strips, approximately 2 cm x 5 cm
- multifibre fabric, strips, approximately 2 cm x 5 cm
- Rubber or plastic gloves

SAFETY PRECAUTIONS

Wear safety goggles at all times in the laboratory Wear rubber or plastic gloved to protect your hands from the dyes. The dyes used in this experiment should be handled with the utmost care since they will dye the skin and clothing.

Be careful to keep the dye powders and solutions off the desk tops. Clean up any spills immediately. Most dye materials are irritants to the skin, eyes, and respiratory system and may be toxic by inhalation and ingestion. In the event of skin contact, wash the affected area well with water. Hydrochloric acid is corrosive and can cause severe burns to the skin along with irritating fumes. Work with the concentrated solution under a hood. In the event of skin contact, wash the affected area well with water.

If the skin is broken or blistered, seek medical assistance. If irritated by the vapours, get fresh air immediately.

Sodium hydroxide is caustic and can cause severe burns to the skin. Do not handle sodium hydroxide pellets with your hands. In the event of skin contact, wash the affected area well with water until there is no soapy feel. If the skin is broken or blistered, seek medical assistance. Sodium nitrite is an irritant to the skin, eyes, and respiratory system. It is toxic by ingestion.

In the event of skin contact, wash the affected area well with water. Sodium nitrite can also be an explosion hazard.

Do not heat this compound dry or in contact with any flammable material. Ammonia vapours are irritants to the eyes and respiratory system. Work with the concentrated solution under a hood. In the event of skin contact, wash the affected area well with water. If irritated by the vapours, get fresh air immediately.

DISPOSAL AND CLEAN UP

Dispose of all materials in the specific containers provided in the laboratory. Dye stains can be removed from glassware by applying a paste of powdered bleach and water, then rinsing. The paste will also remove stains from the hands, which should be washed off with soap and water.

PROCEDURE

Direct Dyeing

Measure 0.1 g of malachite green. Dissolve the malachite green in 200 ml of distilled or deionized water in a 400-mL beaker. Heat the solution to boiling and place one strip each of cotton, wool, silk, and multifibre fabric in the solution.

Allow the cloth strips to remain in the hot solution for 2 to 3 minutes. Remove the strips of cloth with forceps or tongs and rinse thoroughly with water. Are the colours "fast"? Record your results. Save the solution for the next dyeing procedure.

Mordant Dyeing

Measure 0.2 g of tannic acid. Dissolve the tannic acid in 100 ml of distilled or deionized water in a 400-mL beaker. Heat the solution to boiling and place a strip of cotton and multifibre fabric in the solution. Allow the cloth strips to remain in the hot solution for 1 to 2 minutes.

Remove the cloth strips from the tannic acid solution, place them on a paper towel and press the excess solution from them, then immerse in the hot malachite green solution saved from the first procedure. Allow the cloth strips to remain in the hot malachite green solution for 2 minutes. Remove the cloth strips with forceps or tongs and rinse thoroughly with water. Are the colours "fast"? Record your results.

Developed Dyeing

Measure 3 mL of concentrated hydrochloric acid. Add it to a 400-mL beaker. Place the beaker under a fume hood. Measure 0.1 g of p-nitroaniline. Add it to the hydrochloric acid in the 400-mL beaker. Slowly, with stirring, add 10 mL of distilled or deionized water to the mixture. Place the beaker on an ice bath and cool the solution to about 5°C. Measure 0.1 g of sodium nitrite and dissolve it in 5 ml of distilled or deionized water in a small beaker. Add this solution to the cold p-nitroaniline solution. Measure 0.1 g of 2-naphthol (?-naphthol).

Add it to 5 ml of 5per cent sodium hydroxide solution. Immerse a strip of cotton and multifibre fabric in the naphthol solution. Remove the cloth strips and press to remove excess solution. Now immerse the cloth strips in the cold solution previously prepared. After the dye (para red) forms, test for colourfast properties and record your observations.

VAT DYEING

Measure 0.1 g indigo powder in a 125-mL Erlenmeyer flask. Add 0.1 g of sodium hydrosulfite (Na2S2O4), 2 pellets of sodium hydroxide, and 10 mL of water to the flask. Stopper and shake for 2 to 3 minutes. Dilute to about 50 ml with distilled or deionized water.

Holding a cotton strip with forceps, dip the cloth into the dye solution for about 20 seconds. Remove the cloth and expose to air for a few minutes.

Record your observations Holding a multifibre fabric strip with forceps, dip the cloth into the dye solution for about 20 seconds. Remove the cloth and expose to air for a few minutes. Record your observations. Test both cloth strips for colour fastness and record the results.

ALTERNATIVE PROCEDURE

If sodium hydrosulfite is not available, a mixture of 0.2 g indigo, 1.0 g ferrous sulfate, 5 ml water, and 5 ml concentrated ammonium hydroxide can be used. Heat the solution to boiling for a few minutes, allow the ferric hydroxide to precipitate, pour off the solution.

IDENTIFICATION OF AN UNKNOWN FABRIC

Dyeing tests involve the use of T.I.S. Identification stains. These are special mixtures of dyes that colour different fibres different colours used to identify an unknown fabric or fibres.

Fabric Identifaction using T.I.S. Stain No. 1

Solution to boiling. Maintain a hot, but not actively boiling solution. Wet samples of cotton, wool, silk, and multifibre fabric, with distilled or deionied water.

Squeeze out the excess liquid and place the samples in the hot dye bath for 3 to 5 minutes.

Remove the samples and wash out any excess dye. Compare the colour of the cloth samples with the multifibre fabric.

Fabric Identifaction using T.I.S. Stain No. 3A

A second colour test can be conducted using T.I.S. Identification Stain No. 3A. Prepare a 0.05per cent solution of T.I.S. Identification Stain No. 3A. (0.05 g for each 100 mL water) Heat the solution to boiling and add 1 mL 10per cent acetic acid solution. Maintain a hot, but not actively boiling solution.

Wet samples of cotton, wool, silk, and multifibre fabric, with distilled or deionied water. Squeeze out the excess liquid and place the samples in the hot dye bath for 5 minutes.

Remove the samples and wash out any excess dye. Compare the colour of the cloth samples with the multifibre fabric.

NATURAL DYEING OF TEXTILES

Dyeing is an ancient art which predates written records. It was practised during the Bronze age in Europe. Primitive dyeing techniques included sticking plants to fabric or rubbing crushed pigments into cloth. The methods became more sophisticated with time and techniques using natural dyes from crushed fruits, berries and other plants, which were boiled into the fabric and gave light and water fastness (resistance), were developed.

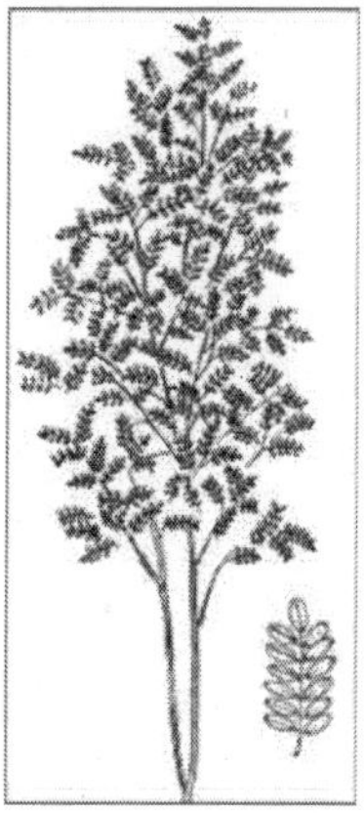

Fig. Indigo.

Some of the well known ancient dyes include madder, a red dye made from the roots of the Rubia tinctorum, blue indigo from the leaves of Indigofera tinctoria, yellow from the stigmas of the saffron plant, and dogwood, an extract of pulp of the dogwood tree. The first use of the blue dye, woad, beloved by the Ancient Britons, may have originated in Palestine where it was found growing wild.

The most famous and highly prized colour through the age was Tyrian purple, noted in the Bible, a dye obtained from the spiny dye-murex shellfish. The Phoenicians prepared it until the seventh century, when Arab conquerors destroyed their dyeing installations in the Levant. A bright red called cochineal was obtained from an insect native to Mexico. All these produced high-quality dark colours. Until the mid-19th century all dyestuffs were made from natural materials, mainly vegetable and animal matter.

Today, dyeing is a complex, specialised science. Nearly all dyestuffs are now produced from synthetic compounds. This means that costs have been greatly reduced and certain application and wear characteristics have been greatly enhanced. But many practitioners of the craft of natural dying (*i.e.* using naturally occurring sources of dye) maintain that natural dyes have a far superior aesthetic quality which is much more pleasing to the eye. On the other hand, many commercial practitioners feel that natural dyes are non-viable on grounds of both quality and economics. In the West, natural dyeing is now practised only as a handcraft, synthetic dyes being used in all commercial applications. Some craft spinners, weavers, and knitters use natural dyes as a particular feature of their work.

In many of the world's developing countries, however, natural dyes can offer not only a rich and varied source of dyestuff, but also the possibility of an income through sustainable harvest and sale of these dye plants. Many dyes are available from tree waste or can be easily grown in market gardens. In areas where synthetic dyes, mordants (fixatives) and other additives are imported and therefore relatively expensive, natural dyes can offer an attractive alternative.

The knowledge required for sourcing and extracting such dyes and mordants is, however, often not available as extensive research work is required to identify suitable plants, minerals, etc. In Zambia for example, there is a wealth of plants available for producing natural dyes, but due to lack of knowledge of the processes involved in harvesting and processing the plants, little use is made of this natural resource.

In some countries, such as India, Nigeria and Liberia, where this research has been carried out, or where there exists a tradition of natural dyeing, natural dyes and mordants are used widely.

TYPES OF TEXTILES SUITABLE FOR DYING

Natural dyes can be used on most types of material or fibre but the level of success in terms of fastness and clarity of colour varies considerably. Users of natural dyes, however, tend to also use natural fibres, and so as suggested, look in more detail at this group.

Natural fibres come mainly from two distinct origins, animal origin or vegetable origin. Fibres from an animal origin include wool, silk, mohair and alpaca, as well as some others which are less well known. All animal fibres are based on proteins. Natural dyes have a strong affinity to fibres of animal origin, especially wool, silk and mohair and the results with these fibres are usually good. Fibres of plant origin include cotton, flax or linen, ramie, jute, hemp and many others. Plant fibres have cellulose as their basic ingredient. Natural dyeing of certain plant based textiles can be less successful than their animal equivalent. Different mordanting techniques are called for with each category. When a blend of fibre of both animal and plant origin is being dyed, then a recipe should be chosen which will accentuate the fibre which is required to be dominant.

EQUIPMENT NEEDED FOR HOME DYEING AND VERY SMALL-SCALE COMMERCIAL DYEING

Most equipment needed for dyeing fabrics at home, or at the very small-scale commercial level, can be found in almost any market place throughout the world.

The following is a list of the equipment requirements and a brief explanation of their use:

- *Heat source*: This can be any type of cooking stove; gas, wood, kerosene, charcoal, electricity. This is used for heating the liquid used during mordanting and dyeing.
- *Pestle and mortar*: Used for milling the natural dye or minerals, where this is called for.
- *Mordanting and dyeing pans*: Stainless steel or enamel pans are the most suitable for dyeing. The size of pan depends upon the quantities of fabric that will be dyed. Do not use pans made from copper, aluminium or iron, unless absolutely necessary, as these metals have properties which can change the colour of the dye.
- *Tirring rods*: Stainless steel or glass rods are best as they can be cleaned and used for different colour dyes. If wooden stirring rods are used then there should be a different spoon for each colour.
- *Thermometer*: This is used to measure the temperature of the liquid during mordanting and dyeing. A long thermometer (to reach the liquid at the bottom of the pan) is preferred, with a range of 0 – 100°C (32 – 210°F).
- *Measuring jugs*: These are used to measure the quantities of liquid called for in the recipe. Sometimes precise quantities are called for.
- *Storage containers*: Used for storing the dyestuffs and mordants. Large glass and plastic jars are ideal. Some mordants and dyes are sensitive to light and should therefore be stored in sealed light-proof containers.
- *Plastic bowls and buckets*: A variety of plastic bowls or buckets of varying sizes are useful when wetting or rinsing fabrics.
- *Strainer*: Used for straining the liquid off the dyestuff in the dyebath.
- *Weighing scales*: Used for obtaining the correct quantities as specified in the recipe. A scales with metric and imperial measurement is useful as conversions from one system to the other are not then needed.
- *Protective equipment*: Gloves for holding hot pans will prevent burns. An apron will protect your clothing. Rubber gloves will prevent skin irritation caused by mordants, and will also prevent you from dyeing your hands. A face mask can cut down the amount of fumes or powder inhaled during the dyeing process.

MORDANTS

Few natural dyes are colour-fast with fibres. Mordants are substances which are used to fix a dye to the fibres. They also improve the take-up quality of the fabric and help improve colour and light-fastness.

The term is derived from the Latin *mordere,* to bite. Some natural dyes, indigo for example, will fix without the aid of a mordant; these dyes are known as 'substantive dyes'. Others dyes, such as madder and weld, have a limited fastness and the colour will fade with washing and exposure to light Traditionally, mordants were found in nature. Wood ash or stale urine may have been used as an alkali mordant, and acids could be found in acidic fruits or rhubarb leaves (which contain oxalic acid), for example.

Nowadays most natural dyers use chemical mordants such as alum, copper sulphate, iron or chrome (there are concerns, however about the toxic nature of chrome and some practitioners recommend that it is not used).

Mordants are prepared in solution, often with the addition of an 'assistant' which improves the fixing of the mordant to the yarn or fibre. The most commonly used mordant is alum, which is usually used with cream of tartar as an additive or assistant. Other mordants are:

- Iron (ferrous sulphate)
- Tin (stannous chloride)
- Chrome (bichromate of potash)
- Copper sulphate
- Tannic acid
- Oxalic acid

Using a different mordant with the same dyestuff can produce different shades, for example;

- *Iron* is used as a 'saddener' and is used to darken colours.
- *Copper sulphate* also darkens but can give shades which are otherwise very difficult to obtain.
- *Tin* brightens colours.
- *Tannic acid,* used traditionally with other mordants, will add brilliancy.
- *Chrome* is good for obtaining yellows.
- *Oxalic acid* is good for extracting blues from berries.
- *Cream of Tartar* is not really a mordant but is used to give a lustre to wool.

Mordants are often poisonous, and in the dye-house they should be kept on a high shelf out of the reach of children. Always use protective clothing when working with mordants and avoid breathing the fumes. The mordant can be added before, during or after the dyeing stage, although most recipes call for mordanting to take place prior to dyeing.

It is best to follow the instructions given in the recipe being used or experiment on a sample before carrying out the final dyeing.

Later in this brief as suggested, explain how the mordant is mixed and used as part of the dyeing process.

These chemical mordants are usually obtained from specialist suppliers or from chemists. Where this is prohibitive, due to location or cost, natural mordants can be used. There are a number of plants and minerals which will yield a suitable mordant, but their availability will be dependent upon your surroundings.

- Some plants, such as mosses and tea, contain a small amount of aluminium. This can be used as a substitute to alum. It is difficult to know, however, how much aluminium will be present and experimentation may be necessary.
- Iron water can be used as a substitute to ferrous sulphate. This can be made simply by adding some rusty nails and a cupful of vinegar to a bucket-full of water and allowing the mixture to sit for a couple of weeks.
- Oak galls or sumach leaves can be used a substitute to tannic acid.
- Rhubarb leaves contain oxalic acid.

NATURAL DYESTUFFS

Dyestuffs and dyeing are as old as textiles themselves. Nature provides a wealth of plants which will yield their colour for the purpose of dyeing, many having been used since antiquity.

As suggested, look at some of these naturally occurring dyes, their source and the colours they produce. Later in the brief as suggested, look at the application of the dyes to textiles.

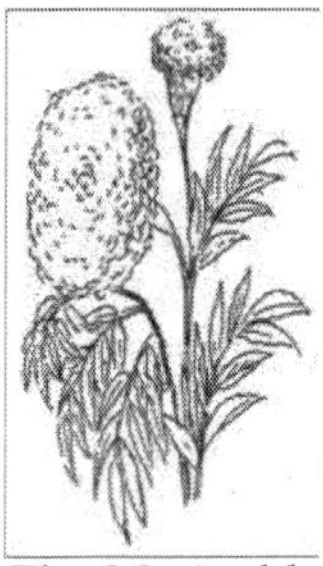

Fig. Marigold.

Almost any organic material will produce a colour when boiled in a dye-bath, but only certain plants will yield a colour that will act as a dye. The plants given in Table are a selection of plants that have stood the test of time, and are used widely and traditionally by natural dyers.

Natural dyes fall into the following categories:

- Leaves and stems
- Twigs and prunings
- Flower heads
- Barks
- Roots
- Outer skins, hulls and husks
- Heartwoods and wood shavings
- Berries and seeds
- Lichens
- Insect dyes

The choice of mordant for a particular plant is dependant upon the material with which it will be used. It is necessary to check a recipe before using a plant, or one can experiment to see what effect a mordant has for a particular application.

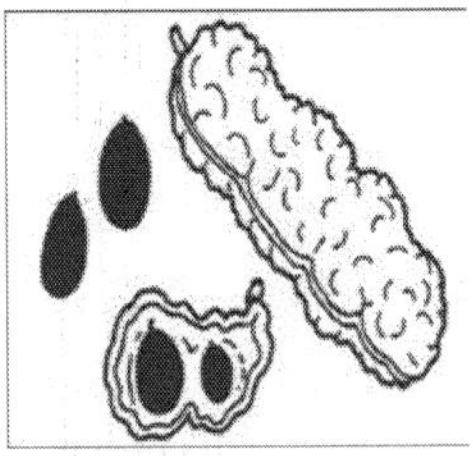

Fig. Groundnuts.

It is recommended that plants be grown specifically for the purpose of dyeing. Harvesting plants from the wild on a non-sustainable basis can endanger the survival of the plant. Many lichens are registered as protected organisms and it is illegal to gather them from the wild. One source of lichen is from timber mills where logged trees are being processed, but ask first!

TESTING DYES

It is always useful and interesting to test the dye which is to be used on a sample of the yarn or fabric to be dyed. The outcome will depend on the fabric, the mordant that has been used and dye that has have been chosen.

Testing is best carried out on a series of marked (for identification) samples, which have been mordanted with a number of different mordants. Tests can be carried out for light, water and washing fastness using simple standard test methods. Box 1 shows a simple test for washing fastness. Similar tests exists for water and light fastness.

Processing of the plant can take one of many forms, but usually takes the form of soaking or boiling the plant to extract the dye. Some plants, such as indigo, need special preparation for use. Some plants will need boiling to extract their dye while others can merely be soaked for extended periods.

PROCESS OF DYEING

Application of the Dye

Dyeing can be carried out at any of the following stages in the textile manufacturing stage:

- The fibres can be dyed before they are spun. *Fibre dyeing* provides a deep penetration of the dye into the fibre, giving even colour and excellent colour-fastness.
- The yarn can be dyed after spinning but before the product is woven or otherwise fabricated. This is called *package dyeing.* Before the fabric is finished, it can be dyed in lengths (*piece dyeing*).This process allows

manufacturers the opportunity to produce fabrics in their natural colours, and then dye them to order.

- In *cross-dyeing,* fabrics of two or more fibres can be dyed so that each fibre accepts a different dyestuff and becomes a different colour, through the use of appropriate dyestuffs for each fibre.

It is essential for the correct identification of the fibre or other fabric to be made before dyeing commences.

METHODS OF DYEING

There are a number of methods of applying dye to a fabric. Although the most common method used for applying natural dyes is the vat method, there are techniques which have been developed to allow patterns to be incorporated during the dying process.

It is worth bearing in mind that using natural dyes is a complex art and the skills required for using natural dyes are learned over many years. Don't be put off if you don't get the desired results at the first attempt!.

Vat Dyeing

In the simplest form of dyeing a textile material is immersed in dye and gradually brought to the boil. Alternatively the fibre is allowed to sit and soak for several hours or days. During this period, agitation is necessary to allow full penetration of the textile by the dyestuff. Depending on the type of fabric and dyestuff used, certain salts or acids may be added to assist absorption of the dye.

The principal difficulty in dyeing mixed yarns and fabrics is to achieve the same colour in both fibres. Cotton fibres may, for instance, absorb dyes rapidly, while the wool fibres will have to be boiled over an extended period to reach the same depth of shade. This could lead to significant damage to the material.

In this case a chemical compound would need to be used to restrain the rate at which the cotton fibre takes up the dyestuff. The amount of dyestuff which is used is usually given in the recipe.

It is usually quoted as a percentage weight of the fabric to be dyed. Box 2 demonstrates a typical, simple recipe for dyeing using natural dyestuff, in this case onion skins.

Batik

Batik is a starch resist-dyeing process, developed on the Island of Java in modern-day Indonesia. Colour is prevented from reaching certain areas of a fabric by covering these areas with molten wax.

The fabric is starched prior to the design being drawn upon it. The wax is applied with a type of cup with a fine pouring spout, usually made of copper. The technique has been developed to a high art form in Indonesia from where it is exported to many parts of the world. Batik paintings, as well as sarongs and lengths of fabric, are produced. When the fabric is dyed, all waxed areas resist the dyestuff. The wax is then removed by placing the fabric in boiling water. For patterns with many colours the same procedure is repeated until the full design is completed.

Tie-dyeing

Tie-dyeing is another popular artisanal dyeing technique. In this resist-dyeing process, waxed thread is tightly tied around the areas chosen to resist the coloured dyestuff, and the fabric is dipped into the dye. The waxed thread is then removed and the fabric dried. This process can be repeated for each colour to be added.

OBTAINING UNUSUAL COLOURS

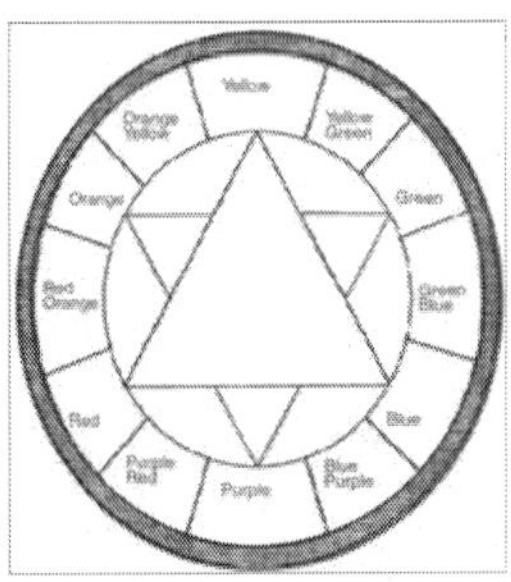

Colours other than those obtained by simply using a single dye, can be obtained by mixing dyes or dyeing a fabric more than once in different dyebaths. The colour triangle is a useful tool in determining the colours needed to produce a required shade.

FABRIC DYEING TIPS

METHODS

- *Washing Machine*: Most convenient.
- *Stovetop*: For very dark colours like black, where the amount of fabric will fit in a large pot.
- *Sink or Bucket*: Suitable for small or delicate items being dyed a light colour.

WASHING MACHINE METHOD

STEP 1

- Take 1 package or a ½ bottle of RIT dye (2 packs or a full bottle will give a full colour particularly for medium, bright or dark colours) and place it in an old sauce pan with a quart of water (1 pint of water per package used).
- Place sauce pan on the stove and heat it up, stirring while it heats. The purpose here is to totally dissolve the dye into the water.

STEP 2

- Fill your washer with hot water (140 degrees is best for dark colours), add the dye/water from your saucepan, plus a splash of vinegar to help set the dye, and turn on the agitator for several minutes- to mix in the dye solution.

STEP 3

- Take a piece of rag and put it in the washer for a few minutes. Pull it out and look at the colour to see if it's what you want. If its too light then add more dissolved dye.
- Never add more than 4 packages or 2 bottles per washer load when dyeing a large amount of fabric.

STEP 4

- Turn machine on again, and let it agitate for another 3 minutes. — This will assure that all the dye is thoroughly mixed in.
- Never pour dye directly on fabric.

STEP 5

- Wet the garments to be dyed and place them loosely into washer and turn machine on.

STEP 6

- Take garments out of machine after the cycle and either line dry or put in dryer.
- If you use the dryer, be sure to wipe it out with a bleach cloth after use to remove any dye.

STEP 7

Clean-Up:

- Fill washer to highest water level.
- Add a small amount of detergent plus a cup of bleach.
- Run a complete cycle.

- Finish by wiping the top of the tub and inside lid with bleach.
- Clean out lint trap.
- Plastic or rubber machine parts may be tinted but won't stain future laundry loads.
- Clean up saucepan and any other items used.

SEWING

Sewing or stitching or tailoring is the fastening of cloth, leather, furs, bark, or other flexible materials, using needle and thread. Its use is nearly universal among human populations and dates back to Paleolithic times (30,000 BCE). Sewing predates the weaving of cloth.

Sewing is used primarily to produce clothing and household furnishings such as curtains, bedclothes, upholstery, and table linens. It is also used for sails, bellows, skin boats, banners, and other items shaped out of flexible materials such as canvas and leather.

Most sewing in the industrial world is done by machines. Pieces of a garment are often first tacked together. The machine has a complex set of gears and arms that pierces thread through the layers of the cloth and semi-securely interlocks the thread.

Some people sew clothes for themselves and their families. More often home sewers sew to repair clothes, such as mending a torn seam or replacing a loose button. A person who sews for a living is known as a seamstress (from seams-mistress) or seamster (from seams-master), dressmaker, tailor, garment worker, or machinist.

"Plain" sewing is done for functional reasons: making or mending clothing or household linens. "Fancy" sewing is primarily decorative, including techniques such as shirring, smocking, embroidery, or quilting.

Sewing is the foundation for many needle arts and crafts, such as applique, canvas work, and patchwork. While sewing is sometimes seen as a semi-skill job, flat sheets of fabric with holes and slits cut into the fabric can curve and fold in complex ways that require a high level of skill and experience to manipulate into a smooth, ripple-free design.

Aligning and orienting patterns printed or woven into the fabric further complicates the design process. Once a clothing designer with these skills has created the initial product, the fabric can then be cut using templates and sewn by manual laborers or machines.

SEAM ALLOWANCE

Seam allowance is the area between the edge of the fabric and the line of stitching. It is usually 1.5 cm away from the edge of the fabric except for the hem, where the seam allowance is usually 2.5 cm or more. This is often the case for standard home dressmaking. Industry seam allowances vary but are usually 0.6 cm.

OCCUPATIONS REQUIRING SEWING

- Bookbinding/Bookbinder
- Shoemaking/Cobbler
- Corsetier
- Draper
- Dressmaker
- Glove/Glover
- Hatter
- Milliner
- Parachute rigger
- Quilting/Quilter
- Sailmaker
- Seamstress
- Tailor
- Taxidermist
- Upholsterer
- Leatherworker

SEWING TOOLS AND ACCESSORIES

Stitching Awl

The Stitch-It-Awl is a hand sewing awl for safely and quickly sewing by hand mainly medium to heavy materials such as sailcloth, canvas, leather, soft luggage, upholstery or webbing. It was developed to solve the problems associated with the use of existing hand sewing devices for heavy materials, such as the needle and leather palm used by sail makers.

Bobbin

A bobbin is the little plastic or metal wheel that fits underneath the needle plate of a sewing machine. Thread runs off the bobbin, gets picked up by the needle's thread and forms the bottom part of a seam. The bobbin is an essential part of a sewing machine. It enables the machine to form a seam from two threads, which makes a stronger, smoother, neater seam. Even the best hand sewer cannot duplicate the look of a machine seam. Bobbins, in one form or another, have been around for years. They have always been a part of the sewing machine, even in the old treadle machine days. Using the term bobbin for this implement probably dates from the days of weaving, since the weft threads of a loom are wound around bobbins. The word bobbin is of indeterminate age and origin, which in English, means it probably dates back at least to Middle English.

The sewing machine bobbin was wound by hand for early machines, but the advent of the electric motor sewing machine meant that the bobbin could be wound on the machine itself. After all, the motor turned, so why not put it to another useful purpose?

Depending on the sewing machine, a bobbin may be wound on top of the machine or underneath the needle plate. Winding a bobbin on top of the sewing machine is not difficult, but the operator must make sure to wind the thread in the right direction. Putting the bobbin on upside down will wind the thread off the bobbin. Some sewing machines require the user to put the bobbin underneath the needle plate, hold the needle thread and turn the hand-wheel until the needle thread "catches" the bobbin thread.

Other machines have a "drop-in" bobbin, so the user only needs to make certain the bobbin is oriented correctly and drop it into the correct place. The machine then picks up the bobbin thread on its own.

A sewer should make sure that the sewing machine's bobbin is of the proper size for the machine. A plastic bobbin should not have cracks or splits. Bobbins are available for most sewing machine models in stores where sewing machines and fabrics are sold, as well as online.

Bodkin

Several different implements go by the name of bodkin, including a weapon, an archaic type of hairpin, and two different sewing notions. As a weapon, a bodkin is a dagger; as a hairpin, it's an ornament with a stiletto shape; and as a sewing tool, it's either an instrument similar to an awl for piercing holes in cloth or a long needle with a large eye and blunt tip for drawing tape, elastic, or ribbon through a casing. The needle-like bodkin is also an item of interest.

It was the tool used to lace stays and bodices. A bodkin was stored in a bodkin case, which looks rather like a flat rectangular version of a lipstick container. Modern bodkins may have several special features. Some needle-like bodkins have a ballpoint tip for ease of movement through the casing. A flat bodkin is a small metal tool that resembles a matchstick with a hole (or holes) in the end for threading.

A third type of bodkin is a piece of wire with one end covered with a plastic tip and the other formed into a loop. A completely different style of bodkin is fashioned like a miniature tongs with gripping teeth on its ends to grasp the material being threaded and a neck ring to tighten and hold it in place. It may prove useful to know that in the absence of a threading bodkin, several other tools can be substituted. A tapestry needle, with its large eye and blunt tip, can serve the purpose of a bodkin by tying the material around the eye so it can't come loose. A large safety pin can also be used by piercing the material to be drawn through the casing, closing the pin, and working it around in the same manner as the bodkin.

Dress Form

A dress form is used to give a three-dimensional view on the substance of clothing that is being sewed. They come in all sizes and shapes for almost every substance of clothing that can be made. When a piece of clothing is made, it can be put on the dress form so one can see how the piece of clothing will turn out. Then one can make alterations upon the clothing after seeing what it looks like on a body.

Scissors

Scissors are hand-operated cutting instruments. They consist of a pair of metal blades, or tangs, connected in such a way that the sharpened edges slide against each other. Scissors are used for cutting various thin materials, such as paper, cardboard, metal foil, thin plastic, cloth, rope and wire. Scissors can also be used to cut hair and food.

Scissors and shears exist in a wide variety of forms depending on their intended uses. Children's scissors, used only on paper, have dull blades and rounded corners to ensure safety. Scissors used to cut hair or fabric must be much sharper. The largest shears used to cut metal or to trim shrubs must have very strong, sharp blades.

Specialized scissors include sewing scissors, which often have one sharp point and one blunt point for intricate cutting of fabric, and nail scissors, which have curved blades for cutting fingernails and toenails. Special kinds of shears include pinking shears, which have notched blades that cut cloth to give it a wavy edge, and thinning shears, which have teeth that cut every second hair strand, rather than every strand giving the illusion of thinner hair.

Tape Measure

A tape measure or measuring tape is a flexible form of ruler. It consists of a ribbon of cloth, plastic, fibre glass, or metal strip with linear-measurement markings. It is a common measuring tool. Its flexibility allows for a measure of great length to be easily carried in pocket or toolkit and permits one to measure around curves or corners. Today it is ubiquitous, even appearing in miniature form as a keychain fob, or novelty item. Surveyors use tape measures in lengths of over 100 m (300+ ft).

Sewing Needle

A sewing needle is a long slender tool with a pointed tip. The first needles were made of bone or wood; modern ones are manufactured from high carbon steel wire, nickel- or gold plated for corrosion resistance. The highest quality embroidery needles are plated with two-thirds platinum and one-thirds titanium alloy. Traditionally, needles have been kept in needle books or needle cases which have become an object of adornment.

A needle for hand sewing has a hole, called the eye, at the non-pointed end to carry thread or cord through the fabric after the pointed end pierces it. Hand sewing needles have different names depending on their purpose.

Needle size is denoted by a number on the packet. The convention for sizing is that the length and thickness of a needle increases as the size number decreases. For example, a size 1 needle will be thicker and longer, while a size 10 will be shorter and finer.

Pattern (Sewing)

In sewing and fashion design, a pattern is an original garment from which other garments of a similar style are copied, or the paper or cardboard templates from which the parts of a garment are traced onto fabric before cutting out and assembling (sometimes called paper patterns). Patternmaking, pattern making or pattern cutting is the art of designing patterns. A custom-fitted basic pattern from which patterns for many different styles can be created is called a sloper or block.

Patterns for Custom Dressmaking

A custom dressmaker frequently employs one of three pattern creation methods. The flat-pattern method begins with the creation of a sloper or block, a basic pattern for a fitted, jewel-neck bodice and narrow skirt, made to the wearer's measurements.

The sloper is usually made of lightweight cardboard or tagboard, without seam allowances or style details. Once the shape of the sloper has been refined by making a series of mock-up garments called toiles (UK) or muslins (US), the final sloper can be used in turn to create patterns for many styles of garments with varying necklines, sleeves, dart placements, and so on.

Although it is also used for women's' clothing, the drafting method is more commonly employed in men's garments and involves drafting a pattern directly onto pattern paper using a variety of straightedges and curves. The pattern draping method is used for more elaborate and unique designs that are hard to obtain through the flat pattern method. This is because it is nearly impossible to account for the way a fabric will drape or hang on the body without an actual 3-dimensional test run.

It involves creating a muslin mock-up pattern by pinning fabric directly on a dress form, then transferring the muslin outline and markings onto a paper pattern or using the muslin as the pattern itself.

Patterns for Home Sewing

Home sewing patterns are generally printed on tissue paper and sold in packets containing sewing instructions and suggestions for fabric and trim. Modern patterns are available in a wide range of prices, sizes, styles, and sewing skill levels, to meet the needs of consumers. Home sewing patterns are graded, that is, redrawn to fit larger and smaller sizes than the original design. Ebenezer Butterick invented the commercially produced graded home sewing pattern in 1863 (based on grading systems used by Victorian tailors), originally selling hand-drawn patterns for men's and boys' clothing.

In 1866, Butterick added patterns for women's clothing, which remains the heart of the home sewing pattern market today. There are some applications today that enable a home sewer to customise a computerized pattern to fit her body measurements and or body shape.

The 3D technology enables the home sewer to see a virtual simulation of the final garment as it will appear on her. This reduces the Time-to-Market as well as the number of muslins/test garments that are needed.

A variation on the theme was evolved by iconic British brand Clothkits. Clothkits devised ingenious cut and sew clothing kits for home sewing that avoided the need for paper patters. Rather than using conventional techniques, Clothkits pre printed fabric with both quirky designs and the pattern lines, to make dressmaking for the novice easier.

Pin

A pin is a device used for fastening objects or material together. It is usually made of steel, or on occasion copper or brass. It is formed by drawing out a thin wire, sharpening the tip, and adding a head. Nails are related, but are typically larger.

Pincushion

A pincushion (or less commonly pin cushion) is a small cushion, typically 3-5 cm across, which is used in sewing to store pins or needles with their heads protruding so as to take hold of them easily, collect them, and keep them tidy.

Pincushions are typically filled tightly with stuffing, so as to hold pins rigidly once placed. The stereotypical design – a tomato with a strawberry attached – seems to have been around forever but, most likely, was introduced during the Victorian Era.Typically, the tomato is filled with wool roving to prevent rust, and the strawberry is filled with an abrasive to clean the pins and keep them sharp.

Pincushions come in all shapes, sizes and forms. Magnetic pincushions have the advantage of being able to easily keep the work area clear of pins, but the drawback that the pins become magnetized easily and an attempt to retrieve one may pick up several.

Rotary Cutter

A rotary cutter is a tool generally used by quilters to cut fabric. It consists of a handle with a circular blade that rotates, thus the tool's name.

Rotary blades can be found in different sizes: usually smaller blades are used to cut around corners, while larger blades are used to cut to straight lines. Several layers of fabric can be cut simultaneously, making it easier to cut out patchwork pieces of the same shape and size than with regular scissors.

In conjunction with a rotary cutting ruler – specially designed to be used with a rotary cutter, and made of thick and resistant plastic – it is also a practical tool for achieving the straight, squared lines that are so important in patchwork. The first rotary cutter was introduced by the Olfa company in 1979.

Seam Ripper

A seam ripper, stitch ripper or stitch unpicker is a small tool used for unpicking stitches. The most common form consists of a handle, shaft and head. The head is usually forked with one side of the fork flattening out and becoming a blade and the other side forming a small point. In some designs the blade side then tapers back to a point to allow easier insertion in tight stitching.

In use the blade is inserted into the seam underneath the thread to be cut. The thread is allowed to slip down into the fork and the tool is then lifted upwards allowing the blade to cut through the thread. Once the seam has been undone in this way the loose ends can be removed and the seam resewn.

Sewing Table

A sewing table or work table is a table or desk used for sewing. Generally is has large amounts of space and a full set of sewing tools. Nearby there will be a chair and a waste bin. A common attachment is a dropleaf to give expanded space. Other attachments can be a cloth bag for storing sewing materials, drawers, or shelves.

Chalk

Chalk is a soft, white, porous sedimentary rock, a form of limestone composed of the mineral calcite. Calcite is calcium carbonate or $CaCO_3$. It forms under relatively deep marine conditions from the gradual accumulation of minute calcite plates (coccoliths) shed from micro-organisms called coccolithophores. It is common to find chert or flint nodules embedded in chalk. Chalk can also refer to other compounds including magnesium silicate and calcium sulfate.

Chalk is resistant to weathering and slumping compared to the clays with which it is usually associated, thus forming tall steep cliffs where chalk ridges meet the sea. Chalk hills, known as chalk downland, usually form where bands of chalk reach the surface at an angle, so forming a scarp slope. Because chalk is porous it can hold a large volume of ground water, providing a natural reservoir that releases water slowly through dry seasons.

Thimble

A thimble is a protective shield worn on the finger or thumb. It is generally used for sewing. The earliest known thimble was Roman and was found at Pompeii. Made of bronze, its creation has been dated to the first century AD. A second Roman thimble was found at Verulamium, present day St Albans, in the UK and can be viewed in the museum there.

United Kingdom Detector Finds Database, thimbles dating to the tenth century have been found in England, and thimbles were in widespread use there by the 1300s. Although there are isolated examples of thimbles made of precious metals — Elizabeth I is said to have given one of her ladies-in-waiting a thimble set with precious stones — the vast majority of metal thimbles were made of brass.

Medieval thimbles were either cast brass or made from hammered sheet. Early centres of thimble production were those places known for brass-working, starting with Nuremburg in the 15th century, and moving to Holland by the 17th.

In 1693, a Dutch thimble manufacturer named John Lofting established a thimble manufactory in Islington, in London, England, expanding British thimble production to new heights.

He later moved his mill to Buckinghamshire to take advantage of water-powered production, resulting in a capacity to produce more than two million thimbles per year. By the end of the 18th century, thimble making had moved to Birmingham, and shifted to the "deep drawing" method of manufacture, which alternated hammering of sheet metals with annealing, and produced a thinner-skinned thimble with a taller shape. At the same time, cheaper sources of silver from the Americas made silver thimbles a popular item for the first time.

Thimbles are usually made from metal, leather, rubber, and wood, and even glass or china. Early thimbles were sometimes made from whale bone, horn, or ivory. Natural sources were also utilized such as Connemara marble, bog oak, or mother of pearl. Rarer works from thimble makers utilized diamonds, sapphires, or rubies.

Advanced thimblemakers enhanced thimbles with semi-precious stones to decorating the apex or along the outer rim. Cabochon adornments are sometimes made of cinnabar, agate, moonstone, or amber. Thimble artists would also utilize enameling, or the Guilloché techniques advanced by Peter Carl Fabergé. Originally, thimbles were used solely for pushing a needle through fabric or leather as it was being sewn. Since then, however, they have gained many other uses.

In the 1800s they were used to measure spirits, which brought rise to the phrase "just a thimbleful". Prostitutes used them in the practice of thimble-knocking where they would tap on a window to announce their presence. Thimble-knocking also refers to the practice of Victorian schoolmistresses who would tap on the heads of unruly pupils with dames thimbles.

Before the 18th century the small dimples on the outside of a thimble were made by hand punching, but in the middle of that century, a machine was invented to do the job. If one finds a thimble with an irregular pattern of dimples, it was likely made before the 1850s.

Another consequence of the mechanization of thimble production is that the shape and the thickness of the metal changed. Early thimbles tend to be quite thick and to have a pronounced dome on the top. The metal on later ones is thinner and the top is flatter. Collecting thimbles became popular in the UK when many companies made special thimbles to commemorate the Great Exhibition held in the Crystal Palace in Hyde Park, London.

In the 19th century, many thimbles were made from silver; however, it was found that silver is too soft a metal and can be easily punctured by most needles. Charles Horner solved the problem by creating thimbles consisting of a steel core covered inside and out by silver, so that they retained their aesthetics but were now more practical and durable. He called his thimble the Dorcas, and these are now popular with collectors. There is a small display of his work in Bankfield Museum, Halifax, England.

Early American thimbles made of whale bone or tooth featuring miniature scrimshaw designs are considered valuable collectibles. Such rare thimbles are prominently featured in a number of New England Whaling Museums. During the First World War, silver thimbles were collected from "those who had nothing to give" by the British government and melted down to buy hospital equipment. In the 1930s and 40s red-topped thimbles were used for advertising.

Leaving a sandalwood thimble in a fabric store was a common practice for keeping moths away. Thimbles have also been used as love-tokens and to commemorate important events. A miniature thimble is one of the tokens in the game of Monopoly. People who collect thimbles are known as digitabulists.

Yarn

Yarn is a long continuous length of interlocked fibres, suitable for use in the production of textiles, sewing, crocheting, knitting, weaving, embroidery and ropemaking. Thread is a type of yarn intended for sewing by hand or machine.

Modern manufactured sewing threads may be finished with wax or other lubricants to withstand the stresses involved in sewing. Embroidery threads are yarns specifically designed for hand or machine embroidery.

Tracing Paper

Tracing paper is a type of translucent paper. It is made by immersing uncut and unloaded paper of good quality in sulphuric acid for a few seconds. The acid converts some of the cellulose into amyloid form having a gelatinous and impermeable character.

When the treated paper is thoroughly washed and dried, the resultant product is much stronger than the original paper. Tracing paper is resistant to oil grease and to a large extent impervious to water and gas.

Tracing paper is named as such for its ability for an artist to trace an image onto it. When tracing paper is placed onto a picture, the picture is easily viewable through the tracing paper. Thus, it becomes easy for the artist to find edges in the picture and trace the image onto the tracing paper. Pure Cellulose fibre is translucent, and it is the air trapped between fibres, that makes paper opaque and looks white. If the fibres are refined and beaten until all the air is taken out, then the resulting sheet will be translucent.

Translucent papers are dense and contain up to 10per cent moisture at 50per cent humidity. This type of paper is roughly 25per cent lighter than regular paper.

The sizing in production will determine whether it is for laser printer or inkjet/offset printing. Tracing paper may be uncoated or coated. Natural tracing paper for laser printing is usually uncoated.

Tracing Wheel

A tracing wheel is an instrument with serrated teeth on a wheel attached to a handle. Used to transfer markings from patterns onto fabric with or without tracing paper, this sewing tool also makes slotted perforations. Such markings might include pleats, darts, buttonholes, notches or placement wikt:lines for appliques or pockets. There are two basic types of tracing wheels available to the modern sewer, one with a serrated edge and one with a smooth edge.

Wax

Traditionally, wax (or beeswax) is a substance secreted by bees and used in constructing their honeycombs.

The term has come to refer more generally to a class of substances with properties similar to beeswax:

- Plastic (malleable) at normal ambient temperatures
- A melting point above approximately 45 °C (113 °F) (which differentiates waxes from fats and oils)
- A relatively low viscosity when melted (unlike many plastics)
- Insoluble in water
- Hydrophobic

Waxes may be natural secretions of plants or animals, artificially produced by purification from natural petroleum or completely synthetic. In addition to beeswax, carnauba (a plant epicuticular wax) and paraffin (a petroleum wax) are commonly encountered waxes which occur naturally. Earwax is an oily substance found in the human ear.

SEWING EQUIPMENT

BAR TACK

Bar tack is a series of hand or machine made stitches used for reinforcing areas of stress on a garment, such as pocket openings, bottom of a fly opening or buttonholes. It consists of a series of close-set zig-zag stitches (machine) or whip-stitches (hand), usually 1/16"-1/8" in width and 1/4"-3/8" in length. In denim jeans, it is often in a contrasting colour, such as orange or white.

BLANKET STITCH

The blanket stitch is a stitch used to reinforce the edge of thick materials. Depending on circumstances, it may also be called a whip stitch or a crochet stitch. It is defined as "A decorative stitch used to finish an unhemmed blanket. The stitch can be seen on both sides of the blanket."

This stitch has long been both an application by hand and as a machine sewn stitch. When done by hand, it is sometimes considered a crochet stitch. When done by machine, it is called a whip stitch or, sometimes, a Merrow Machine Company Crochet Stitch, after the first sewing machine that was used to sew a blanket stich.

This machine was produced and patented by the Merrow Machine Company in 1877. The defining characteristic of the crochet machine is its ability to sew with yarn and stitch thick goods with a consistent overlock edge. From 1877-1925 the machine evolved dramatically, and consequently so did the capacity of manufacturers to produce goods with the whip stitch.

To this day, Merrow retains a patent on the mechanism that creates the blanket stitch, renewing it with each redesign, which has occurred hundreds of times since its introduction in 1877. The most recent patent renewal occurred in 2007. In the early 1990s Chinese manufacturers copied the design of the original crochet sewing machines; Merrow is currently involved in litigation with several Chinese firms who are accused of violating Merrow's intellectual property rights.

BUTTONHOLE STITCH

Buttonhole stitch and the related blanket stitch are hand-sewing stitches used in tailoring, embroidery, and needle lace-making. Buttonhole stitches catch a loop of the thread on the surface of the fabric and needle is returned to the back of the fabric at a right angle to the original start of the thread.

The finished stitch in some ways resembles a letter "L" depending on the spacing of the stitches. For buttonholes the stitches are tightly packed together and for blanket edges they are more spaced out.

The properties of this stitch make it ideal for preventing raveling of woven fabric. Buttonhole stitches are structurally similar to featherstitches.

Applications

In addition to reinforcing buttonholes and preventing cut fabric from raveling, buttonhole stitches are used to make stems in crewel embroidery, to make sewn eyelets, to attach applique to ground fabric, and as couching stitches. Buttonhole stitch scallops, usually raised or padded by rows of straight or chain stitches, were a popular edging in the 19th century. Buttonhole stitches are also used in cutwork, including Broderie Anglaise, and form the basis for many forms of needlelace.

Variants

Examples of buttonhole or blanket stitches include:

- Blanket stitch
- Buttonhole stitch
- Closed buttonhole stitch, in which the tops of the stitch touch to form triangles
- Crossed buttonhole stitch, in which the tops of the stitch cross
- Detached buttonhole stitch, in which rows of buttonhole stitches are worked to form a "floating" filling stitch
- Buttonhole shading, in which rows of buttonhole stitch are sewn in related colours to give a naturalistic shaded effect
- *Buttonhole stitches combined with knots include*:
  - Top knotted buttonhole stitch
  - German knotted buttonhole stitch
  - Tailor's buttonhole stitch
  - Armenian edging stitch

Buttonhole bars are parallel rows of thread laid across an open space in lace or cutwork and then completely covered with closely space buttonhole stitches.

CHAIN STITCH

Chain stitch is a sewing and embroidery technique in which a series of looped stitches form a chain-like pattern. Chain stitch is an ancient craft - examples of surviving Chinese chain stitch embroidery worked in silk thread have been dated to the Warring States period (5th-3rd century BC). Handmade chain stitch embroidery does not require that the needle pass through more than one layer of fabric.

For this reason the stitch is an effective surface embellishment near seams on finished fabric. Because chain stitches can form flowing, curved lines, they are used in many surface embroidery styles that mimic "drawing" in thread. Chain stitches are also used in making tambour lace, needlelace, macramé and crochet.

CROSS-STITCH

Cross-stitch is a popular form of counted-thread embroidery in which X-shaped stitches in a tiled, raster-like pattern are used to form a picture. Cross-stitch is usually executed on easily countable evenweave fabric called aida cloth. The stitcher counts the threads in each direction so that the stitches are of uniform size and appearance. This form of cross-stitch is also called counted cross-stitch in order to distinguish it from other forms of cross-stitch. Sometimes cross-stitch is done on designs printed on the fabric (stamped cross-stitch); the stitcher simply stitches over the printed pattern.

DARNING

Darning is a sewing technique for repairing holes or worn areas in fabric or knitting using needle and thread alone. It is often done by hand, but it is also possible to darn with a sewing machine. Hand darning employs the darning stitch, a simple running stitch in which the thread is "woven" in rows along the grain of the fabric, with the stitcher reversing direction at the end of each row, and then filling in the framework thus created, as if weaving.

Darning is a traditional method for repairing fabric damage or holes that do not run along a seam, and where patching is impractical or would create discomfort for the wearer, such as on the heel of a sock.

Darning also refers to any of several needlework techniques that are worked using darning stitches:

- Pattern darning is a type of embroidery that uses parallel rows of straight stitches of different lengths to create a geometric design.
- Net darning, also called filet lace, is a 19th century technique using stitching on a mesh foundation fabric to imitate lace.
- Needle weaving is a drawn thread work embroidery technique that involves darning patterns into barelaid warp or weft threads.

LOCKSTITCH

A lockstitch is the most common mechanical stitch made by a sewing machine. The lockstitch uses two threads, an upper and a lower. The upper thread runs from a spool kept on a spindle on top of or next to the machine, through a tension mechanism, a take-up arm, and finally through the hole in the needle.

The lower thread is wound onto a bobbin, which is inserted into a case in the lower section of the machine.

To make one stitch, the machine lowers the threaded needle through the cloth into the bobbin area, where a hook catches the upper thread at the point just after it goes through the needle.

The hook mechanism carries the upper thread entirely around the bobbin case, so that it has made one wrap of the bobbin thread.

Then the take-up arm pulls the excess upper thread (from the bobbin area) back to the top forming the lockstitch ideally in the centre of the thickness of the material, the tension mechanism prevents the thread from being pulled from the spool side, the needle is pulled out of the cloth, and the feed dogs pull the cloth back one stitch length, the cycle is repeated as the machine turns mechanically.Lockstitch is so named because the two threads, upper and lower, "lock" together in the hole in the fabric which they pass through.

The term "single needle stitching", often found on dress shirt labels, refers to lockstitch, as opposed to chain stitch which unravels easily and is usually used on lower quality garments.

OVERLOCK

An overlock stitch sews over the edge of one or two pieces of cloth for edging, hemming or seaming. Usually an overlock sewing machine will cut the edges of the cloth as they are fed through (such machines are called 'sergers'), though some are made without cutters.

The inclusion of automated cutters allows overlock machines to create finished seams easily and quickly. An overlock sewing machine differs from a lockstitch sewing machine in that it uses loopers fed by multiple thread cones rather than a bobbin. Loopers serve to create thread loops that pass from the needle thread to the edges of the fabric so that the edges of the fabric are contained within the seam. Overlock sewing machines usually run at high speeds, from 1000 to 9000 rpm, and most are used in industry for edging, hemming and seaming a variety of fabrics and products. Overlock stitches are extremely versatile, as they can be used for decoration, reinforcement, or construction.

Overlocking is also referred to as "overedging", "merrowing" or "serging". Though "serging" technically refers to overlocking with cutters, in practice the four terms are used interchangeably.

PAD STITCH

Pad stitches are types of running stitches, made by taking small stitches perpendicular to the line of stitching. Pad stitches secure two or more layers of fabric together and give the layers more firmness; smaller/denser stitches create more firmness. Pad stitches may also enforce an overall curvature of the layers. Tailors pad stitch a jacket's lapel and undercollar to give them more firmness and to help them maintain their curvature.

The line of stitching usually runs parallel the direction of the most important curve of the layers. For example, pad stitches in a suit's lapel run parallel to the lapel's roll line; pad stitches in the undercollar of a tailored jacket run parallel to the collar back edge.

RUNNING STITCH

The running stitch or straight stitch is the basic stitch in hand-sewing and embroidery, on which all other forms of sewing are based. The stitch is worked by passing the needle in and out of the fabric. Running stitches may be of varying length, but typically more thread is visible on the top of the sewing than on the underside. Running stitches are used in hand-sewing and tailoring to sew basic seams, in hand patchwork to assemble pieces, and in quilting to hold the fabric layers and batting or wadding in place. Loosely spaced rows of short running stitches are used to support padded satin stitch.

Running stitches are a component of many traditional embroidery styles, including kantha of India and Bangladesh, and Japanese sashiko quilting.

TACK (SEWING)

In sewing, to tack or baste is to make quick, temporary stitching intended to be removed.

Tacking is used in a variety of ways:

- To temporarily hold a seam or trim in place until it can be permanently sewn, usually with a long running stitch made by hand or machine called a tacking stitch or basting stitch.
- To temporarily attach a lace collar, ruffles, or other trim to clothing so that the attached substance may be removed easily for cleaning or to be worn with a different garment. For this purpose, tacking stitches are sewn by hand in such a way that they are almost invisible from the outside of the garment.
- To transfer pattern markings to fabric, or to otherwise mark the point where two pieces of fabric are to be joined. A special loose looped stitch used for this purpose is called a tack or tailor's tack.
- A basting stitch is essentially a straight stitch, sewn with long stitches and unfinished ends. The basting stitch is used for temporarily holding sandwiched pieces of fabric in place. The stitch is removed after the piece is finished. Often used in quilting or embroidery.

TOPSTITCH

Topstitching is a sewing technique. It is used most often on garment edges such as necklines and hems, where it helps facings to stay in place and gives a crisp edge.

Decorative topstitching is designed to show, and may be done in a fancy thread or with a special type of stitch. Otherwise, topstitching is generally done using a straight stitch with a thread that matches the fashion fabric.

LAUNDRY

Whether you wash your clothes at home or send them to a laundry, you expect them to come out clean and crisp, looking and feeling as they did when they were new. The dirt, oils, perspiration, and stains should all have vanished, leaving the fabrics and their colours completely intact. Moreover, the clothes should retain their shapes, sizes, structures, and surface textures. With so many expectations, it's no wonder that the word "miracle" appears so frequently in advertisements for laundry detergents.

Achieving these many goals is something of a balancing act. The chemicals that contaminate clothes aren't always so different from those that give them their structures and colours. Trying to remove one chemical while leaving the other isn't easy and washday in the nineteenth century was hardly a treat for the garments being cleaned. However, in recent years laundering has developed from a simple art to an advanced technology.

The miracles promised by the detergent commercials are almost reality. We'll examine physical and chemical mechanisms that make those miracles possible.

Questions to Think About: How do soaps and detergents help to clean clothes? What is the difference between a soap and a detergent? What is hard water and how do you "soften" it? Why does soap form soap scum in hard water? How do detergents keep soil from redepositing on the clothes? Is dry cleaning really "dry"? Why do some clothes wrinkle and shrink when you wash them in water but not when you dry clean them? How does bleach remove stains? Why do fabric softeners increase the volume and fluffiness of towels? Why do fabric softeners reduce static cling? How do "brighteners" make clothes appear whiter than white?

Experiments to Do: You don't have to do laundry to know that soaps and detergents help to remove oil and grease from just about anything. Spread a little oil on a rag or your hand and try to wash it off with water. You'll find that the oil is difficult to remove with water because it doesn't dissolve in water—oil and water don't mix. But if you add a little soap or detergent to the water and try washing again, you'll find that the oil is carried away in the water. The soap or detergent molecules surround tiny droplets of oil and allow the water to remove them.

Soaps and detergents are remarkable materials. They are "at home" in both water and oil and they assist water in handling oil.

SOAP

One of the most difficult problems in laundering clothes is how to remove all of the different soils in a single operation. Some soils consist of polar molecules, those that have electric charges or electrically charged regions, while others consist of nonpolar molecules, those that are effectively neutral throughout.

These two types of soils are so different from one another that a liquid that dissolves one is unlikely to dissolve the other. To make things worse, there are also soils that don't dissolve well in anything, or at least not in anything that you could imagine putting on your clothes. Getting all of these soils out of the clothes without harming the clothes is what laundry is all about.

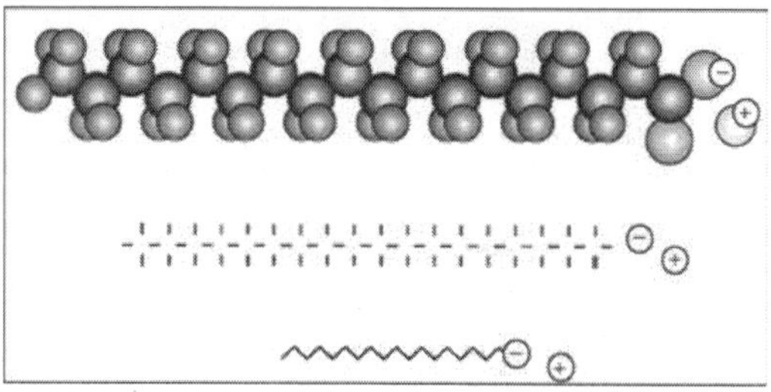

Fig. Soap is a Peculiar Salt.

Polar soil molecules include salts from perspiration and ground dirt. These salts generally dissolve in water, where they become ions that are carried away in shells of water molecules. Because they are basically at home in water, these polar soils are described as hydrophilic. Because carbohydrates such as sugar have electrically charged regions and form hydrogen bonds with water, they dissolve easily in water and are thus also hydrophilic.

Soap is a Peculiar Salt. Its Negative Ion Consists of a Long Hydrophobic (Water Avoiding) Hydrocarbon Chain Attached to a Hydrophilic (Water Loving) Carboxylate Group. A Nearby Positive Ion, Usually soDium (Na), Balances the Negative Charge of the Carboxylate group.

Nonpolar soil molecules include oils, fats, and waxes from skin, foods, and plants. These oily molecules tend to dissolve in nonpolar solvents such as gasoline or kerosene. Because they can't form hydrogen bond with water molecules, they don't bind well with water and are essentially insoluble in it.

These nonpolar soils are described as hydrophobic (water avoiding). You could launder your clothes by first washing them in water to dissolve and remove hydrophilic soils and then laundering them in gasoline to dissolve and remove hydrophobic soils. But this would take a long time and would be very hard on the fabric.

After the process was over, your clothes would have aged considerably, yet some of the soils would still remain. Cleaning clothes requires something more than water and gasoline. That's why we use soaps, detergents, bleaches, and brighteners in our laundry.

Soap is a peculiar type of salt. Like all salts, soap contains a mixture of positively and negatively charged ions. There is nothing special about the positive ions, which are usually just sodium or potassium atoms that are missing an electron. What makes soap so unusual and so effective at cleaning is its negative ions. The negative ions in soap have the negative charge located at one end of a very long molecule. The other end of the molecule is an uncharged hydrocarbon chain such as those encountered in oil molecules.

The negative soap ion is so long that its two ends operate independently. Its charged end is polar and hydrophilic. Water molecules cling to this end's electric charge and try to carry it into solution. But the soap ion's hydrocarbon end is nonpolar and hydrophobic. This end of the soap molecule is expelled from water but binds nicely to oil molecules.

The one half of a soap ion is at home in water and the other half is at home in oil. The hydrophilic end is attracted to water while the hydrophobic oil end is attracted to oil. This split affinity causes soap ions to accumulate at interfaces between water and oil. The negative ions spontaneously orient themselves at such an interface with their electrically charged ends in the water and their hydrocarbon ends in the oil. The positive ions hover around in the water near the interface to keep everything electrically neutral.

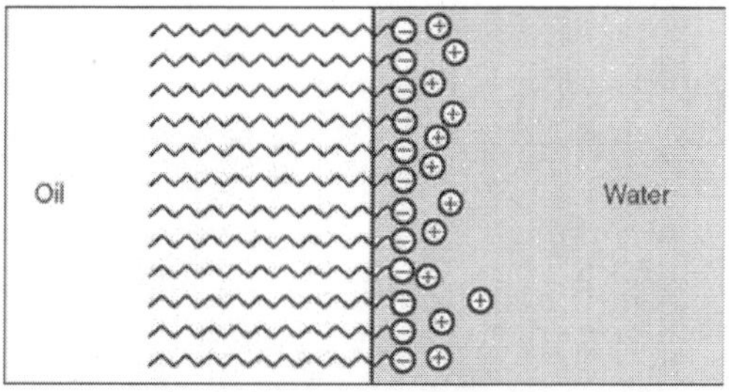

Fig. Soap Negative Ions Move Spontaneously to Interfaces Between Water and Oil. Their Hydrophobic ends Project into the Oil and their Hydrophilic ends Project into the Water.

This tendency for soap ions to order themselves at interfaces is an example of *self-organizing behaviour*. While mixtures of table salt and water are random and homogeneous, mixtures of soap and water are not. Even when there's no oil present, soap ions migrate to water's surface because individual soap ions don't mix freely with the water. Since water molecules don't bond well to the nonpolar hydrocarbon chains, the water molecules push the soap ions to the water's surface. A tiny amount of soap added to a bowl of water soon creates an ultra-thin layer of soap ions on the surface of the water—a layer that's only a single molecule thick. The soap ions arrange themselves with their polar ends in the water and their nonpolar ends in the air. The uppermost water molecules in the bowl are then able to hydrogen bond to the soap ions above them and don't pull together as strongly as they would if they had only air above them. Thus the water molecules contribute little to the liquid's surface tension.

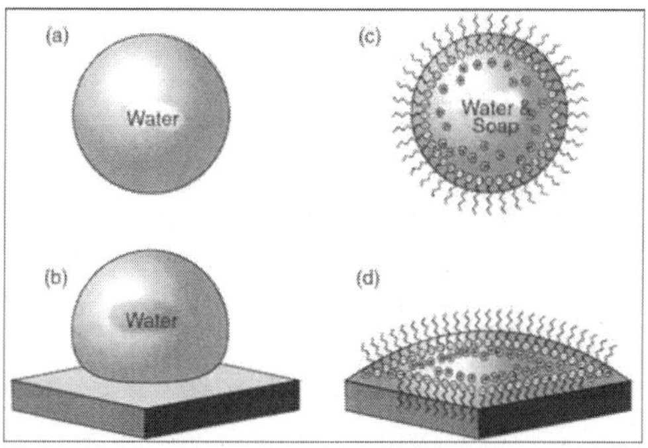

(*a*) Surface Tension in Pure Water Causes its Droplets to be Spherical. (*b*) These Water Droplets Remain Almost Spherical on many Surfaces. (*c*) Soap Ions Coat the Outside of a Water Droplet and Dramatically Reduce the Surface Tension. (*d*) The Soapy Droplet is able to Spread out more Easily and wets many Surfaces Completely. The hydrocarbon chains of the soap ions now form the uppermost layer in the liquid.

With nothing above them to stick to, these chains pull together and create surface tension. However they attract one another with van der Waals forces, not hydrogen bonds, and create a surface tension only about 30per cent that of water molecules. The soap's presence in the water significantly reduces its surface tension. This reduced surface tension is soap's first contribution to the laundering process. Pure water keeps to itself, beading up on any surface that doesn't bind strongly to water molecules. Surface tension makes falling water droplets spherical and they remain almost spherical on oily, hydrophobic surfaces. But adding just a tiny bit of soap to the water reduces each droplet's surface tension and allows it to wet the surface. A soapy droplet spreads outward because van der Waals forces attracting the droplet to the surface are strong enough to stretch it out into a flat puddle. In effect, soapy water is "wetter" than pure water.

Soapy water doesn't bead up on fabrics; it soaks right in. When you are cleaning clothes and want the water to wet every fibre in the fabric, you add soap to the water. Because soap helps water to wet surfaces, it's a wetting agent. It's also a surfactant or *surfaceactive agent* because of its tendency to modify the properties of surfaces or interfaces.

There are other kinds of surfactants, but soaps and soap-like materials are the most important group. However, not all soap molecules make it to the water's surface. If you put lots of soap in the water, or the surface is far away, the soap ions assemble themselves into spherical structures called micelles and remain inside the water. In these micelles, all of the soap ions are oriented with their charged, polar ends pointing outward and their uncharged, nonpolar ends pointing inward. The water molecules stick to the micelles' polar outsides and carry the micelles about. As usual, the positive soap ions hover about nearby to keep everything electrically neutral.

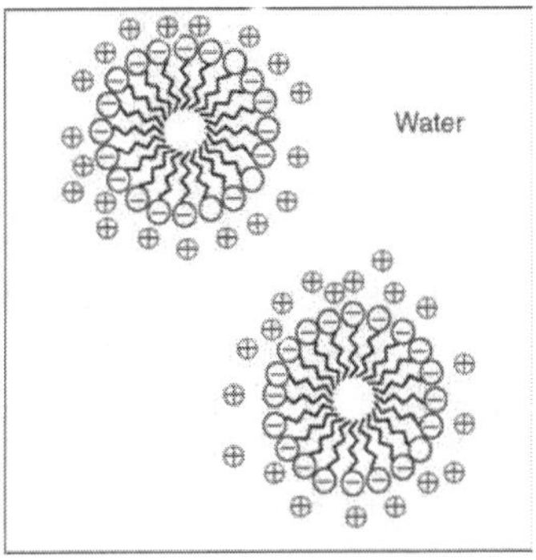

Fig. In Water, Negative Soap Ions from Spherical Micelles. The Hydrophobic Chains from the Centres of these micelles and tend to Accumulate Oily Soil Moledules.

These micelles are soap's second contribution to the laundering process. They tend to trap and collect oily soil molecules. The inside of a micelle is a nonpolar environment and ideal for oil molecules. When an oil molecule bumps into a micelle, the water pushes it into the centre of the micelle and there it remains. The micelles in soapy water move randomly, collecting any oil molecules they encounter in their travels. With a little thermal or mechanical agitation, micelles can even pluck oil molecules from the surfaces of fabrics.

Naturally, this is helpful when you are doing laundry. Little by little, the oily soils in the clothes become trapped in micelles in the water. Soap also helps to remove insoluble debris from clothes. Micelles form around dust particles and help the water to carry these particles away. Since most dust particles don't dissolve in any liquids, soap micelles are essential to their removal from clothing.

Since soap micelles are composed of negatively charged soap ions, they are negatively charged objects and tend to repel one another in the water. They remain separate and mobile and are easily washed down the drain along with their contents. In fact, most fabrics also become negatively charged in water. Their fibres include weakly attached hydrogen atoms that are carried away as positive ions by the water. The fibres are left with negative charges and they tend to repel the negatively charged soap micelles. This repulsion prevents soils from redepositing on the clothes.

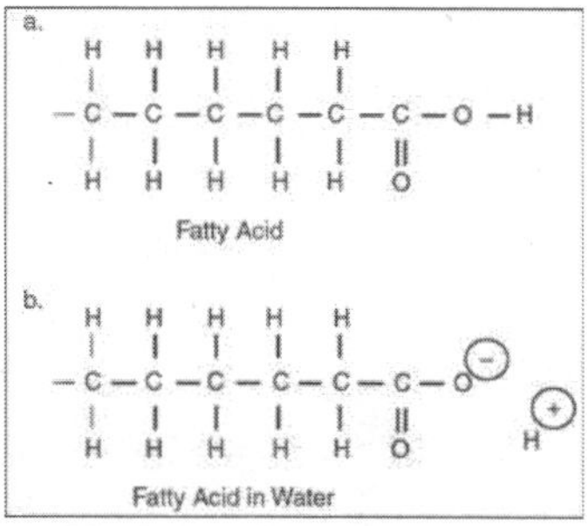

(a) Fatty Acid is a Long Hydrocarbon Chain Ending in a carboxylate Group. (b) In Water, the Carboxylate group's HydrogenAtom is Carried away by water Molecules as a Positive Ion, Leaving the rest of the Molecule Negatively Charged.

This arrangement, soap micelles in water, is a stable emulsion. Unlike a simple mixture of oil and water, it doesn't separate when you let it sit. A surfactant that helps to form and stabilize emulsions is called an emulsifier. Emulsifiers are particularly important in food preparation, where egg yolks, lecithin, and various gums are used to make mayonnaise, chocolate, and other foods smooth and creamy.

Soap is clearly useful in laundering clothes, but where does it come from and what is its structure? Soap is made from naturally occurring oils and fats. Each molecule of oil or fat consists of three fatty acid molecules bound to a glycerin molecule.

A fatty acid molecule resembles a paraffin or olefin molecule, with its long chain of carbon atoms surrounded by hydrogen atoms. But the fatty acid molecule has a special arrangement of carbon, oxygen, and hydrogen atoms, a *carboxylate group*, at one end that makes it an organic acid.

An acid is a molecule that can easily lose a positively charged hydrogen ion when it is mixed with water. One of the hydrogen atoms at the special end of the fatty acid falls off easily because the adjacent oxygen atom has largely removed its electron.

Oxygen and hydrogen form a covalent bond, with a pair of electrons between them, but the oxygen atom attracts the pair of electrons more strongly than the hydrogen atom does. As a result, the hydrogen atom's nucleus is relatively exposed and is easily carried away by passing water molecules. This loss leaves a negatively charged fatty acid ion.

In an oil or fat, these three fatty acid molecules are not ionized. Instead, they have reacted with a glycerin molecule like three large ships docking at a small port. The glycerin molecule has a chain of three carbon atoms and each of these carbon atoms plays host to one of the fatty acids. The resulting structure is called a triglyceride.

Assembled in this manner, the triglyceride is nonpolar and virtually insoluble in water. It looks and feels like petroleum oil because both have the same long hydrocarbon chains. However triglycerides are digestible while petroleum oils are not.

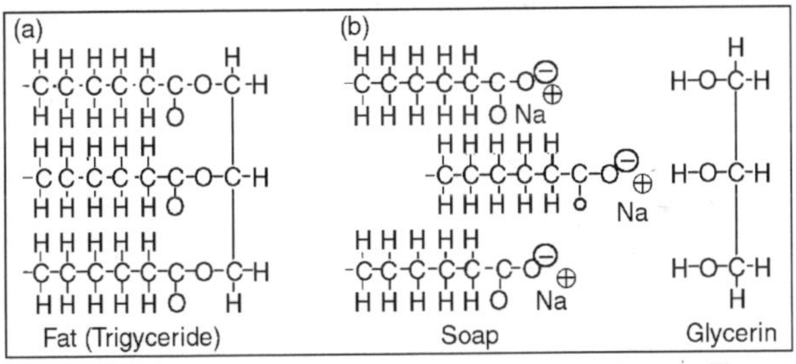

(*a*) A Fat Molecule or Triglyceride Consists of Three Fatty Acids Bonded to a Glycerin Molecule. When the Triglyceride Reacts with Sodium Hydroxide (lye), the Fatty Acids Break free of the Glycerin and Produce a Mixture of Soap and Glycerin (*b*).

Triglycerides composed entirely of paraffin-like fatty acids are called saturated fats because they have only single covalent bonds and as many hydrogen atoms as possible. Such molecules experiences strong van der Waals forces, forming fats that remain solid at relatively high temperatures. These fats are found in animals and tropical plants such as palms and coconuts.

Triglycerides containing olefin-like fatty acids are called unsaturated fats because they have double bonds that reduce their hydrogen atoms count. Double bonds stiffen the hydrocarbon chains and prevent them from bonding as strongly to one another.

Oils contain these molecules melt at relatively low temperatures and are found in fish and temperate plants such as soybeans and corn. Unfortunately, people find the less healthy saturated fats more tasty and satisfying than the unsaturated fats. Converting unsaturated fats to saturated fats, a process called hydrogenation, is commonly used to stiffen oils for use in foods such as margarine and candy.

Sodium hydroxide is a salt consisting of positive sodium ions and negative hydroxyl ions (a hydrogen and an oxygen atom together) and it rapidly dissolves into independent ions when you put it in water. When you mix fat, water, and lye together, the hydroxyl ions from the lye attack the fat molecules and remove the fatty acids from the glycerin as negative ions.

Soon the water is filled with glycerin molecules, negative fatty acid ions, and positive sodium ions. When the reaction is complete and most of the water is removed, the result is soap.

The glycerin may or may not be removed. The hardness of the soap depends on the fats from which it was made. Saturated fats produce hard bar soaps while unsaturated fats produce soft liquid soaps. Soft hand soaps often include the glycerin.

While most modern soaps are made with lye and are thus sodium salts, earlier soaps where made with potassium hydroxide obtained from wood ash and lime and were potassium salts.

WATER SOFTENING

Unfortunately, laundering clothes isn't quite this easy. While soap is wonderful at removing oils and fats from fabric, it has problems in hard water.

Hard water is any water with more than about 120 milligrams of positively charged calcium and magnesium ions per litre. These two metal ions, and a few others, bind with the negative soap ions and form insoluble soap scums that deposit themselves on sinks, showers, bathtubs, washing machines, and clothing.

If you try to launder clothes with soap in hard water, you are in for a messy surprise. The problem occurs because calcium and magnesium ions behave differently from the ions of sodium and potassium normally found in soap. Sodium and potassium atoms each have one more electron than they need to complete an electronic shell—a quantum physical structure that is particularly stable.

That extra electron is relatively easily removed, creating a positively charged ion that is easily drawn into solution in water. Water is so strongly attracted to sodium ions that almost every sodium salt in existence dissolves in water. Sodium's fantastic solubility explains why there is so much sodium in seawater.

Potassium ions are almost as soluble as sodium ions. Salts consisting of sodium or potassium positive ions and soap negative ions are extremely soluble in water. But calcium and magnesium atoms have two more electrons than they need to complete an electronic shell.

They give up those two electrons somewhat reluctantly to form positively charged ions and aren't particularly soluble in water. While some calcium and magnesium salts are modestly soluble in water, calcium and magnesium soap salts are not.

When you put soap in hard water, the positive calcium and magnesium ions in the water combine with the negative soap ions and quickly form insoluble salts. These salts are pasty solids that cling to everything. If you want to do laundry in a place where the water contains substantial amounts of dissolved minerals, you must either remove the calcium and magnesium ions from the water or replace the soap with something else.

Actually, you often do both. Removing the calcium and magnesium ions is the first option. This step is called *water softening* and is done routinely in most industrial laundries long before the water enters the washing equipment.

There are several different ways to soften water, but the most interesting plan and the one used most often in houses is called ion exchange. The water passes through an ion exchange material that replaces the calcium and magnesium ions with sodium ions.

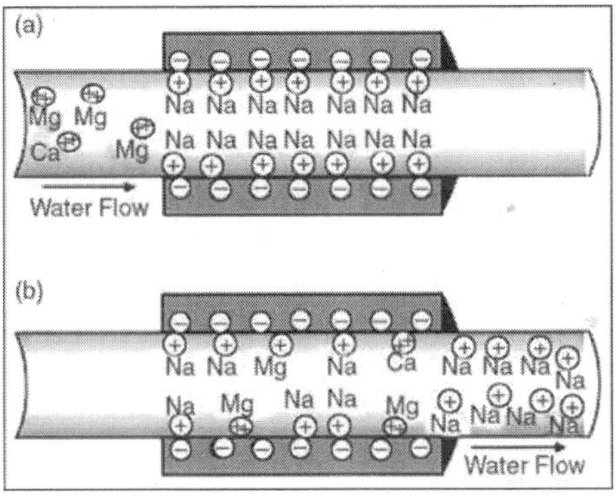

(a) Fresh ion Exchange Water Softener Contains Sodium Ions, Located near Negatively Charged Sites in a Resin or Zeolite Ceramic. (b) As Water Containing Magnesium (Mg) and Calcium (Ca) Ions Passes Through the Softener and the Megnesium and Calcium Ions remain Behind.

The ion exchange material is a special ceramic (zeolite) or plastic resin with many negatively charged regions in its porous structure. To keep the material electrically neutral, a positive ion is located near each negative region. The negative regions are part of the material, so they can't go anywhere, but the positive ions are mobile.

When the ion exchange material is fresh, nearly all of the positive ions inside it are sodium ions. As hard water flows through the material, the sodium ions are gradually replaced by calcium and magnesium ions. Since the sodium ions are more soluble in water than the calcium and magnesium ions, the sodium ions tend to enter the water and the calcium and magnesium ions tend to leave it. Each calcium or magnesium ion that sticks to the ion exchange material releases two sodium ions, which leave the water softener in the water itself.

While the water leaving an ion exchange water softener still contains dissolved ions, they are sodium ions rather than calcium or magnesium ions.

The sodium ions cause no trouble when laundering clothes or washing your skin, but people who are on low sodium diets should avoid softened water. Moreover, you shouldn't use softened water in a steam iron.

When all of the sodium ions in the ion exchange material have been replaced by calcium or magnesium ions, the water softener stops softening the water. To regenerate the ion exchange material, you must flush it with very concentrated salt water. The many sodium ions in the salt water dislodge most of the calcium and magnesium ions and return the ion exchange material to its original condition—it is once more full of sodium ions and ready to soften water.

Because many homes don't have water softeners, most laundry soaps soften the water themselves. They contain chemicals called builders that bind to the calcium and magnesium ions and keeping them away from the soap ions. The most effective of these builders is sodium tripolyphos-phate and, at one time, most household detergents contained large amounts of it.

However, phosphates encourage the growth of algae, threatening the ecologies of rivers, lakes, and bays, and have been banned from detergents in many regions. Builders such as sodium carbonate, citric acid, and sodium citrate are now often used instead.

Another building technique used in some products is to incorporate small zeolite ceramic particles directly in the detergent. These builder particles exchange sodium ions for calcium and magnesium ions and soften the water directly in the washer. They fall to the bottom of the washer and are rinsed away.

DETERGENTS

But water softening is only half the solution. Because it's hard to eliminate all of the calcium and magnesium from water, the manufacturers also eliminate the soap from the laundry powder. That's right—most laundry detergents aren't soap at all.

Instead, they are synthetic detergents that are structurally related to natural soap but aren't quite the same. Actually, soap is a type of detergent. Detergents are a broad class of molecules that stabilize mixtures of oil and water. There are many other kinds of molecules that can perform this task and thus many types of detergents. The most common laundry detergents are the linear alkylbenzenesulfonates. These petroleum products are sodium salts, just like most soaps, but the structures of the negative ions are different.

A negative soap ion is a long hydrocarbon chain attached to a negatively charged carboxylate group. The detergent ion is also a long hydrocarbon chain, attached to an aromatic or benzene ring, attached to a negatively charged sulfonate group. The sulfonate group involves one sulfur atom and three oxygen atoms.

The two important parts of the detergent molecule are the charged head and the long nonpolar tail. The charged end is a *sulfonate group* in which a sulfur atom attaches to four other atoms: a carbon atom and three oxygen atoms. This arrangement is roughly tetrahedral in shape. But sulfur normally attaches to only two atoms, so how is this arrangement possible?

First, the sulfur atom forms normal covalent bonds with the carbon atom and with one of the oxygen atoms. That oxygen atom has an extra electron, making it a negatively charged ion that can only form a single covalent bond.

Second, the sulfur atom allows each of the two other oxygen atoms to share a pair of its electrons. Because these shared electrons can travel between both atoms, their wavelengths increase and their kinetic energies decrease. In this manner, the oxygen atoms become attached to the sulfur atom. The overall result is a negatively charged structure that's easily carried about by water molecules.

The detergent molecule's long nonpolar tail is essentially an unbranched paraffin chain, also referred to as a *linear alkyl group*. This chain provides the oily tail of the detergent molecule. While early alkylbenzenesulfonate detergents included branched paraffin chains, these proved to be less biodegradable than the linear versions. Bacteria can metabolize long linear chains because those chains are common in animals and plants, but branched chains are rare in nature and bacteria are unprepared for them. To keep detergent foam out of streams and lakes, manufacturers have learned to produce purely linear detergent molecules.

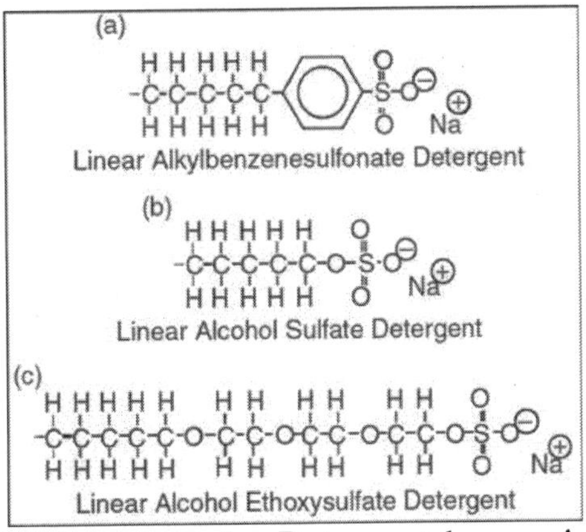

(a) The most Common Laundry Detergent has an Aromatic or Benzene Ring Connecting a Long Hydrophobic Hydrocarbon Chain to a Hydrophilic Sulfonate Group. Common Shampoo Detergents Connect the Chain to the Sulfonate With Either (b) An Oxygen Atom or (c) A string of Oxygen and Carbon Atoms.

The last piece of the detergent molecule is the aromatic or benzene ring. This ring is a vestige of the manufacturing process. It's much easier to attached the linear alkyl tail and the sulfonate head separately to an aromatic ring than it is to attach them directly to one another. Unfortunately, the ring actually reduces the biodegradability of the molecule somewhat. There are other detergents, such as linear alcohol sulfates and linear alcohol ethoxysulfates, in which the aromatic ring is replaced by an oxygen atom or a string of oxygen and carbon atoms.

The most common linear alcohol sulfate is sodium lauryl sulfate, with 14 carbon atoms in its hydrophobic chain. Sodium laureth sulfate is a common linear alcohol ethoxysulfate, also with 14 carbon atoms in its chain. These detergents are derived from tropical oils and often used in shampoos and dishwashing liquids.

Since calcium and magnesium ions don't cause these sulfonate or sulfate detergents to form insoluble detergent scums, why do laundry detergents still worry about softening the water? Unfortunately, calcium and magnesium ions interfere with the micelles, making it difficult for them to extract soil from fabric and keep it suspended in water. Because each calcium or magnesium ion has twice the positive charge of a sodium or potassium ion, these highly charged ions approach the micelles closely and partially neutralize their surfaces.

Since these neutralized micelles don't repel one another or the fabric well, they do a poor job of cleaning clothes. That's why laundries and laundry detergent still work to remove the calcium and magnesium ions.

Before leaving detergents, it's worth noting that not all detergents are negative ions (anions). It's also possible to construct detergent molecules that are positive ions (cations) and even ones that aren't ions at all. However, cationic detergents and surfactants aren't used in laundry detergents because they tend to stick to fabric—we'll discuss their use as fabric softeners later on. But nonionic detergents are often used to launder clothes.

The only requirement for a detergent is that its molecules stabilize a mixture of oil and water. Nonionic surfactant molecules don't have an electric charge, but they do have a hydrophilic end and a hydrophobic end. Like most detergents, the hydrophobic end is just a long hydrocarbon chain.

But the hydrophilic end is also a long chain, consisting of oxygen and carbon molecules attached one after the next and decorated with hydrogen atoms. The oxygen atoms in this special chain form hydrogen bonds with water molecules, giving that end of the molecule its hydrophilic character. These nonionic molecules form micelles and are very effective at removing grease.

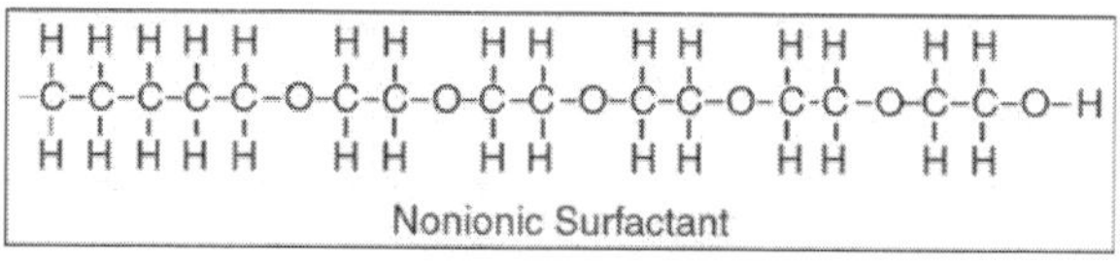

Fig. Nonionic Surfactants have a Hydrophobic Hydrocarbon Chain (on the left) Attached to a Hydrophilic Chain Containing Oxygen Atoms (on the right).

Nonionic surfactants are unaffected by hard water and are actually better than anionic detergents at removing some soils—they are particularly good at removing skin oils from synthetic fabrics. However, they aren't salts and exist either as liquids or waxy solids. As a result, nonionic surfactants are difficult to formulate into powdered detergents but are common in liquid detergents.

BLEACHES AND ENZYMES

Not all soils can be removed from fabrics with detergent and water. Molecules that form covalent bonds with the fabric create stains that can only be eliminated with bleaches or enzymes. Bleaches act to destroy the colouration of stain molecules or to cut them free from the fabric. Enzymes act to dice up large stain molecules into smaller fragments that can be washed away.

With a little luck, these steps can be taken without destroying the fabric or its colour. Unlike soaps and detergents, bleaches react chemically with the soil molecules. They are particularly aggressive at converting double bonds to single bonds by attaching oxygen and chlorine molecules to the two atoms involved. Double bonds often give organic molecules their colours so this sort of rearrangement tends to make them colourless.

Just as atoms absorb and emit photons of light that are characteristic of their electronic states, so molecules absorb and emit photons that are characteristic of their electronic states. Each electron in a molecule is sensitive to passing electromagnetic radiation and responds to photons that can transfer it to some unoccupied state with more energy. If the electron finds such a photon, it may undergo a radiative transition to the excited electronic state and absorb the photon. The electron will eventually return to its original state, converting the extra energy into thermal energy, but the photon will be gone forever. If a particular molecule contains electrons that can absorb photons of visible light in this manner, it will appear coloured.

In most single covalent bonds, the two electrons are bound so tightly between the two nuclei that any transition to a new electronic state requires more energy than a photon of visible light can provide. Only ultraviolet light can cause radiative transitions in these electrons. Molecules based entirely on single covalent bonds are normally unaffected by visible light and are thus colourless.

However, the outer electrons in a double covalent bond aren't so tightly bound and can be transferred to other electronic states relatively easily—a photon of visible light may well be able to cause the transfer. A double bond that absorbs blue photons from passing light appears yellow. One that absorbs red photons appears cyan. The usual rules of subtractive colour apply.

Just how much energy it takes to cause this transfer depends on the chemical nature and environment of the double bond. Most important are the two atoms joined by the double bond. In addition to double bonds involving a pair of carbon atoms, there are also carbon-oxygen, carbon-nitrogen, nitrogen-oxygen, and nitrogen-nitrogen double bonds. These double bonds are often coloured, particularly the latter two. Groups of atoms that give rise to colour in molecules are called chromophores.

However the exact colour of a double bond is determined by the detailed structure of the molecule around it. Since all of the electrons in that molecule affect one another through electrostatic forces and the Pauli exclusion principle, the whole molecular structure contributes to the colour of the electrons in the double bond itself. A subtle change in a molecule's structure may change its colour from red to orange. That's how organic dye manufacturers construct rich pallets of colours from a small number of different chromophores.

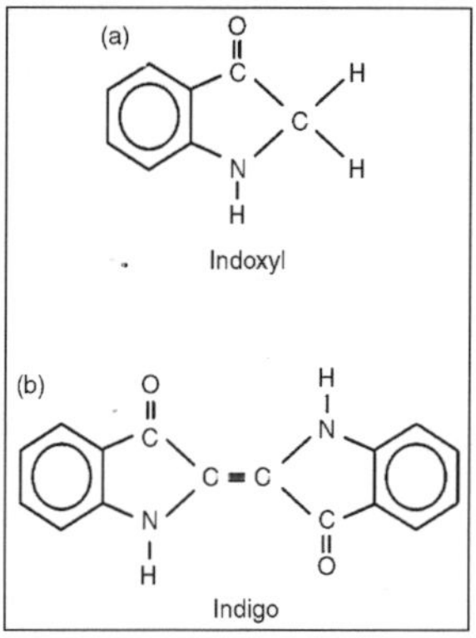

(*a*) Indoxyl is a Colourless, Water-soluble Chemical Obtained from a Fermented Plant Extract. When Indoxyl is Exposed to Oxygen in the Air, it Reacts Pairwise to form Water-insoluble Indigo (*b*), The Blue Dye used in Blue Jeans.

The Double Bonded Carbon Atoms at the Centre of this Molecule are the Chromophore and the Rest of the Molecule Determines the Precise Colour of the dye. Bleaches and Ultraviolet Light can Destroy the Double bond, Giving Blue Jeans a Faded Look.

Coloured molecules are wonderful if you are an artist, but you don't want extraneous ones attached to your clothes. That's where bleach comes in. Bleach attacks double bonds, destroying the chromophores in the stain molecules. The molecules may remain on the fabric but they no longer absorb visible light.

The two major classes of bleaches are chlorine bleaches and oxygen bleaches. The chlorine bleaches tend to put chlorine and oxygen atoms on the two atoms involved in a double bond. The double bond vanishes as one of its atom binds to a chlorine atom and its other atom binds to the oxygen atom of a hydroxyl group (OH).

Unfortunately, chlorine bleaches are so effective at attacking chemicals that they damage the clothes, too. Sometimes they destroy the chromophores in dye molecules, turning coloured fabric white. Other times they modify the dye molecules and change the fabric's colour. But chlorine bleaches also damage natural fibres themselves, breaking up their molecules and weakening the fabric. While chlorine bleach may succeed in cutting stain molecules free from your clothing, it may also cut holes in the clothing itself.

Oxygen bleaches used hydrogen peroxide to attack double bonds. A hydrogen peroxide molecule is a water molecule with an extra oxygen atom inserted between the oxygen atom and one of the hydrogen atoms. This molecule decomposes in water and its fragments, either ions or free radicals, attack double bonds. Once again, they destroy chromophores and decolourize stains.

Hydrogen peroxide is less reactive than chlorine bleach and causes less fabric damage. It's also less damaging to commercial dye molecules than chlorine bleach. However, since hydrogen peroxide is often used to bleach hair, oxygen bleaches can obviously destroy the colours in some natural fibres.

Hydrogen peroxide itself is rather unstable and is often generated right in the washer by the decomposition of another molecule, sodium perborate. This decomposition occurs only above about 50 °C, so bleaching must be done in hot water. Some laundry detergents contain activators that help sodium perborate decompose in cooler water but it still works best in hot water.

Enzymes are biological catalysts. Your body uses a great many different enzymes to catalyze various chemical reactions that would otherwise rarely occur at body temperature or that might proceed along the wrong paths without help. Enzymes help to construct molecules, to take them apart, or to rearrange their components.

The enzymes used most often in detergents are those that degrade proteins. Protein molecules cling tightly to fabrics, are insoluble in water, and prevent detergents from penetrating to the fibres. They act as binders for other molecules, creating stains that are hard to remove. Familiar proteinaceous stains include blood, milk, eggs, and gravy. The most effective way to remove these stains is by taking the protein molecules apart. This decomposition is done by *proteolytic enzymes*—enzymes that catalyze reactions between water and protein.

In these reactions, protein molecules are broken up and water molecule fragments cap the severed ends. In time, proteolytic enzymes can dice up long protein molecules into their constituent parts: *amino acids* and short sequences of amino acids called *peptides*. The stain falls apart and is carried away by the water and detergents. Meat tenderizers operate in a similar fashion, using papain—a proteolytic enzyme extracted from unripe papaya—to degrade protein in meat before cooking.

However, proteolytic enzymes may have an effect on the people who use them. You certainly don't want the protein in your body decomposed while you do laundry or while you wear freshly laundered clothes. Although studies seem to indicate that the enzymes in household detergents pose no serious health threat, they're used sparingly in detergents to avoid any possible adverse effects.

BRIGHTENERS AND FABRIC SOFTENERS

Not all laundry chemicals disappear down the drain when the wash is done. Brighteners and fabric softeners do their jobs by remaining on the clothes long after they leave the drier. Brighteners affect the appearances of the clothes while fabric softeners affect their feels. With age, white fabrics such as cotton tend to absorb more and more blue light and begin to look yellow. Cleaning and bleaching do little to reduce this effect. In fact, bleached fabric molecules tend to appear slightly yellow themselves. The old fashioned solution to the yellowing problems was to adding *bluing* to the wash.

This blue dye absorbed red and green light, balancing the blue absorption of the fabric itself so that the fabric appeared colourless. To mask the yellowing of age, bluing darkened the whole fabric to a light gray—the amount of light reflected by the fabric was noticeably less than that striking it.

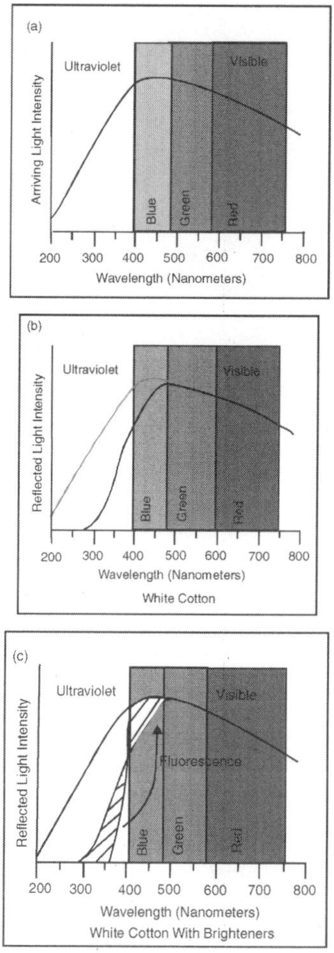

When White Cotton Cloth is Exposed to Sunlight (*a*), It Reflects a Little less Blue Light than it should (*b*) and Appears Slightly Yellow. This Yellowness Increases with age. But when Fluorescent Brighteners are Added to the Cotton (*c*), they Convert Ultraviolet Light into Blue Light and make the Cloth Appear Dazzlingly White.

Instead of using bluing, virtually all modern laundry detergents add chemicals that optically brighten the fabric. These brighteners are actually fluorescent dyes, designed to absorb ultraviolet light and emit bluish-white visible light. Instead of absorbing red and green light to balance the white appearance of a fabric, the brighteners reintroduce the missing blue light.

They work best in sunlight, which is rich in ultraviolet light. A brightener molecule absorbs a photon of ultraviolet light and reemits it as a photon of blue light. The energy not reemerging from the molecule in the second photon is converted through vibrations into internal energy in the clothes.

When clothes are washed in these fluorescent dyes, the dye molecules stick to the fabric to create brightened fabric. We can't see the ultraviolet light that the brightened fabric absorbs but we can see the bluish light that it emits. With the missing blue light restored by this fluorescence processes, the brightened fabric appears dazzlingly white.

In fact, it may emit more blue light than it is exposed to, making it effectively "whiter than white." In a room illuminated only by ultraviolet light, the brighteners give clothes an eerie violet glow. Fabric softeners are also chemicals that remain on fabrics after laundering. They are primarily cationic surfactants called *quaternary ammonium compounds*. These compounds are based on the positive ammonium ion, which is itself based on a positive nitrogen ion.

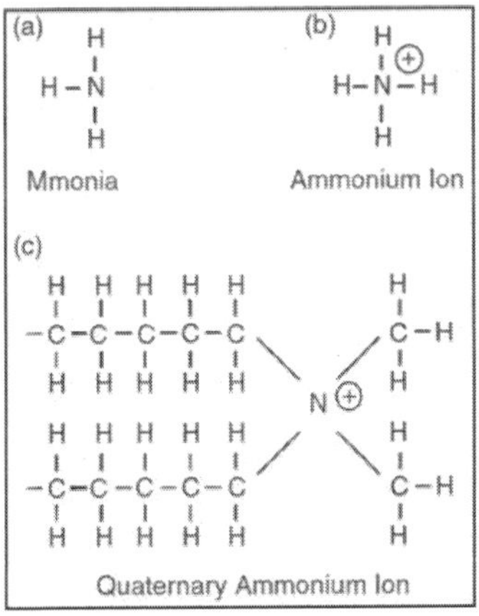

A normal nitrogen atom has five valence electrons and must share three of these to reach the four pairs needed to complete an electronic shell. If it shares those electrons with three hydrogen atoms, it forms an ammonia molecule. But if it's missing an electron, the nitrogen atom must share four electrons to complete its shell and can actually bind to four hydrogen atoms. In that case, it forms a positive ammonium ion. In quaternary ammonium compounds, a positive nitrogen ion forms covalent bonds with four other atoms. However, these atoms aren't necessarily hydrogen atoms. In fabric softeners, the central nitrogen ion binds to four hydrocarbon chains. Two of these chains are short, only one carbon atom long, while the other two chains may contain as many as 18 carbon atoms. These long chains are hydrophobic and have the same oily character as most lubricants.

This oily character is what gives these compounds their fabric softening ability. When you apply the softener to wet fabric, its positively charged surfactant ions are drawn towards the negatively charged fibres and stick to them strongly. While anionic surfactants are repelled by wet fabric and help to clean it, cationic surfactants are attracted to wet fabric and help to soften it.

(a) An Ammonia Molecule is Three Hydrogen Atoms Bound to a Nitrogen Atom. (b) An Ammonium ion is four Hydrogen Atoms Bound to a Positively Charged Nitrogen Ion. (c) The Quaternary Ammonium Ion used in fabric Softeners is Formed by Replacing the four Hydrogen Atoms with Hydrocarbon Chain. The surfactant molecules stick to the fabric with their long hydrophobic chains pointing outward. These molecules decorate every fibre in every thread of the clothing, giving them all an oily coating. The hydrocarbon chains lubricate the fabric so that each fibre slides easily within a thread and each thread slides easily within the fabric. This lubrication enhances the flexibility of the fabric and makes it feel softer and more flexible. Fabric softeners also make fabric surfaces slightly hydrophobic, so that they dry more easily in the spin dry cycle of a washing machine.

In this cycle, the clothes travel rapidly around in a circle, always accelerating towards the centre of the circle and experiencing huge inward forces from the washer's metal drum. Water's inertia causes it to lag behind the accelerating clothes and it leaves the drum through perforations. By making the fabric slightly hydrophobic, the fabric softener helps the clothes to shed water as they spin, so that they don't have to spend as much time in a hot drier later on.

Fabric softeners also raise the nap on cotton terry towels. Cotton fibres are normally hydrophilic and cling tightly to water droplets. As water droplets dry up, they shrink and pull the cotton fibres towards one another. By the time an untreated towel is dry, its fibres have been crushed together by these forces and it has little nap.

But a towel that has been coated by quaternary ammonium compounds is hydrophobic enough that the water droplets can't pull its fibres together as they dry. The nap remains loose and thick, giving the towel a fluffy appearance and feel. Unfortunately, this same hydrophobic coating slightly reduces the towel's absorbency—a real problem for cotton diapers. To keep it under control, don't use too much fabric softener.

Despite their hydrophobic chains, quaternary ammonium compounds actually attract a few water molecules to the surface of the fabric

They are hygroscopic, meaning that they attract water molecules directly out of the air. Since water conducts electricity very weakly, fabric that has been treated with fabric softeners is very slightly conducting. This conductivity reduces the accumulation of static electricity on the fabric and eliminates static cling.

In a drier, untreated clothes rub against one another and sliding friction transfers electric charge from one region of fabric to another. Large charge imbalances are created and the clothes leave the dry clinging to one another with electrostatic forces. However, treated clothes are lubricated in the drier and experience weaker frictional forces. They transfer less electric charge as they tumble and the small charge imbalances that are created quickly dissipate through the moisture attracted by the fabric softener.

Quaternary ammonium compounds are also used in conditioners and shampoos to soften hair and reduce static electricity problems—they will coat and lubricate just about anything. They are actually bactericidal because they coat bacteria and smother them. These compounds also deactivate some of the enzymes in bacteria and upset their metabolisms. Some antiseptic throat lozenges and mouth washes use quaternary ammonium compounds to kill germs.

Unfortunately, the positive charges of cationic quaternary ammonium compounds make them relatively incompatible with the negative charges of anionic detergents. When they're present together in the water, these two types of ions attract one another and may clump together. This clumping is avoided by keeping the two types of surfactants separate, which is why softeners are usually added during the rinse cycle, in the drier, or in a separate conditioner when washing your hair. However, some detergent and shampoo formulators have successfully combined cationic softeners and anionic detergents.

DETERGENT ADDITIVES

Formulated detergents contain a number of important components that work together to clean clothes. We've already examined the anionic and nonionic detergents (surfactants), the builders (water softeners), the bleaches, and the brighteners. But there are also foam stabilizers, corrosion inhibitors, soil redeposition inhibitors, and processing agents.

Foam stabilizers are there to control bubble formation. These chemicals can either enhance or suppress foaming. Believe it or not, foam is unrelated to a detergent's ability to clean clothes. The same goes for shampoos and dishwashing detergents. However, the amount of foam a detergent produces may influence its use. If the detergent foams excessively, you may think the detergent is more powerful than it is and cut back on the amount you use. As a result, you may not use enough to clean your clothes properly. If the detergent doesn't foam much, you may think that it isn't working and buy another brand. So the detergent and shampoo manufactures carefully control the foaminess of their products.

Air bubbles don't last long in pure water because water's surface tension causes them to tear. The final layers of water molecules on the bubble's outer and inner surfaces pull together so strongly that any tiny defect immediately initiates a rip that lets the air out of the bubble. By reducing water's surface tension, soaps and detergents remove its tendency to rip and stabilize air bubbles.

But how long each air bubble lasts depends on many features of the mixture and not on its ability to clean things. Some surfactant molecules make particularly stable and long lasting bubbles while other molecules deliberately introduce defects that pop the bubbles. Methyl silicone polymers ("methicones") are particularly effective at weakening bubbles so that they tear and collapse. These polymers are common in antifoam additives and are even included in some antiacid tablets.

Foam boosters are common in detergents and shampoos that are used by hand, where foam is regarded as a sign of effectiveness. Antifoaming agents are often used in washing and dishwashing machine detergents where you don't see the foam anyway and foam interferes with the machine's operation. Corrosion inhibitors are important in detergent because the ions in detergent would otherwise quickly rust the steel in a washing machine.

Rusting is an electrochemical reaction of the type explored in the supplement on batteries. In normal rusting, the iron in steel is attacked by negatively charged hydroxyl ions. However other negatively charged ions, including detergent ions, can also attack iron and rust it. So detergents include corrosion inhibitors. These compounds are usually sodium silicates. They form thin glassy coatings on the washer parts and inhibit rusting.

Soil redeposition inhibitors enhance the negative charge of wet fabric fibres. Some fabrics, particularly synthetic ones, don't acquire a strong negative charge in water. They need this electrostatic charge to keep the negatively charged detergent micelles from redepositing their soils on the fabric. So detergents include carboxymethyl cellulose, which attaches itself to the fibres and adds to their negative charge.

Finally, processing agents simply give the detergents the right structures in their boxes or bottles. Sodium sulfate helps to bulk up powdered detergent and make it pour easily. Sodium xylene sulfonate helps to keep all of the components of very concentrated liquid detergents in solution.

DRY CLEANING

Washing clothes in water isn't always a good idea. Fibres such as cotton, wool, silk, and rayon, are very hydrophilic and soak up water molecules like sponges. These fibres form hydrogen bonds with water molecules at various sites on their molecules and accumulate large quantities of water.

This water takes up space and causes the fibres to swell. Cotton, wool, and silk fibres increase by about 1per cent in length and about 15per cent in thickness. Rayon expands even more, by 3per cent in length and about 25per cent in thickness. This swelling distorts the fabric and changes its structure. When the fabric eventually dries, it may have shrunk or wrinkled.

To avoid damage caused by this cycle of expansion and contraction, you can send your clothes to be dry-cleaned. Dry cleaning takes place in a nonpolar solvent. Since this solvent doesn't form hydrogen bonds, it's only weakly attracted to the fibres by van der Waals forces and doesn't cause them to swell. The clothes don't lose their shapes.

The solvents used in dry cleaning have evolved over the years since petroleum oils were first found to remove stains. Early dry cleaning was done with gasoline, resulting in many dramatic fires. In 1928, a less flammable solvent became available. The Stoddard solvent, named for the president of the National Institute of Dry cleaning, W. J. Stoddard, is less volatile than gasoline because it contains larger hydrocarbon molecules. It's obtained by distilling crude oil and its vapour will not ignite in air at temperatures below 38 °C.

Nonetheless, Stoddard solvent is still dangerous during hot air drying so nonflammable nonpolar solvents have largely replaced it. The most common solvent in dry cleaning is now perchloroethylene.

Its molecule consists of a pair of carbon molecules connected by a double bond and each attached to two chlorine atoms. The chlorine atoms bind so strongly to the carbon atoms that the molecule doesn't react with oxygen and forms a nonflammable liquid.

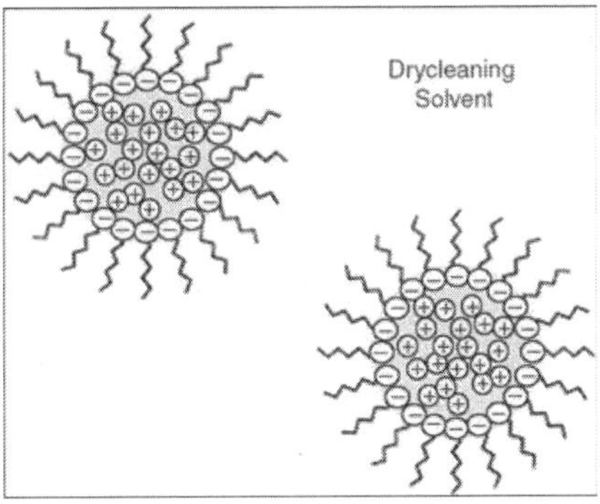

Fig. Detergents from Inside-out Micelles in dry Cleaning Solvent. The Polar Hydrophilic ends of the Molecules project Inward, Towards a tiny Drop of Water. The Nonpolar Hydrophobic end Project Potward into the Solvent.

When you put clothes in either Stoddard solvent or perchloroethylene, the oily soils dissolve. These nonpolar solvents attract the oily molecules with van der Waals forces and carry them away.

Chlorinated solvents clean better than hydrocarbons because they bind more strongly to oily soils. Chlorine atoms are more polarizable than hydrogen atoms and produce stronger van der Waals forces, which is why perchloroethylene doesn't boil until it is heated to 121 °C. However, these nonpolar solvents are unable to dissolve salts and other polar soils.

They are also poor at removing insoluble soils such as dust. To help in removing these other soils, dry cleaning solvents include detergents and a little water. The detergents form inside-out micelles in the nonpolar solvents, arranged with their nonpolar ends on the outside and their polar ends on the inside.

Each micelle surrounds a tiny droplet of water. Just as in water cleaning, detergents help to carry away substances that aren't soluble in the principal cleaning liquid. The water in the dry cleaning mixture is carefully adjusted so that the clothes neither gain nor lose moisture during the cleaning process.

In air, water molecules are continually leaving and returning to the clothing and an equilibrium is reached. At this equilibrium, the water molecules still move back and forth but the moisture in the clothing doesn't change significantly. The actual moisture level in the fabric then depends only on the relative humidity of the air, which is typically about 70per cent in a dry cleaning shop.

The same leaving and returning process takes place in the dry cleaning solvent. Water molecules move back and forth between the fabric and the solvent and establish an equilibrium.

Like air, the dry cleaning solvent has a relative humidity and a dry cleaner tries to maintain this relative humidity at the same value as the air in the shop. That way, the fabrics don't accumulate too many water molecules and swell, nor do they lose too many water molecules and dry out.

But the polar soils leave the fabrics, become trapped in the detergent micelles, and never return. With the help of detergents, nonpolar dry cleaning solvents carry away nonpolar, polar, and insoluble soils from clothes without affecting the structure of the cloth. The dry cleaner then removes the solvent from the clothes by spinning them and drying them in hot air. Because solvents are expensive and environmentally damaging, dry cleaners collect the solvents for reused. They do this by filtering and distilling the liquid solvents and by condensing the gaseous solvent molecules onto chilled surfaces. When this type of solvent recycling is done effectively, a dry cleaner can operate for a long time on the same supply of solvent.

6

Textile Fabric Computing

Fabric computing or unified computing involves the creation of a computing fabric consisting of interconnected nodes that look like a 'weave' or a 'fabric' when viewed collectively from a distance. Usually this refers to a consolidated high-performance computing system consisting of loosely coupled storage, networking and parallel processing functions linked by high bandwidth interconnects (such as 10 Gigabit Ethernet and InfiniBand) but the term has also been used to describe platforms like the Azure Services Platform and grid computing in general (where the common theme is interconnected nodes that appear as a single logical unit).

The fundamental components of *fabrics* are "nodes" (processor(s), memory, and/or peripherals) and "links" (functional connection between nodes). While the term "fabric" has also been used in association with storage area networks and switched fabric networking, the introduction of compute resources provides a complete "unified" computing system.

Other terms used to describe such fabrics include "unified fabric", "data centre fabric" and "unified data centre fabric". According to Ian Foster, director of the Computation Institute at the Argonne National Laboratory and University of Chicago, "grid computing 'fabrics' are now poised to become the underpinning for next-generation enterprise IT architectures and be used by a much greater part of many organizations." IBM, TIBCO, Brocade, Cisco, HP, Egenera, Avaya and Xsigo Systems currently manufacture computing fabric equipment.

HISTORY

While the term has been in use since the mid to late 1990s the growth of cloud computing and Cisco's evangelism of *unified data centre fabrics* followed by *unified computing* (an evolutionarydata centre architecture whereby blade servers are integrated or *unified* with supporting network and storage infrastructure) starting March 2009 has renewed interest in the technology. Other companies offering unified or fabric computing systems include Liquid Computing Corporation and Egenera. There have been mixed reactions to Cisco's architecture, particularly from rivals who claim that these proprietary systems will lock out other vendors. Analysts claim that this "ambitious new direction" is "a big risk" as companies like IBM and HP who have previously partnered with Cisco on data centre projects (accounting for $2–3bn of Cisco's annual revenue) are now competing with them.

Key Characteristics

The main advantages of *fabrics* are that a massive concurrent processing combined with a huge, tightly-coupled address space makes it possible to solve huge computing problems (such as those presented by delivery of cloud computing services) and that they are both scalable and able to be dynamically reconfigured. Challenges include a non-linearly degrading performance curve, whereby adding resources does not linearly increase performance which is a common problem with parallel computing and maintaining security.

TEXTILE FABRICS

However, a knowledge of the characteristics of each kind of fibre helps a good deal, as every fibre has certain inherent qualities, which cannot be wholly reproduced in any imitation fibre. For example, wool has its own features and no other fibre has identica1 ones.

The look and even the touch of pure wool flannel may be copied in the cotton flannelette and an inexperienced purchaser may find it difficult to distinguish the wool from the cotton. Fabrics made with newer blends, synthetic fibres are also not easy to judge merely by the appearance or touch. To identify such fabrics, the help of other tests, even of chemical tests is needed.

FEATURES OF FIBRES

Natural Fibres

Man-made Fibres

These fibres do not take place in fibre form but have been turned into it by man, by breaking down from their original form and reassembled into various sort of structure, *e.g.,* rayons.

Cellulose-Fibres

These fibres have as their origin cellulose which is the chief matter of the plant cells. Cellulose is a complex compound made up of carbon, hydrogen and oxygen with the molecular formula of $(C_6H_{10}O_6)n$.

Cotton comprises about 91 per cent and hemp rame and flax contain similarly large amounts of cellulose. Cellulose is very sensitive to the action of mineral acids, and oxidising agents.

However, it is very resistant to alkalies, including strong caustic alkalies at high temperature and pressure. Cellulosic fibres are low in resiliency so the fabrics wrinkle easily. Due to the high absorbency of the fibre they are comfortable for summer wear.

Fig. Cellulose Fibre

ANIMAL-FIBRES

Wool and silk are obtained from animals and are, therefore, known as animal fibres. They have protein as one of the chief constituents and are made up of carbon, hydrogen and nitrogen. Wool contains sulphur besides the above elements. Both are destroyed by concentrated mineral acids but the specific action of the dilute acids is not very harmful on either. Thus, dilute acids are the basis of dyeing and finishing processes for the materials made from these fibres. Wool as well as silk burn with the special odour of burning proteins like burning of hair, dals or milk.

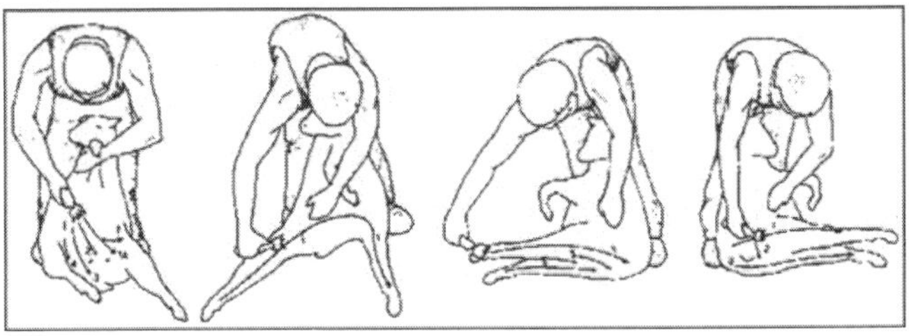

Fig. Animal-Fibres

Alkalies have a harmful effect on both. Animal fibres are very resilient and the wrinkles go out between wearings. However, they are bad conductors of heat and build up static electricity in cold and dry weather. Wool loses about 40 per cent of its strength when wet and silk, 15 per cent. Wool has a natural crimp which supports to increase its elasticity and strength. It has a springiness or resilience which no other natural fibre possesses. commonly speaking, woollen fabrics are soft to the touch and present a fuzzy appearance.

Worsteds are an exception, as their yarns are tightly twisted and firm. Wool, mixed with cotton, produces in the fabric a distinct hardness and heaviness about the material, which further tends to soil readily, fade and to wrinkle. Silk has a luxurious soft texture, and a deep lustre. The sheen is not loud or bright as in man-made fabrics.

This fibre is strong and retains most of its strength even when wet. It possesses elasticity and does not wear out soon. It is the lightest of all the fibres. Man-made fabrics may similiar to pure silk in appearance but are not so light. Silks weighted with metallic salts are still heavier and produce a sharp crease on the line of the fold. There are two kinds of man-made fibres: thermo-plastic and non-thermo-plastic.

Thermo-plastic Fibres

These are those that soften with heat and thus become liable or if the heat is sufficient high, will melt. These fibres are resistant to friction and wrinkling.

Fig. Thermo-plastic Fibres

They have low moisture content which makes them an easy to care fabric, however, they are not easy to dye. Due to this property they build up static electricity in cold and dry weather like animal fibres.

Non-Thermo-plastic Fibres

Which neither soften or melt with the application of heat but will scorch or burn if the temperature is very high. As a group they are soft, absorbent, pliable and comfortable to wear. They do not accumulate static electricity. They, except for mineral fibres, may be cared much as the natural fibres.

Fibres of Mineral Nature

These are inorganic and used for fireproof fabrics. Asbestos is practically the only natural mineral fibre. It is utilised in fabrics intended for make dress, fire-proof curtains and screens, and for man uses. It is also used for floor and table mats. Mineral fibres are really drawn threads from metals, *e.g.,* gold, silver in tissues and brocades, and even baser metals are utilised in fabrics like tinsel. Thus, their origin is inorganic matter. They have an unusually bright sheen. Cotton is limp and has a dull surface because of to its fibre structure.

It lacks the lustre and the natural creaminess of linen. It is quite inflammable, and soils and crushes easily. It is heavier than comparable fabrics of other fibre content. It is brittle and is elastic when dry; but, when moist, its strength and elasticity increases. Mercerized cotton is stronger, smoother and lustrous, hence it soils less readily than cotton. Linen is smooth and cool to the touch as compared to cotton. Fine linen has a lustre almost equal to that of silk. It has crispness and is strong enough for hard wear. Moisture soaks through and spreads over linen fabrics much more quickly than in the case of cotton fabrics.

Linen, when torn, has straight, glossy fibre-ends of unequal length. The fibre-ends of torn cotton curl and are lustreless. Linen yarn when not twisted, a more or less parallel arrangement of glossy, individual fibre. Rayon is produced to replace pure silk and so has a similar appearance, but with a better luster generally; though sometimes dull rayons are also produced. The rayon fibre if subtly examined, will be found to be coarser and heavier than a silk fibre. Rayon realised stiffer to the touch and is less elastic. It breaks more easily than silk, thus showing less strength. It loses strength when wet and regains it when dry.

FIBRE CLASSIFICATION

Textile fibres are normally broken down into two main classes, natural and man-made fibres. All fibres which come from natural sources (animals, plants, etc.) and do not require fibre formation or reformation are classed as natural fibres. Natural fibres include the protein fibres such as wool and silk, the cellulose fibres such as cotton and linen, and the mineral fibre asbestos. Man-made fibres are fibres in which either the basic chemical units have been formed by chemical synthesis followed by fibre formation or the polymers from natural sources have been dissolved and regenerated after passage through a spinneret to form fibres.

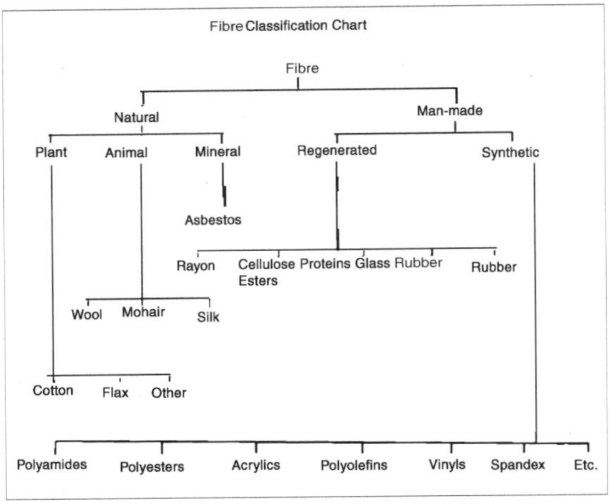

Fig. Classification of Natural and Man-made Fibres.

Those fibres made by chemical synthesis are often called synthetic fibres, while fibres regenerated from natural polymer sources are called regenerated fibres or natural polymer fibres. In other words, all synthetic fibres and regenerated fibres are man-made fibres, since man is involved in the actual fibre formation process.

In contrast, fibres from natural sources are provided by nature in ready-made form. The synthetic man-made fibres include the polyamides (nylon), polyesters, acrylics, polyolefins, vinyls, and elastomeric fibres, while the regenerated fibres include rayon, the cellulose acetates, the regenerated proteins, glass and rubber fibres.

Another method of classifying fibres would be according to chemical structure without regard of the origin of the fibre and its starting materials. In this manner all fibres of similar chemical structure would be classed together. The natural man-made fibre classification given in figure does this to a certain extent. In this way, all fibres having the basic cellulosic unit in their structures would be grouped together rather than separated into natural and man-made fibres.

This book essentially presents the fibres in groups of similar basic chemical structure, with two exceptions. In one case the elastomeric fibres have been grouped together due to their exceptional physical property, high extensibility and recovery.

FIBRE PROPERTIES

There are several primary properties necessary for a polymeric material to make an adequate fibre:

- Fibre length to width ratio,
- Fibre uniformity,
- Fibre strength and flexibility,
- Fibre extensibility and elasticity, and
- Fibre cohesiveness.

Certain other fibre properties increase its value and desirability in its intended end-use but are not necessary properties essential to make a fibre. Such secondary properties include moisture absorption characteristics, fibre resiliency, abrasion resistance, density, luster, chemical resistance, thermal characteristics, and flammability.

FIBRE LENGTH TO WIDTH RATIO

Fibrous materials must have sufficient length so that they can be made into twisted yarns. In addition, the width of the fibre (the diameter of the cross section) must be much less than the overall length of the fibre, and usually the fibre diameter should be 1/100 of the length of the fibre. The fibre may be "infinitely" long, as found with continuous filament fibres, or as short as 0.5 inches (1.3 em), as found in staple fibres. Most natural fibres are staple fibres, whereas man-made fibres come in either staple or filament form depending on processing prior to yarn formation.

FIBRE UNIFORMITY

Fibres suitable for processing into yarns and fabrics must be fairly uniform in shape and size. Without sufficient uniformity of dimensions and properties in a given set of fibres to be twisted into yarn. The actual formation of the yarn may be impossible or the resulting yarn may be weak, rough, and irregular in size and shape and unsuitable for textile usage. Natural fibres must be sorted and graded to assure fibre uniformity, whereas synthetic fibres may be "tailored" by cutting into appropriate uniform lengths to give a proper degree of fibre uniformity.

FIBRE STRENGTH AND FLEXIBILITY

A fibre or yarn made from the fibre must possess sufficient strength to be processed into a textile fabric or other textile article. The resulting textile must have sufficient strength to provide adequate durability during end-use. Many experts consider single fibre strength of 5 grams per denier to be necessary for a fibre suitable in most textile applications, although certain fibres with strengths as low as 1.0 gram per denier have been found suitable for some applications.

The strength of a single fibre is called the tenacity, defined as the force per unit linear density necessary to break a known unit of that fibre. The breaking tenacity of a fibre may be expressed in grams per denier (g/d) or grams per tex (g/tex). Both denier and tex are units of linear density (mass per unit of fibre length) and are defined as the number of grams of fibre measuring 9000 meters and 1000 meters, respectively. As a result, the denier of a fibre or yarn will always be 9 times the tex of the same fibre. Since tenacities of fibres or yarns are obtained by dividing the force by denier or tex, the tenacity of a fibre in grams per denier will be 1/9 that of the fibre tenacity in grams per tex. As a result of the adaption of the International System of Units. The appropriate length unit for breaking tenacity becomes kilometer (km) of breaking length or Newtons per tex (N/tex) and will be equivalent in value to g/tex. The strength of a fibre yarn or fabric can be expressed in terms of force per unit area, and when expressed in this way the term is tensile strength. The most common unit used in the past for tensile strength has been pounds force per square inch or grams force per square centimeter.

In 51 units, the pounds force per square inch x 6.895 will become kilopascals (kPa) and grams force per square centimeter x 9.807 will become megapascals (MPa). A fibre must be sufficiently flexible to go through repeated bending without significant strength deterioration or breakage of the fibre. Without adequate flexibility, it would be impossible to convert fibres into yarns and fabrics, since flexing and bending of the individual fibres is a necessary part of this conversion. In addition, individual fibres in a textile will be subjected to considerable bending and flexing during enduse.

FIBRE EXTENSIBILITY AND ELASTICITY

An individual fibre must be able to undergo slight extensions in length (less than 5 per cent) without breakage of the fibre. At the same time the fibre must be able to almost completely recover following slight fibre deformation. In other words, the extension deformation of the fibre must be nearly elastic. These properties are important because the individual fibres in textiles are often subjected to sudden stresses, and the textile must be able to give and recover without significant overall deformation of the textile.

FIBRE COHESIVENESS

Fibres must be capable of adhering to one another when spun into a yarn. The cohesiveness of the fibre may be due to the shape and contour of the individual fibre s or the nature of the surface of the fibres.

In addition, long-filament fibres by virtue of their length can be twisted together to give stability without true cohesion between fibres. Often the term "spinning quality" is used to state the overall attractiveness of fibres for one another.

MOISTURE ABSORPTION AND DESORPTION

Most fibres tend to absorb moisture (water vapor) when in contact with the atmosphere. The amount of water absorbed by the textile fibre will depend on the chemical and physical structure and properties of the fibre, as well as the temperature and humidity of the surroundings.

The percentage absorption of water vapor by a fibre is often expressed as its moisture regain. The regain is determined by weighing a dry fibre, then placing it in a room set to standard temperature and humidity (21° ± 1° C and 65 per cent relative humidity [RH] are commonly used). From these measurements, the percentage moisture regain of the fibre is determined:

$$\text{Percentage Regain} = \frac{\text{Conditioned weight} - \text{Dry Weight}}{\text{Dry Weight}} \times 100\%$$

Percentage moisture content of a fibre is the percentage of the total weight of the fibre which is due to the moisture present, and is obtained from the following formula:

$$\text{Percentage moisture content } \frac{\text{Conditioned weight} = \text{Dry weight}}{\text{Conditioned weight}} \times 100\%$$

The percentage moisture content will always be the smaller of the two values. Fibres vary greatly in their regain, with hydrophobic (water-repelling) fibres having regains near zero and hydrophilic (water-seeking) fibres like cotton, rayon, and wool having regains as high as 15 per cent at 21°C and 65 per cent RH. The ability of fibres to absorb high regains of water affects the basic properties of the fibre in end-use. Absorbent fibres are able to absorb large amounts of water before they feel wet, an important factor where absorption of perspiration is necessary. Fibres with high regains will be easier to process, finish, and dye in aqueous solutions, but will dry more slowly.

The low regain found for many man-made fibres makes them quick drying, a distinct advantage in certain applications. Fibres with high regains are often desirable because they provide a "breathable" fabric which can conduct moisture from the body to the outside atmosphere readily, due to their favourable moisture absorption-desorption properties. The tensile properties of fibres as well as their dimensional properties are known to be affected by moisture.

FIBRE RESILIENCY AND ABRASION RESISTANCE

The ability of a fibre to absorb shock and recover from deformation and to be generally resistant to abrasion forces is important to its end-use and wear characteristics. In consumer use, fibres in fabrics are often placed under stress through compression, bending, and twisting (torsion) forces under a variety of temperature and humidity conditions. If the fibres within the fabric possess good elastic recovery properties from such deformative actions, the fibre has good resiliency and better overall appearance in end-use. For example, cotton and wool show poor wrinkle recovery under hot moist conditions, whereas polyester exhibits good recovery from deformation as a result of its high resiliency.

Resistance of a fibre to damage when mobile forces or stresses come in contact with fibre structures is referred to as abrasion resistance. If a fibre is able to effectively absorb and dissipate these forces without damage, the fibre will show good abrasion resistance. The toughness and hardness of the fibre is related to its chemical and physical structure and morphology of the fibre and will influence the abrasion of the fibre. A rigid, brittle fibre such as glass, which is unable to dissipate the forces of abrasive action, results in fibre damage and breakage, whereas a tough but more plastic fibre such as polyester shows better resistance to abrasion forces. Finishes can affect fibre properties including resiliency and abrasion resistance.

LUSTER

Luster refers to the degree of light that is reflected from the surface of *a* fibre or the degree of gloss or sheen that the fibre possesses. The inherent chemical and physical structure and shape of the possesses. The inherent chemical and physical structure and shape of the fibre *can* affect the relative luster of the fibre. With natural fibres the luster of the fibre is dependent on the morphological form that nature gives the fibre, although the relative luster can be changed by chemical and/or physical treatment of the fibre as found in processes such *as* mercerisation of cotton.

Man-made fibres can vary in luster from bright to dull depending on the amount of delusterant added to the fibre. Oelusterants such *as* titanium dioxide tend to scatter and absorb 1ight, thereby making the fibre appear duller. The desirability of luster for *a* given fibre application will vary and is often dependent on the intended end-use of the fibre in *a* fabric or garment form and on current fashion trends.

RESISTANCE TO CHEMICALS IN THE ENVIRONMENT

A textile fibre to be useful must have reasonable resistance to chemicals it comes in contact with in its environment during use and maintenance. It should have resistance to oxidation by oxygen and other gases in the air, particularly in the presence of light, and be resistant to attack by microorganisms and other biological agents. Many fibres undergo light-induced reactions, and fibres from natural sources are susceptible to biological attack, but such deficiencies can be minimized by treatment with appropriate finishes. Textile fibres come in contact with a large range of chemical agents on laundering and dry cleaning and must be resistant from attack under such conditions.

OENSITY

The density of a fibre is related to its inherent chemical structure and the packing of the molecular chains within that structure. The density of a fibre will have a noticeable effect on its aesthetic appeal and its usefulness in given applications. For example, glass and silk fabrics of the same denier would have noticeable differences in weight due to their broad differences in density. Fishnets of polypropylene fibres are of great utility because their density is less than that of water. Oensities are usually expressed in units of grams per cubic centimeter, but in 51 units will be expressed as kilograms per cubic meter, which gives a value 1000 times larger.

THERMAL AND FLAMMABILITY CHARACTERISTICS

Fibres used in textiles must be resistant to wet and dry heat, must not ignite readily when coming in contact with a flame, and ideally should self-extinguish when the flame is removed. Heat stability is particularly important to a fibre during dyeing and finishing of the textile and during cleaning and general maintenance by the consumer. Textile fibres for the most part are made up of organic polymeric materials containing carbon and burn on ignition from a flame or other propagating source. The chemical structure of a fibre establishes its overall flammability characteristics, and appropriate textile finishes can reduce the degree of flammability. A number of Federal, state, and local statutes eliminate the most dangerous flammable fabrics from the marketplace.

PRIMARY FIBRE PROPERTIES FROM AN ENGINEERING PERSPECTIVE

The textile and polymer engineer must consider a number of criteria essential for formation, fabrication, and assembly of fibres into textile substrates. Often the criteria used will be similar to those set forth above concerning end-use properties.

Ideally a textile fibre should have the following properties:

- A melting and/or decomposition point above 220°C.
- A tensile strength of 5 g/denier or greater.
- Elongation at break above 10 per cent with reversible elongation up to 5 per cent strain.
- A moisture absorptivity of 2 per cent-5 per cent moisture uptake.
- Combined moisture regain and air entrapment capability.
- High abrasion resistance.
- Resistance to attack, swelling, or solution in solvents, acids, and bases.
- Self-extinguishing when removed from a flame.

FIBRE FORMATION AND MORPHOLOGY

Fibre morphology refers to the form and structure of a fibre, including the molecular arrangement of individual molecules and groups of molecules within the fibre. Most fibres are organic materials derived from carbon combined with other atoms such as oxygen, nitrogen, and halogens. The basic building blocks that organic materials form as covalently-bonded organic compounds are called monomers.

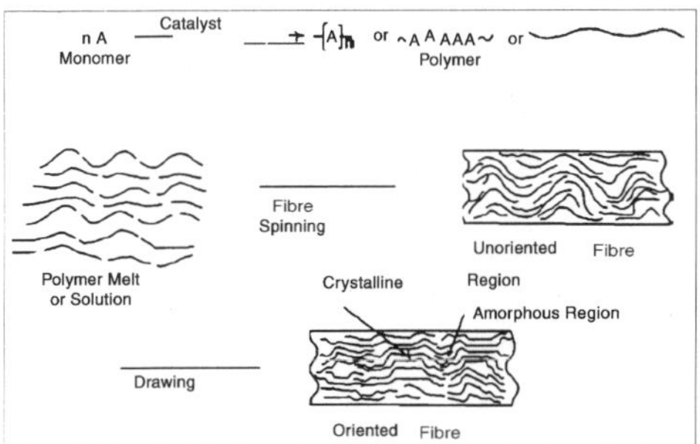

Fig. Polymerisation Sequence and Fibre Formation.

Covalent bonds involve the sharing of electrons between adjacent atoms within the monomer, and the structure tons between adjacent atoms within the monomer, and the structure of the monomer is determined by the type, location, and nature of bonding of atoms within the monomer and by the nature of covalent bonding between atoms. Monomers react or condense to form long-chain molecules called polymers made up of a given number (n) of monomer units which are the basic building unit of fibres. On formation into fibres and orientation by natural or mechanical means the polymeric molecules possess ordered crystalline and non-ordered amorphous areas, depending on the nature of the polymer and the relative packing of molecules within the fibre. For a monomer A the sequence of events to fibre formation and orientation would appear as shown in figure.

Polymers with repeating units of the same monomer (An) would be referred to as homopolymers. If a second unit B is introduced into the basic structure, structures which are copolymers are formed with structures as outlined in figure.

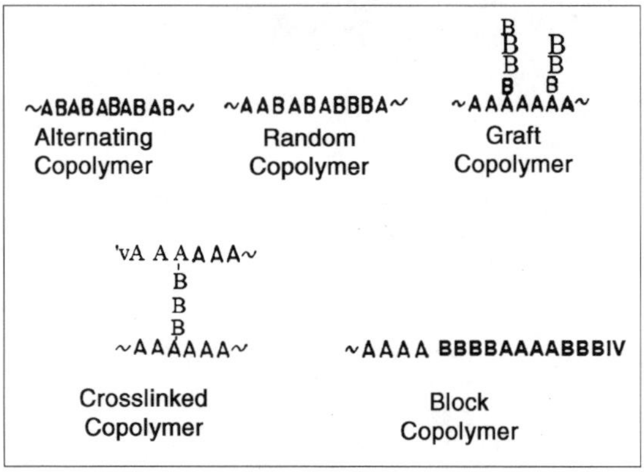

Fig. Copolymer Structures.

POLYMER FORMATION

Synthetic polymers used to form fibres are often classified on the basis of their mechanism of polymerisation—step growth (condensation) or chain growth (addition) polymerisation. Step growth polymerisation involves multifunctional monomers which undergo successive condensation with a second monomer or with itself to form a dimer, which in turn condenses with another dimer to form a tetramer, etc., usually with loss of a small molecule such as water. Chain growth involves the instantaneous growth of a long molecular chain from unsaturated monomer units, followed by initiation of a second chain, etc. The two methods are outlined schematically:

$$\text{Step Growth:} \quad nA \rightarrow \frac{n}{2} AA \rightarrow AAAA \rightarrow$$

$$\text{Chain Growth:} \quad nA \rightarrow (A)n \rightarrow (A)n \qquad nA \rightarrow (A)n$$

The average number of monomer repeating units in a polymer chain (n) is often referred to also as the degree of polymerisation, DP. The DP must exceed an average 20 units in most cases to give a polymer of sufficient molecular size to have desirable fibre-forming properties. The overall breadth of distribution of molecular chain lengths in the polymer will affect the ultimate properties of the fibres, with wide polymer size distributions leading to an overall reduction of fibre properties. Although the polymers from natural fibres and regenerated natural fibres do not undergo polymerisation by the mechanisms found for synthetic fibres, most natural polymers have characteristic repeating units and high degrees of polymerisation and are related to step growth polymers. Basic polymeric structures for the major fibres are given in figure.

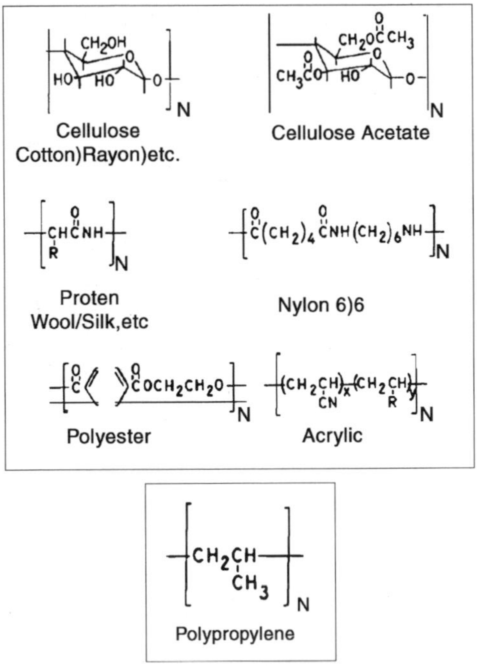

Fig. Basic Polymeric Structures for Major Fibres.

FIBRE SPINNING

Although natural fibres come in a morphological form determined by nature, regenerated and synthetic man-made fibres can be "tailor-made" depending on the shape and dimensions of the orifice (spinning jet) that the polymer is forced through to form the fibre. There are several methods used to spin a fibre from its polymer, including melt, dry, wet, emulsion, and suspension spinning. Melt spinning is the least complex of the methods. The polymer from which the fibre is made is melted and then forced through a spinneret and into air to cause solidification and fibre formation. Dry and wet spinning processes involve dissolving the fibre-forming polymer in an appropriate solvent, followed by passing a concentrated solution (20 per cent- 50 per cent polymer) through the spinneret and into dry air to evaporate the solvent in the case of dry spinning or into a coagulating bath to cause precipitation or regeneration of the polymer in fibre form in the case of wet spinning.

There is a net contraction of the spinning solution on loss of solvent. If a skin of polymer is formed on the fibre followed by diffusion of the remainder of the solvent from the core of the forming fibre, the cross-section of the fibre as it contracts may collapse to form an irregular popcorn like cross section.

Emulsion spinning is used only for those fibre-forming polymers that are insoluble. Polymer is mixed with a surface-active agent (detergent) and possibly a solvent and then mixed at high speeds with water to form an emulsion of the polymer. The polymer is passed through the spinneret and into a coagulating bath to form the fibre. In suspension spinning, the polymer is swollen and suspended in a swelling solvent. The swollen suspended polymer is forced through the spinneret into dry hot air to drive off solvent or into a wet non-solvent bath to cause the fibre to form through coagulation.

The spinning process can be divided into three steps:

1. Flow of spinning fluid within and through the spinneret under high stress and sheer.
2. Exit of fluid from the spinneret with relief of stress and an increase in volume (ballooning of flow).
3. Elongation of the fluid jet as it is subjected to tensile force as it cools and solidifies with orientation of molecular structure within the fibre.

Common cross sections of man-made fibres include:

- Round,
- Trilobal,
- Pentalobal,
- Dog-bone, and
- Crescent shapes.

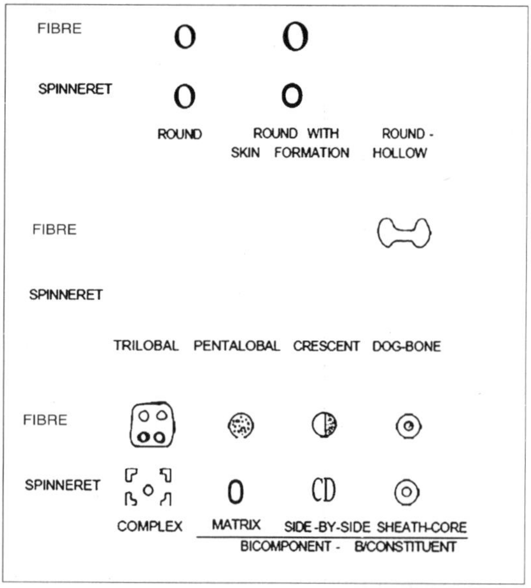

Fig. Fibre Cross Sections.

When two polymers are used in fibre formation as in bicomponent or biconstituent fibres, the two components can be arranged in a matrix, side-by-side, or sheath-core configuration. Round cross sections are also found where skin formation has caused fibre contraction and puckering (as with rayons) has occurred or where the spinneret shape has provided a hollow fibre. Complex fibre cross-sectional shapes with special properties are also used.

FIBRE DRAWING AND MORPHOLOGY

On drawing and orientation the man-made fibres become smaller in diameter and more crystalline, and imperfections in the fibre morphology are improved somewhat. Side-by-side bi component or bi constituent fibres on drawing become wavy and bulky. In natural fibres the orientation of the molecules within the fibre is determined by the biological source during the growth and maturity process of the fibre. The form and structure of polymer molecules with relation to each other within the fibre will depend on the relative alignment of the molecules in relationship to one another. Those areas where the polymer chains are closely aligned and packed close together are crystalline areas within the fibre, whereas those areas where there is essentially no molecular alignment are referred to as amorphous areas. Dyes and finishes can penetrate the amorphous portion of the fibre, but not the ordered crystalline portion.

CROSS-SECTION

A number of theories exist concerning the arrangement of crystalline and amorphous areas within a fibre. Individual crystal line areas in a fibre are often referred to as microfibrils. Microfibrils can associate into larger crystalline groups, which are called fibrils or micelles. Microfibrils are 30-100 Å in length, whereas fibrils and micelles are usually 200-600 Å in length. This compares to the individual molecular chains, which vary from 300 to 1500 Å in length and which are usually part of both crystal line and amorphous areas of the polymer and therefore give continuity and association of the various crystal line and amorphous areas within the fibre.

A number of theories have been developed to explain the interconnection of crystal line and amorphous areas in the fibre and include such concepts as fringed micelles or fringed fibrils, molecular chain folding, and extended chain concepts. The amorphous areas within a fibre will be relatively loosely packed and associated with each other, and spaces or voids will appear due to discontinuities within the structure. Figure outlines the various aspects of internal fibre morphology with regard to polymer chains.

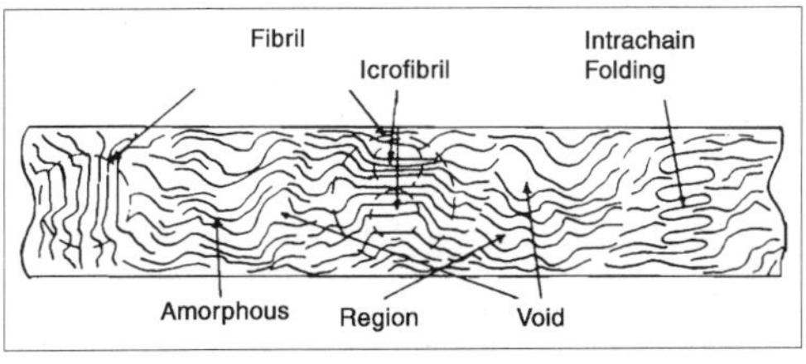

Fig. Aspects of Internal Fibre Morphology.

The force s that keep crystalline areas together within a fibre include chemical bonds (covalent, ionic) as well as secondary bonds. Cova lent bonds result from sharing of electrons between atoms, such as found in carboncarbon, carbon-oxygen, and carbon-nitrogen bonding, within organic compounds. Covalent bonds joining adjacent polymer chains are referred to as crosslinks. Ionic bonding occurs when molecules donate or accept electrons from each other, a s when a metal salt reacts with acid side chains on a po1ymer within a fibre. Chemical bonds a re much stronger than secondary bonds formed between polymer chains, but the total associative force between polymer chains can be large, since a very large number of such bonds may occur between adjacent polymer chains.

Hydrogen bonds are the strongest of the secondary bonds and occur between electropositive hydrogen atoms and electronegative atoms such as oxygen, nitrogen, and halogen s on opposing polymer chains. Nylon, protein, and cellulosic fibres are capable of extensive hydrogen bonding. Van der Waals interactions between polymer chains occur when clouds of electrons from each chain come in close proximity, thereby promoting a small attractive force between chains.

the more extended the cloud of electrons, the stronger the van der Waals interaction will be. Covalent bonded materials will show some uneven distribution of electron density over the molecule due to the differing electronegativity of the atoms and electron distribution over the molecule to form dipoles. Dipoles on adjacent polymer chains of opposite charge and close proximity are attracted to each other and promote secondary bonding. When a synthetic fibre is stretched or drawn, the molecules in mo s t cases will orient themselves in crystalline areas parallel to the fibre axis, although crystalline areas in some chain-folded polymers such as polypropylene can be aligned vertical to the fibre axis. The degree of crystallinity will be affected by the total forces available for chain interaction, the distance between parallel chains, and the similarity and uniformity of adjacent chains. The structure and arrangement of individual polymer chains also affects the morphology of the fibre. Also, configurations or optical isomers of polymers can have very different physical and chemical properties.

BULKING, TEXTURISING, AND STAPLE FORMATION

Thermoplastic man-made fibre s can be permanently heat-set after drawing and or ientation. The fibre will possess structural integrity and will not shrink up to that setting temperature.

Al so, thermoplastic fibres or yarns from this fibre s can be texturised to give three dimensional loft and bulkiness:

1. Through fibre deformation and setting at or near their softening temperature,

2. Through air entanglement, or
3. Through differential setting within fibres or yarns.

Table. Texturising Methods.

Heat-setting Techniques	**Air Entanglement**	**Differential Setting**
False Twist	Air Jet	Bicomoponent-
Knife edge		biconstituent
Stuffer box		Fibre orientation
Gear crimping		Heat shrinkage of
Autowist		thermoplastic fibres
Knit-de-knit		in a blend

Schematic representations of these methods are given in figure.

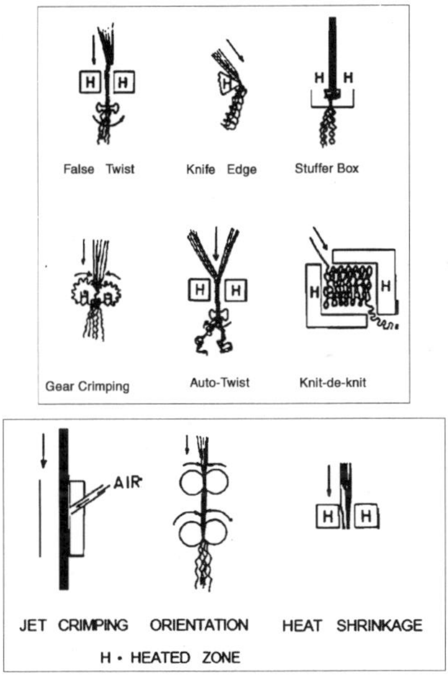

Fig. Texturising Methods.

Heat-Setting Techniques

The false twist heat-setting technique is extremely rapid, inexpensive, and most widely used. The filament fibre tow is brought in contact with a high-speed spindle running vertical to the moving tow. This action results in a high twist in the tow up to the spindle. The twisted tow is heated near its softening point before passing the spindle and then cooled and untwisted to give a wavy bulky yarn.

In knife edge texturising, filament tows or yarns are passed over a heated knife edge under tension. The fibres near the knife edge are changed in overall orientation in relation to the unheated yarns or portion of the filaments away from the knife edge, thereby causing bulking of the yarn. In stuffer box texturising, the filament tow is fed into a heated box, causing the tow to double up against itself. On removal, the cooled tow retains the zigzag configuration caused by the process. In gear crimping, the tow is passed between heated intermeshing gears. On cooling the fibres retain the shape induced by the heated gears. In auto-twisting, two tows or yarns are twisted together and then heat set. On untwisting the yarns have equal but opposite twists, which causes a spiral bulking of the yarn. In the knit-de-knit process a yarn is fill knitted, heat set, cooled, and deknitted to give a bulked yarn retaining the shape and curvature of the knit.

Air Entanglement

In air entanglement texturising, a fibre tow is loosely fed into and through a restricted space and a high-speed air jet is impinged on the fibres at a 45° angle. The loose fibres within the tow are looped to give a texturised effect.

Differential Setting

Heat shrinkage techniques cause a bulking of fibre tows containing different fibres through heating one component of the blend sufficiently to cause heat shrinkage of the fibre and compaction, contraction, and bulking. Side-by-side bicomponent and biconstituent fibres recover different degrees on each side from fibre stretching and causing a waving, crimping, or bulking of the fibre.

STAPLE FORMATION

Continuous filaments can be cut into staple by wet or dry cutting techniques. In wet cutting, the wet spun fibre is cut to uniform lengths right after spinning, while dry cutting involves partial cutting, debonding, and shuffling of the dry tow to form a sliver. Before the filament or staple is used in yarn spinning, spin finishes are added to give lubricity and antistatic characteristics to the fibres and to provide a greater degree of fibre cohesiveness. Such finishes are usually mixtures including such materials as fatty acid esters, mineral oils, synthetic esters, silicones, cationic amines, phosphate esters, emulsifiers, and/or non-ionic surfactants. Spin finishes are formulated to be oxidation resistant, to be easily removed by scouring, to give a controlled viscosity, to be stable to corrosion, to resist odour and colour formation, and to be non-volatile and readily emulsifiable.

STRUCTURE-PROPERTY RELATIONSHIPS

The basic chemical and morphological structure of polymers in a fibre determine the fundamental properties of a fabric made from that fibre. Although physical and chemical treatments and changes in yarn and fabric formation parameters can alter the fabric properties to some degree, the basic properties of the fabric result from physical and chemical properties inherent to the structure of the polymer making up the fibre. From these basic properties, the end-use characteristics of the fibre are determined. To that end, in the following stages as suggested, describe the various textile fibres in terms of their basic structural properties, followed by physical and chemical properties, and finally the end-use characteristics inherent to constructions made from the fibre. Initially the name and general information for a given fibre is set forth followed by an outline of the structural properties, including information about chemical structure of the polymer, degree of polymerisation, and arrangement of molecular chains within the fibre.

Physical properties include mechanical (tensile) and environmental properties of the fibre, whereas the effect of common chemicals and chemically induced processes on the fibre is listed under chemical properties. The end-use properties are then listed and include properties coming inherently from the structural, physical, and chemical properties of the fibre as well as end-use properties that involve evaluation of performance, subjective aspects, and aesthetics of the fabrics. Where possible, the interrelationships of these properties are presented.

TECHNICAL TECHNOLOGY USES IN FIBRES AND TEXTILES

A number of definitions have been used to describe the term 'technical textiles' with respect to their intended use, functional ability and their non-aesthetic or decorative requirements. However, none of these carefully chosen words include the fundamental fibre elements, technical or otherwise, which make up the technical textile structures. The omission of the word 'fibre' may indeed be deliberate as most technical textile products are made from conventional fibres that are already well established. In fact over 90per cent of all fibres used in the technical sector are of the conventional type. Specially developed fibres for use in technical textiles are often expensive to produce and have limited applications.

Historically, utilisation of fibres in technical capacities dates back to the early Egyptians and Chinese who used papyrus mats to reinforce and consolidate the foundations respectively of the pyramids and the Buddhist temples. However, their serious use in modern civil engineering projects only began after the floods of 1953 in The Netherlands in which many people lost their lives. The event initiated the famous Delta works project in which for the first time synthetic fibres were written into the vast construction programme. Since then, geotextiles in particular have matured into important and indispensable multifunctional materials. Use of silk in semitechnical applications also goes back a long way to the lightweight warriors of the Mongolian armies, who did not only wear silk next to their skin for comfort but also to reduce penetration of incoming arrows and enable their subsequent removal with minimal injury. Use of silk in wound dressing and open cuts in web and fabric form also dates back to the early Chinese and Egyptians.

In light of extensive utilization of conventional fibres in the technical sector, this chapter initially attempts to discuss fibres under this category highlighting their importance and the scope of their versatility. The discussion covers concisely an outline of fibre backgrounds, chemical compositions and their salient characteristics. It then introduces other fibres which have been specially developed to perform under extreme stress and/or temperature; ultrafine and novel fibres are also discussed. Finally, the chapter concludes by identifying areas of application and the roles that selected fibres play in fulfilling their intended purpose. The complete range of fibres available to the end-user and some of their mechanical properties.

HIGH MODULUS ORGANIC FIBRES

Keller's much-published contribution to the understanding of crystal growth in 1957 and confirmation of the tendency of polymers to form folded-chain crystals, provided the insight and inspiration for development of high strength, high modulus organic fibres that would surpass conventional fibres. During crystallisation, long chains of molecules fold back on themselves to form folded-chain crystals which only partly unfold during normal drawing. In the Netherlands in the 1970s DSM developed a super drawing technique known as *gel spinning,* which uses dilute solutions of ultra-high molecular weight polymers such as polyethylene to unfold the chains further and thus increase both tensile strength and fibre modulus. Ultra-high molecular weight polyethylene (UHMWPE) fibres, Dyneema or Spectra, are today the strongest fibres known, with tensile moduli in excess of 70 GN $m^{\sim 2}$. Weight for weight this fibre genus is claimed to be 15 times stronger than steel and twice as strong as aromatic polyamides such as Kevlar. It is also low in density, chemically inert and abrasion resistant. It, however, melts at around 150 °C and thermally degrades at 350 °C which restrict its use to low temperature applications.

To achieve even better performance characteristics at higher temperatures, other means of fibre production were explored in the 1960s. One successful approach eventually led to the advent of liquid crystalline polymers. These are based on polymerisation of long stiff molecules such as *para*-phenylene terephthalamide achieving molecular weights averaging to around 20 000. The influence of the stiff aromatic rings, together with hydrogen-bonding crosslinks, combines the best features of both the polyamides and the polyesters in an extended-chain configuration.

Molecular orientation of these fibres is brought about by capillary shear along the flow of the polymer as it exits from the spinneret thus overriding the need for subsequent drawing. Kevlar by DuPont and Twaron by Akzo (now

Acordis) were the first of such fibres to appear in the early 1970s. There now exists a series of first, second and third generations of *para*-aramids. Kevlar HT for instance, which has 20per cent higher tenacity and Kevlar HM which has 40per cent higher modulus than the original Kevlar 29 are largely utilised in the composite and the aerospace industries.

Para-aramids generally have high glass transition temperatures nearing 370 °C and do not melt or burn easily, but carbonise at and above 425 °C. All aramid fibres are however prone to photodegradation and need protection against the sun when used out of doors. These will be discussed later. Other high tenacity and high modulus fibres include the isotropically spun Tech-nora (Teijin) and Supara, based upon *para*-aramid copolymers, with slightly lower maximum strength and modulus values than Kevlar. Several other melt-spinnable liquid crystalline polymers are also available.

Combustion-resistant Organic Fibres

The fibres were developed following earlier observations that aromatic polymer backbones yielded improved tensile and heat resistance compared with conventional fibres. However, if the polymer chains have lower symmetry and order, then polymer tractability and textile fibre characteristics are improved. Solvent-spun Nomex and Conex were the first so-called *meta*-aramids made from poly(meta-phenylene isophthalamide) and were produced by DuPont in 1962 and by Teijin in 1972, respectively. The *meta*-phenylene isophthalamide molecule is identical in all but the position of its —NH— and —CO— groups to *para*-phenylene terephthalamide (or Kevlar) molecules. The different positioning of these groups in *meta*-aramids creates a zig-zag molecular structure that prevents it from full crystallisation thus accounting for its relatively poorer tensile properties. However, Nomex is particularly well known for its resistance to combustion, high decomposition temperature prior to melting and high limited oxygen index (LOI), the minimum amount of oxygen required to induce ignition.

Melt-spinnable aromatic fibres with chains containing paraphenylene rings, like polyether ether ketone (PEEK), polyether ketone (PEK) and poly(*p*-phenylene sulphide) (PPS), also have high melting points but, since their melting points occur prior to their decomposition temperature, they are unsuitable for fire-retardant applications. However, their good chemical resistance renders them suitable for low temperature filtration and other corrosive environments. The polyheterocyclic fibre, polybenzimidazole or PBI, produced by Hoechst-Celanese has an even higher LOI than the aramids.

It has excellent resistance to both heat and chemical agents but remains rather expensive. P84, initially produced by Lenzing and now produced by Inspec Fibres, USA, comprises polyimide groups that yield reasonably high resistance to fire and chemical attack. The acrylic copolymer-based fibre produced by

Acordis known as Inidex (although now no longer produced) unlike many aramid fibres has high resistance to UV (ultraviolet) radiation and a fairly high LOI at the expense of much reduced tenacity and rather low long-term exposure resistance to heat.

Oxidised acrylic fibre, best known as Panox (SGL, UK) is another crosslinked, high combustion-resistant material produced by combined oxidation and pyrolysis of acrylic fibres at around 300 °C. Although black in appearance, it is not classed as carbon fibre and preserves much of its original non-carbonaceous structure. In addition, it does not have the graphitic or turbostratic structure of carbon fibres. It retains a sufficiently high extension-at-break to be subjected to quite normal textile processes for yarn and fabric formation. Panox has a very high LOI of 0.55. Fully carbonised, or carbon fibres with even higher tensile properties, are achieved by full carbonisation of these oxidised precursor fibres or by the melt spinning of liquid crystalline mesophase pitch followed by their oxidation and pyrolysis.

Ultra-fine and Novelty Fibres

Ultra-fine or microfibres were developed partly because of improved precision in engineering techniques and better production controls, and partly because of the need for lightweight, soft waterproof fabrics that eliminate the more conventional coating or lamination processes. As yet there are no universal definitions of microfibres. *Textile Terms and Definitions* simply describes them as fibres or filaments with linear densities of approximately 1.0dtex or less. Others have used such terms as fine, extra-fine and micro-fine corresponding to linear densities ranging from 3.0dtex to less than 0.1dtex. They are usually made from polyester and nylon polymers, but other polymers are now being made into microfibres. The Japanese first introduced microfibres in an attempt to reproduce silk-like properties with the addition of enhanced durability. They are produced by at least three established methods including island-in-sea, split process and melt spinning techniques and appear under brand names such as Mitrelle, Setila, Micrell, Tactel and so on. Once in woven fabric form their fine diameter and tight weave allows up to 30000 filaments cm^2, making them impermeable to water droplets whilst allowing air and moisture vapour circulation.

They can be further processed to enhance other characteristics such as peach-skin and leather-like appearances. The split technique of production imparts sharp-angled edges within the fibre surface, which act as gentle abraders when made into wiping cloths that are used in the optical and precision microelectronic industries. Microfibres are also used to make bacteria barrier fabrics in the medical industries. Their combined effect of low diameter and compact packing also allows efficient and more economical dyeing and finishing. Finally, constant pressure to achieve and develop even more novel applications of fibres has led to a number of other and, as yet, niche fibrous products. In principle, the new ideas usually strive to combine basic functional properties of a textile material with special needs or attractive effects.

For example, Solar-Aloha, developed by Descente and Unitika in Japan, absorbs light of less than 2|jm wavelength and converts it to heat owing to its zirconium carbide content. Winter sports equipment made from these materials use the cold winter sun to capture more than 90per cent of this incident energy to keep the wearer warm. Another interesting material gives rise to thermochromic fabrics made by Toray which have a uniform coating of microcapsules containing heat-sensitive dyes that change colour at 5 °C intervals over a temperature range of -40 °C to 80 °C creating 'fun' and special effects.

Cripy 65 is a scented fibre produced by Mitsubishi Rayon (R) who have enclosed a fragrant essence in isolated cavities along the length of hollow polyester fibres. The scent is gradually released to give a consistent and pleasant aroma. Pillows and bed linen made from these materials are claimed to improve sleep and sleeping disorders. The effect can also be achieved by printing or padding microcapsules containing perfumes into fabrics which subsequently burst and release the perfume. With careful handling, garments made from these materials are said to maintain this property for up to two years. Infrared-emitting and bacteria-repelling fibres are some of the other emerging novel fibres.

High Performance Inorganic Fibres

Any fibre that consists of organic chemical units, where carbon is linked to hydrogen and possibly also to other elements, will decompose below about 500°C and cease to have long-term stability at considerably lower temperatures. For use at high temperatures it is therefore necessary to turn to inorganic fibres and fibres that consist essentially of carbon. Glass, asbestos and more recently carbon are three well-known inorganic fibres that have been extensively used for many of their unique characteristics.

Use of glass as a fibre apparently dates back to the ancient Syrian and Egyptian civilizations which used them for making clothes and dresses. However, the very high modulus or stiffness displayed by these fibres means that they are quite brittle and can easily be damaged by surface marks and defects. They are, therefore, best utilised by embedding in matrix forms where the fibres are fully protected.

Epoxy resins, polyester and other polymers, as well as cement, have commonly been used both to protect and to make use of their contributory strength. Glass-reinforced boat hulls and car bodies, to name but two application areas of such composites, reduce overall weight and cost of fabrication as well as eliminating the traditional problems of rotting wood and rusting metals associated with traditional materials. Their good resistance to heat and very high melting points have also enabled them to be used as effective insulating materials.

Asbestos is a generic name for a variety of crystalline silicates that occur naturally in some rocks. The fibres that are extracted have all the textile-like properties of fineness, strength, flexibility and more importantly, unlike conventional fibres, good resistance to heat with high decomposition temperatures of around 550 °C. Their use as a reinforcement material has been found in clay-rich prehistoric cooking pots discovered in Finland. In more recent times they have been extensively used to reinforce brittle matrices such as cement sheeting, pipes, plastics and also as heat insulators. However, with the discovery of their carcinogenic hazards, their use has gradually declined and alternative fibres have been developed to replace them totally.

High purity, pyrolysed acrylic-based fibres are classified as carbon fibres. The removal of impurities enhances carbon content and prevents the nucleation and growth of graphite crystals which are responsible for loss of strength in these fibres. Carbon fibres with different structures are also made from mesophase pitch. The graphite planes in PAN-based fibres are arranged parallel to the fibre axis rather than perpendicular as is the case with pitch-based carbon fibres. Their high strength and modulus combined with relatively low extensibility means that they are best utilised in association with epoxy or melt-spinnable aromatic resins as composites.

Increasing demand in the defence and aerospace industries for even better performance under extreme conditions led, within the last quarter of the 20th century, to yet another range of new and rather expensive metal oxides, boron and silicon-based fibres. These are often referred to as ceramic fibres and will now be briefly discussed. Aluminosilicate compounds are mixtures of aluminium oxide (Al_2O) and silicon oxide (SiO_2); their resistance to temperature depends on the mixing ratio of the two oxides. High aluminium oxide content increases their temperature tolerance from a low of 1250 °C to a maximum of 1400 °C.

However, despite their high temperature resistance, these fibres are not used in high stress applications owing to their tendency to creep at high temperatures. Their prime applications are in insulation of furnaces and replacement of asbestos fibres in friction materials, gaskets and packings. Both aluminium oxide or alumina fibres and silicon oxide or silica fibres are also produced. Pure boron fibres are too brittle to handle but they can be coated on tungsten or carbon cores.

Their complex manufacturing process makes them rather expensive. Their prime application is in lightweight, high strength and high modulus composites such as racket frames and aircraft parts. Boron fibre use is limited by their thickness, their relatively poor stability in metal matrices and their gradual loss of strength with increasing temperature. Boron nitrides (BN) are primarily used in the electronic industry where they perform both as electrical insulators and thermal conductors.

The most outstanding property of silicon carbide (SiC) is the ability to function in oxidizing conditions up to 1800 °C with little loss of mechanical properties. Silicon carbide exceeds carbon fibre in its greater resistance to oxidation at high temperatures, its higher compressive strength and better electrical resistance. SiC fibres containing carbon, however, lose some tensile properties at the expense of gaining better electrical conductivity.

Finally, many of the inorganic fibres so far referred to are now also produced in microcrystalline or whisker form and not in the more normal textile-fibre form. Whiskers have extremely good mechanical properties and their tensile strength is usually three to four times those of most reinforcing fibres. They are, however, costly to produce and their inclusion into composite structures is both difficult and cumbersome.

CONVENTIONAL AND NATURAL FIBRES

Cotton accounts for half of the world's consumption of fibres and is likely to remain so owing to many of its innate properties and for economical reasons that will not be discussed here. Cotton is made of long chains of natural cellulose containing carbon, hydrogen and oxygen otherwise known as polysaccharides. The length of the chains determines the ultimate strength of the fibre. An average of 10000 cel-lulosic repeat or monomeric units make up the individual cellulose chains which are about 2|jm in length. The linear molecules combine into microfibrils and are held together by strong intermolecular forces to form the cotton fibre. The unique physical and aesthetic properties of the fibre, combined with its natural generation and biodegradability, are reasons for its universal appeal and popularity. Chemical treatments such as Proban and Pyrovatex are two examples of the type of durable finishes that can be applied to make cotton fire retardant. High moisture absorbency, high wet modulus and good handle are some of the more important properties of cotton fibre.

Wool, despite its limited availability and high cost, is the second most important natural fibre. It is made of protein: a mixture of chemically linked amino acids which are also the natural constituents of all living organisms. Keratin or the protein in the wool fibre has a helical rather than folded chain structure with strong inter- and intrachain hydrogen bonding which are believed to be responsible for many of its unique characteristics. Geographical location, the breeding habits of the animals, and climatic conditions are some of the additional variables responsible for its properties.

The overall high extensibility of wool, its natural waviness and ability to trap air has a coordinated effect of comfort and warmth, which also make it an ideal insulating material. The sophisticated dual morphology of wool produces the characteristic crimp which has also been an inspiration for the development of some highly technical synthetic fibres. Wool is inherently fire retardant, but further improvements can be achieved by a number of fire-retardant treatments. Zirconium- and titanium-treated wool is one such example which is now universally referred to as Zirpro (IWS) wool.

Flax, jute, hemp and ramie, to name but a few of the best fibres, have traditionally taken a secondary role in terms of consumption and functional requirements. They are relatively coarse and durable, and flax has traditionally been used for linen making. Jute, ramie and to a lesser extent other fibres have received attention within the geotextile sector of the fibre markets which seeks to combine the need for temporary to short-term usage with biodegradability, taking into account the regional availability of the fibres.

Silk is another protein-based fibre produced naturally by the silkworm, *Bombyx Mori* or other varieties of moth. Silk is structurally similar to wool with a slightly different combination of amino acids which make up the protein or the fibroin, as it is more appropriately known. Silk is the only naturally and commercially produced continuous filament fibre which has high tenacity, high lustre and good dimensional stability. Silk has been and will remain a luxury quality fibre with a special place in the fibre market. However, its properties of biocompatibility and gradual disintegration, as in sutures, have long been recognised in medical textiles.

Synthetic Fibres

All synthetic fibres originate from coal or oil. The first synthetic fibre that appeared on the world market in 1939 was nylon 6.6. It was produced by DuPont and gained rapid public approval. A series of nylons commonly referred to as polyamides now exists in which the amide linkage is the common factor. Nylon 6.6 and nylon 6 are most popular in fibre form. They are melt extruded in a variety of cross-sectional shapes and drawn to achieve the desired tenacity. They are well known for their high extensibility, good recovery, dimensional stability and relatively low moisture absorbency. Nylon 6.6 in particular soon became a popular household carpet fibre and was developed into fibres commonly known as Antron manufactured by DuPont. Antron fibres of various generations use additives, varied cross-sectional shapes and modified surface characteristics to enhance the aesthetic and visual appeal of carpets as well as improving their resilience and dissipating static charges.

Nylon was later surpassed by the even more popular fibre known as polyester, first introduced as Dacron by DuPont in 1951. Polyester is today the second most used fibre after cotton and far ahead of other synthetics both in terms of production and consumption. Polyethylene terephthalate or polyester is made by condensation polymerisation of ethylene glycol and terephthalic acid followed by melt extrusion and drawing. It can be used in either continuous form or as short staple of varying lengths.

The popularity of polyester largely stems from its easycare characteristics, durability and compatibility with cotton in blends. Its very low moisture absorbency, resilience and good dimensional stability are additional qualities. Many manufacturers across the world produce polyester under different commercial names with almost tailor-made properties. A high glass transition temperature of around 70 °C with good resistance to heat and chemical degradation also qualifies this polymer for most technical textile applications. These will be discussed in greater detail later. Flame-retardant Trevira CS and Trevira high tenacity, both polyesters developed and marketed by Trevira GmbH in Germany, are examples of the many varieties available today.

Wool-like properties are shown by polyacrylic fibres which are produced by the polymerisation of acrylonitrile using the addition route into polyacrylonitrile. They can then be spun into fibres by dry or wet spinning methods. Orlon was produced by DuPont. It had a distinctive dumbbell shaped cross-section and was extruded by the dry process in which the solvent is evaporated off. Acrilan produced by Mon-santo and Courtelle produced by Acordis are spun by the wet extrusion technique and have near circular cross-sections. Acrylic fibres now also appear in bicomponent form with wool-like characteristics.

Chemically modified acrylics, principally the modacrylics, include chlorine atoms in their molecular structure which are responsible for their low burning behaviour and, unlike acrylics, have the ability to self-extinguish once the source of ignition has been removed. A selected fibre in terms of fibre chemistry is Oasis, a superabsorbent fibre made by the collaborative efforts of Acordis and Allied Colloids, based on crosslinking copolymers of acrylic acid. This fibre is claimed to absorb moisture many times its own weight and holds it even under pressure. Its application in hygiene and medical care in different forms is being pursued.

Polyolefin fibres include both polyethylene and polypropylene made by addition polymerisation of ethylene and propylene and subsequent melt extrusion, respectively.

Polyethylene has moderate physical properties with a low melting temperature of about 110 °C for its low density form and about 140 °C for its high density form which severely restricts its application in low temperature applications.

Polypropylene has better mechanical properties and can withstand temperatures of up to 140 °C before melting at about 170 °C. Both polymers have a density less than that of water which allows them to float as ropes, nets and other similar applications. The availability, low cost and good resistance to acid and alkaline environments of polypropylene has greatly influenced its growth and substantial use in geotextile applications.

Finally, elastane fibres are synthetic-based elastomeric polymers with at least 85per cent segmented polyurethane in their structure. They have rubber-like properties, which means they can be extended up to six or more times their original length. They are used in combination with most natural and synthetic fibres in knitted and woven materials. They were initially produced by DuPont in 1959 under the now well-known trademark of Lycra.

Regenerated Fibres

Viscose rayon was the result of the human race's first attempts to mimic nature in producing silk-like continuous fibres through an orifice. Cellulose from wood pulp is the main constituent of this novel system, started commercially in the early 1920s. Thin sheets of cellulose are treated with sodium hydroxide and aged to allow molecular chain breakage.

Further treatment with carbon disulphide, dissolution in dilute sodium hydroxide and ageing produces a viscous liquid, the viscose dope, which is then extruded into an acid bath. he continuous filaments that finally emerge are washed, dried and can be cut to staple lengths. The shorter cellulose molecules in viscose and their partial crystallisation accounts for its rather inferior physical properties relative to cotton. Further development and refinement of the manufacturing technique have created a whole range of fibres with improved properties.

High tenacity and high wet modulus viscose compare in all but appearance to cotton in both dry and wet conditions. Chemically altered regenerated cellulose di- and triacetates do not burn like cotton and viscose to leave a fluffy black ash, but melt and drip instead. This characteristic enables them to be shaped or textured to enhance their visual and aesthetic appeal. Hollow viscose modifications give enhanced bulk and moisture absorbency and have an improved cotton-like feel. Fire-retardant (FR) viscose was first introduced in the 1960s. A major example is produced by Lenzing in Austria by incorporating organophosphorous compounds into the spinning dope prior to extrusion. The additive is reasonably stable and has no chemical interaction with the cellulose molecules. It is also unaffected by bleaching, washing, dry cleaning, dyeing and finishing processes.

Early in the 1990s Kemira (now Sateri Fibres) of Finland introduced an alternative version of FR viscose known as Visil in which polysilicic acid is present. The fibre chars upon heating leaving a silica residue. Lyocell, is the latest addition to this series of fibres, commercially known as Tencel (Acordis), has all the conventional properties of viscose in addition to its much praised environmentally friendly production method. The solvent used is based on non-toxic JV-methyl morpholine oxide used in a recyclable closed loop system, which unlike the viscose process avoids discharge of waste. Highly absorbent derivatives of Tencel, known as Hydrocell are establishing a foothold in wound dressing and other medical-related areas of textiles.

CIVIL AND AGRICULTURAL ENGINEERING

Natural fibres such as flax, jute and ramie can be used for most temporary applications where, for instance, soil erosion is the problem. The geotextiles made from these natural polymers help to prevent the erosion of soils by allowing vegetative growth and their subsequent root establishment. Once the purpose is served, the geotextile material gradually disintegrates into the soil. In most medium to long term applications however, where physical and chemical durability and dimensional stabilities are of prime concern, synthetic fibres are preferred.

There are currently at least six synthetic polymers considered suitable for this purpose; they include:

- Polypropylene
- Polyester
- Polyethylene
- Polyvinyl chloride
- Polyamide
- Aramids.

Polypropylene is by far the most utilised geotextile, followed by polyester, the other three trailing behind with polyamide as the least used synthetic polymer. *Para*-aramids are only used where very high creep resistance and tolerance to prolonged heating are required.

Generally, geosynthetics must have lifetimes which are defined and relate to their function. Durability, therefore, is mainly determined by the resistance offered by the component fibre and its assembled structure to the degrading species that causes a reduction in the tensile and mechanical properties. At least three main degrading mechanisms have been identified that ultimately determine the durability and life of the contending polymer; they include:

- Physical degradation
- Chemical degradation
- Biological degradation.

Physical degradation is usually sustained during transport or installation in one form or another. Initiation of cracks is normally followed by their subsequent propagation under environmental or normal stress. In the first instance, polymer susceptibility to physical degradation depends on such factors as the weight of the geotextile; generally lightweight or thin geotextiles suffer larger strength losses than thick and heavy materials. Secondly, woven geotextiles suffer slightly larger strength losses than nonwovens, owing to their greater stiffness and, finally, the generic nature of the polymer itself plays an important role.

Polyester, for instance, suffers greater strength losses than polyolefin fibres owing to its high glass transition temperature of around 70 °C. Under normal soil conditions, the polymer is relatively brittle and therefore susceptible to physical damage. Polypropylene with its glass transition averaging at about -10 °C is much more pliable under these conditions but suffers from the rather critical phenomenon of extension with time or creep.

Chemical degradation is the next mechanism by which chemical agencies, often in combination with ultraviolet light or/and heat, attack the polymer whilst in use or being stored. The presence of energy starts off a self-destructing chain of events that ultimately renders the polymer ineffective. Ultraviolet radiation normally attacks the surface of the polymer and initiates chain breakage or scission which leads to embrittlement and eventual failure of the polymer.

Generally, chemical degradation is a function of polymer type, thickness and availability of stabilizers. The type, quantity, and distribution of stabilizers also controls the degree of resistance to degradation. Polyolefins are particularly susceptible to ultraviolet degradation and need protection using light-stabilising additives. Oxidation due to heat operates similarly by weakening the polymer thus causing rapid degradation. A range of antioxidants are often included in the polymer during manufacture and processing to minimise this harmful effect.

Biological degradation can result from at least three types of microbiological attack: direct enzymatic attack, chemical production by microorganisms which may react destructively with the polymer, and attack on the additives within the polymer. High molecular weight polymers are much more immune to biological attack than low molecular weight polymers because microorganisms cannot easily locate the molecular chain endings. However, some microorganisms can permeate less digestible polymers in order to gain access to food.

Automotive and Aeronautics

Mechanical functionality has increasingly become an almost secondary requirement for travel safety, weight efficiency, comfort and material durability of the transporting medium. From bicycles to spacecraft, fibres in one form or another fulfil these important requirements. Carbon fibre reinforcement of the frame of a bicycle ridden in the 1992 Olympics was the first of its kind to allow a comfortable win ahead of its competitors. The one-piece, light, fibre-reinforced composite structure and design of the bike have since become an earmark of this industry.

In the simplest terms, composites utilise unique fibre properties such as strength, stiffness and elasticity whilst incorporating the compression resistance, and torsion and bending characteristics of the matrix used. Glass fibres have been traditionally used for the manufacture of boat hulls and car bodies. UHMWPE, aramids and carbon-reinforced composites in a variety of matrix materials are now routinely produced and utilised in low to very high temperature applications. Reinforcement of tyres is another area where the transport industry has benefited from fibres and where rapid temperature change and changing weather conditions demand effective response and durability.

High tenacity viscose and polyester yarns built into the internal structure of tyres now address these needs adequately. In addition to all the obvious upholstery materials used in the interior of cars, trains and aeroplanes, fibres are now used in one form or another in such parts as the engine components, fan belts, brake pads, door panels, seat skeletons, seat belts and air bags. Fire-retardant additives or inherently fire-retardant fibres are today standard requirements in the public sector use of all transport in order to improve safety.

Protection and Defence

In textile terms, protection and defence can be a passive response in which the textile product receives and absorbs the impending impact or energy in order to protect the underlying structure. Ballistic garments are obvious examples, where the assembled fibrous material is deliberately designed to slow down and reduce the penetration of an incoming projectile. But they may also have a more active role, where the fibres show positive response by generating char or protective gases, shrinking or expanding to prevent penetration of moisture or vapour and so on. In each scenario, the protection of the underlying structure is the common objective.

In the early days, leather and metal mesh garments were used to protect the body against sword and spear attacks, but with the passage of time development of new materials occurred and, for example, early in the 1940s, nylon-based flack jackets were introduced. They were a considerable improvement on the leather and metal garments but were still rather heavy and uncomfortable to wear. With the advent of *para*-aramids in the early 1970s, advanced fibres were for the first time used to make much more acceptable protective gear. In these garments, Kevlar or Twaron continuous filament yarns are woven into tight structures and assembled in a multilayer form to provide maximum protection. Their high tenacity and good energy absorption combined with high thermal stability enables these garments to receive and neutralise a range of projectiles from low calibre handguns, that is 0.22 to 0.44 inch (5.6-11.2mm) to military bullets within the 5.56-7.62mm range. In the latter case, the fabric will require facing with ceramic tiles or other hard materials to blunt the tips of metal-jacket spitzer-pointed bullets.

To strike a better balance between garment weight, comfort and protection an even stronger and lighter fibre based on ultra-high molecular weight polyethylene was developed and used in the early 1990s, which immediately reduced average garment weight by 15per cent. UHMWPE, commercially known as Dyneema (DSM) and in composite form as Spectra Shield by Allied Signal, is now used to make cut-resistant gloves and helmets, as well as a wide range of protective garments. However, unlike Kevlar, with a fairly low melting temperature of 150 °C, it is best suited to low temperature applications.

Fire is another means by which fatal and non-fatal injuries can be sustained. No textile-based material can withstand the power and ferocity of a fire for sustained periods of greater than 10min or so. However, if the fire ignition point could be increased or the organic structure converted to a carbonaceous char replica and its spreading rate delayed, then there may just be enough time gained to save an otherwise lost life. This is in essence the nature and objective of all fire-retardant compounds. Such treatments give fibres the positive role of forming char, reducing the emission of combustible volatile gases and effectively starving the fire of oxygen or providing a barrier to underlying surfaces.

In contrast, water-repellent fabrics may shed water by preventing water droplets from physically passing through them owing to their fine fibre dimensions and tight weaves, as is the case with ventile fabrics or perforated barrier inlays such as Goretex, whose tiny holes allow water vapour to pass through them but not water droplets. Wax and chemically treated fabrics reduce water/fabric surface tension by allowing water to roll off but do not allow permeation of moisture and air, so compromising comfort.

FABRICS OF PLAIN WEAVE NATURE

Wool

Homespun, flannel crepe.

Silk

Crepe de chine, chiffon, shantung, crepe georgette, taffetta and voile.

Cotton

Calico, cambric, canvas, cheese cloth, chintz, cretonne, crepe, cotton, flannelette, long cloth, gingham, muslin, organdy, seersucker and voile.

Linen

Cambric, dress linen, handkerchief linen and towelling.

Rayon

Chifton,.crepe, geogette, seersucker, organdy, taffetta and voile.

Variation of Plain Weave

Basket Weave

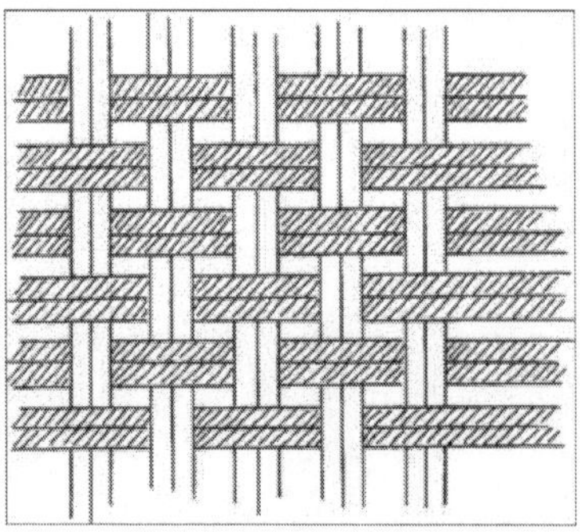

Fig. Basket Weave

Basket weave is a balanced weave. The fabrics with basket weave have flenth and if the count is not very highlite fabric is even porous and pliable.

However, fabrics with arrangements like 3x3, 4x4, 6x6 snag easily. In this two or more yarns in both warp and filling are treated as one, and interlaced as in plain weave. This weave is utilised in materials for sport. Basket weave. coats and suits. This is a relatively loose weave and therefore, the fabrics are more likely to shrink.

Rib Weave

Fig. Rib Weave

This is a variation of plain weave. In this, heavier yarns are utilised in the warp than those in the weft, and this produces a ribbed effect. Sometimes, the order is reversed and the heavier yarn is used in the weft. Ego eve, faille gross grain bengalene, ottoman, broad cloths poplin and iabralco are some of the instances.

Twill Weave

Fig. Twill Weave

The simplest form of the twill weave is made by throwing the filling yarn over two warp yarns, then under one, over two, under one and so 'on, with a progression of one. This weave needs at least three shifts of the loom to complete the unit of the design and three harnesses. In this weave, either the warp or the filling may be utilised to form the face of the cloth. The progression, which produces the diagonal effect of the twill may be to the right or to the left. It may be plain or varied which makes the cloth decorative. The common variation in the twill weave is herring bone. In this, the diagonal effect runs in one direction a few rows and them in the opposite direction.

The whole pattern is thus repeated. Another variation is made for a giamon Q pattern. Varifions are also introduced by using yarns of various sizes, qualities and colours. It will weave has fewer points of interlacings than plain weave. So it allows closer picking of warp yarns to produce heavier fabrics which results in longer wear. Twill weave produces strong material because of the tightly twisted yarns which are utilised to bring out the diagonal effect, and the compactness of its construction.

Twill weave fabrics are very precious because of their elaborate construction, but they are strong and stand hard and long wear. This weave is generally used in wool and cotton fabrics where durability is prime need. Twill weave fabrics do not-shew dirt or dust as much as the fabric woven in plain weaves do, and are, therefore, more appropriate for dresses, men's shirts and suits and children's garments. The side on which the diagonal effect is more prominent is the right side *of* the cloth. Bet when twill weave fabrics are finished by 'napping', the napped side is the right-side.

Fabrics of Twill Weave Nature

Wool

Covert cloth, broad cloth, cashmere; flannel, gabardine, tweed, serge, worsted

Cotton

Covert cloth denim, drill, gabardine,.Jean, khaki serge.

Linen

Table linen, towels, drills and ticking.

Silk

Twill, serge, Surat.

Sateen and Satin

Satin Weaveing

This weave differs from the twin weave as in this the interlacing of the warp and filling yarns not regular. The filling yarn is passed over several warp threads. The floasts in sateen are generally shorter than those in the stin weave. Other variations may be introduced by throwing long or short floats or by throwing long or short floats upon the surface. Then again, a combination of satin and sateen weaves may be utilised to produce elaborate or figured designs.

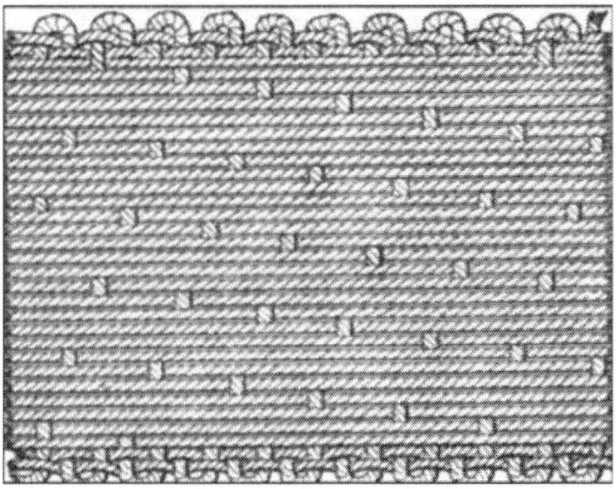

Fig. Satin Weaveing

The satin and sateen weaves need more shafts in weaving them are needed in the stain or twill weaves and, therefor fabrics of the former weaves are very expensive. The Satin or, Sateen weave fabrics are more decorative than serviceable. These are durable only if they are not subjected to excessive or hard use. Stain or Sateens with short floats are more durable. To differentiate between satin and sateen weaves fingers are passed over the cloth to condition the direction of the floats. If the fingers run-smoohtly and smoothly along the length olthe fabric, thus indicating the presence of the warp floats, the weave is satin. This weave is most commonly utilised in the silks and rayons. Satins are extremely lustrous and smooth the sateen weave will have floats running on the cross side of the fabric. This weave is utilised in a few mercerised cotton fabrics. These, fabrics are not as lustrous or smooth as satin.

Sateen and Satin Fabrics

Dress satin, slipper satin, satin crepe, satin georgettee, satin canton, satin twill, sateen, venetian sateen.

Weaves of Novelty Nature

Figure Weaving

Several fabrics have various kinds of designs woven into them. There are two methods of making these figure weaves —Jacquard and Dobby. A Jacquard loom is essential for more elaborate patterns buti Dobby attachment is sutpcient for simple and small geometrical designs.

Dobby Weave

Only a few geometrical designs are possible in this weave, such as the diamond patterns. The Dobby attachment consists of a chain of parochial strips of wood hung on the top of the loom.

Fig. Dobby Weave

Pegs are inserted in each of the strips to connote the pattern. These strips of wood control the movements of the harness to form a shed. Another chain of strips of wood controls the shuttle, and thus the designs are woven in the fabric. The Dobby attachment can control no more than twenty-five harness. This is the reason, for the limited number of designs. Instances of Dobby weave are the "Birds eye" used in diapers and sharkskin suiting Huckaback or huck towtling of small diamond designs are also made in the Dobby weave

Jacquard Loom Weaveing

The Jacquard loom was invented by a French man, Joseph Jacquard, in 1801. The intricate arrangement of this loom weaves the most beautiful and exhaustive designs in the fabrics. In this loom every individual warp is controlled independently and not in series as in ordinary looms. In the place of harnesses, a series. of oblong punched cards called 'Jacquard card' control the raising of the warps. As several cards as there are picks are used in the design. The designs are first worked out on a graph paper in points. Then these points of the designs are punched on the cards in the same order and are rotated over a cylinder on the top of the loom. Long cords which hold fine steel wires are also attached to another set of horizontal wires knownas 'needles' on the" top of the loom.

Each of the steel wires has an eye through which a warp yarn is threaded. In working the loom, the needles press forward against a Jacquard Card. The needles which go through the punched holes in the card pull some cords which in their turn pull the steel wires and so the warps are-faised to form a shed simultaneousdly.

The shuttle passes through throwing in the filling yarn. The cylinder Which holds the cards, turns slightly and another card israised. This again forms a contact with the needles, a new shred is formed, and the filling yarn is passed through. This process is continued till all the cards have been used once. The pattern is thus completed. The process is then repeated till the whole length of the cloth is woven.

The Jacquard weave fabrics are fairly expensive because of:

- The elaborate process used for this weave,
- The time and skill required for preparing the cards. The actual weave is a combination of two or more kinds of fundamental weaves or a variation of such combinations.

For instance, in damask the design is woven in sateen weave with filling floats, and the ground of the cloth is in satin weave with warp floats. In brocades, the design may be in the twill weave and the background in satin weave. The face and back of the fabrics in figured weaves may display the same design (as in the case of huckaback) or in the reverse as in damask, the design may be totally different on either side as in brocades. The durability of the figured fabrics relies on the length of the floats, and the type of the yarn. Long floats increase the sheen and softness of the fabric but affect its wearing quality.

TREATING FABRICS WITH CHEMICAL REAGENTS

These comprise a process of treating the fabrics with chemical reagents which change either the appearance or the,intrinsic properties of the cloth, *e.g.*. mercerising changes their appearance of the cloth, whereas crease resistant finishes or. fire proofing changes its properties.

Process of Mercerising

This process provides a high degree of lustre tottons through the chemical action of caustic soda. John Mercer, a calico printer in England, first discovered in 1844, that the application of a strong solution of caustic soda made cotton cloth transparent. The process comprises of impregnating the yarn or the fabric with cold caustic soda solution, applied under tension to reduce shrinkage, and to increase the lustre. The process removes the twist in the fibre and thus led to the fibre to become smooth and cylindrical which produces a silky effect. Mercerised cotton absorbs dyes very soon.

Process of Fire-Proof Finish

A simple method which comprises treatment with boric acid and borax. This, however, dissolves in water. Another method is the treatment of the fabric with chlorinated compounds, like Vinyle Chloride or chlorinated rubber and antimony oxide. This solution covers the fabric with a thin, non-inflammable film which renders the fabric fire-proof.

The Process of Crease Resistant Finish

The process is used mostly for cotton because cotton, owing to its natural inelasticity, wrinkles badly. In this process the fabrics are impregnated with a solution of synthetic resin, live:

- Phenol formaldehyde,
- Urea formaldehyde
- Acrylic resins, and dried by a very high temperature in a moist atmosphere.

This forms a clear insoluble resin in the fibre; which improves the resilience of the fabric. Urea formaldehyde is colourless and can be utilised for white and light coloured fabrics but phenol formaldehyde being dark coloured is used for dark coloured fabrics only.

Process of Creping

Creping by the chemical process is the treatment of fabrics with caustic soda. The soda paste is used to fabrics in a definite design of stripes or figures. The parts to which the paste is applied shrink leaving the other parts unshrunk; thus the effect of a puckered or creped material will be produced. Another method is to apply a paste of a substance which opposed the effect of caustic soda to the fabric in a definite design, and then place the fabric in a solution of caustic soda.

The untreated places will shrink and produce the crepe effect. This method is more lasting than the mechanical one. The crepe produced in weaving is of course very different to the crepe effect produced by these finishes.

Process of Dyeing

Dyeing may also be called a finishing process as it colours the fabric and so adds to its eauty. Dyeing is a very ancient Indian art which the other countries learnt from the Indians.

In the earliest stages, the fabrics were coloured with the juices of flowers, fruits, stems, leaves and barks of plants and trees. Later on, dye stuffs from vegetable and mineral sources came into utilisation. The chemical dyes are the discovery of relatively recent times. There are a number of methods utilised for dyeing the fabrics at various stages.

Process of Raw Stock-dyeing

Fibres of wool, cotton, linen and waste silk are dyed. The stock of fibres are rotated in the dye bath, then removed and dried.

Process of Piece-dyeing

This is the process utilised for dyeing materity woven from any fibre. The penetration of colour is not quite so thorough as in the preceding forms of dying. The fabrics art wrapped around rollers which are placed in the dye bath and left there until they are saturated with the colour. Cotton fabrics 'require to be held under tension while passing through the dye. The other fabrics are not held taut.

Process of Stuh-dyeing

This method is brought into use in dyeing wool fibres after these have been combed. The skeins of wool slivers are hung on rods. The rods are then placed on the top of the dye bath. The rods are made to rotate, so that the slubbings keep constantly moving through the dye bath. These are then rinsed and dried.

Process of Skein-dyeing

Yarns of almost all fibres are dyed. Cotton and linen are generally dyed at the yarn stage. But the penetration of colour in skein-dyeing is much better. Skeins of yarns are hung on rods, which are placed in a dye bath. These rods rotate causing the skeins to 'circulate through the dye bath. This method is known as skein-dyeing.

Cross Dyeing

Fabrics woven with mixed yarns, like cotton, wool, or acetate rayon and viscose rayon, are dyed by this method. The affinity of dye-stuffs differs according to the two sorts of yarns mixed in weaving. thus, the fabrics are to be dyed in two successive dyebaths. In one, the dye suitable for one kind of yarn is used and in the other is utilised the dye suitable for the other yarn. Sometimes, a mixed fabric is dyed in one and the same dye bath and so each one of the yarn takes its own particular and. Various shade of cot our, or one takes the colour and the other remains colourless.

Process Tie and Dye

This is a kind of resist dying, which was practised as far back as the sixteenth century in India *o*r probably even earlier. It is a simple method and yet man, fascinating and intricate designs are produced in the fabric, as for instance in the famous 'Patola' of Gujarat. By this method either the yarn is dyed before the fabric is woven as in *'Paldani'* or the woven material is dyed as in *'Bandani'* saris of Gujarat, Kathiawar and Rajasthan.

The parts of the fabric on the yarn which are to oppose the dye are tied with strings or narrow strips of cloth. The strings or the sirips are waxed to enhance the dye-resistance and so get better results. The skeins of the yarn or the material is then dipped in the dye bath. Then these are rinsed, dried and the strings removed. This method of dying produces beautiful patterns. Work of 'Bandani' this is used by women in India to dye their saris and scarves or 'dupattas' at home.

Process of Batik

This is another form to resist dyeing which produces patterns like those in prints. This also was first practised in India. The Javanese then took this art to their country, and now 'Batik' dyeing is a big industry in Java. The method is as follows: A full sized design is drawn or traced on the material. Hot, melted wax is applied on the lines of the design with a paint brush. The size of the brush depends on the thickness of the lines of the design.

After coating the lipes with wax, the fabric is immersed in cold water to harden the wax and to wet the fabric. The dye bath is then prepared. Cold water dye or dyes which car be utilised at low temperature are used for"Batik' dyeing. If the dye bath is hot, the wax applied on the design will melt and the process will be a failure.

The dye-bath vessel must be lane enough to permit the entire fabric to be moved freely without any pressure or friction which might cause the wax lines to crack and thus spoil the design.

The material is kept in the dye bath long sufficient for the cloth to be thoroughly impregnated with the dye. The cloth is then rinsed and dried and the waf is removed. A 'Batik' in which more than one colour and an intricate design is used needs very great care and more than one operation is necessary as in the case of multi-coloured print.

Practice of Printing

Printing also was practised in India thousands of years before the Christian Era. The designs were either drawn by a brush or were stamped on the material by wooden blocks carved or cut in relief. A simple method of printing utilised at first comprised dipping the carved block in a paste of colour and then stamping the fabrics with it in a definite pattern.

This method is still utilised extensively. Later, a screen made of fine wires (wire-cloth) or silk bolting cloth came into use. In machine printing engraved copper rollers, or cylinders are utilised. The dyes used in printing are the same as those used for dyeing but these are utilised in the form of pastes. If mordants are necessary, the fabrics are treated with mordant solution before being printed.

Practice of Block Printing

This is follows is in India even today to get beautiful patterns. Blocks are made of wood or wood and lino. The design is carved on lino which is generally cut to a thickness of ¼ inch. This cut piece of lino is stuck on to a wooden piece of the same size. In India, several printers use only wood blocks on which the design has been carved.

The paste of colour is poured on trays lined with cloth. The material to be printed is spread on a padded table and held taut and smooth. Then the block carrying pate of one colour is stamped on the material and allowed to dry. Then another block carrying paste of various colour is stamped over it to form the multi-coloured patterns. The process is repeated over the entire fabric-surface which is to be printed.

Process of Screen Printing

Silk bolting cloth or wire cloth is stretched over a wooden frame. Over this, a line is drawn of such portions of the design as are to take one colour. The sphere of the screen which are to be utilised to prevent the spreading over of the colour are filled with water proof varnish, enamel or some other insoluble filter The fabric to be printed is then stretched on a table which is padded and covered with oil-cloth.

The wooden frame with the screen is then placed in the material, and the paste of colour is brushed over its surface and then lightly pressed. The colour is permitted to dry. Then another frame carrying a several colour and different outline is utilised and the process continues till the design is complete. A separate screen is required for the application of each colour in the design. The size of the screen used is generally 4 ft. by 6 ft.

Machine Printing

A printing machine comprises several copper cylinders or rollers which are engraved. The actual printing of the design on the cloth by this method takes very little time but the engraving of the design on the rollers means hard and careful work lasting many days. The roller is as wide as the cloth. The number of rollers needed depend on the number of colours utilised in the design as one roller can print only one colour.

MANUFACTURE OF THE INDIVIDUAL FIBRES

A considerable lag always exists between the marketing of a new product and the publication of specific details about its manufacture. Of the five fibres under consideration, excellent accounts of the production steps are available for nylon 66 only, a product which has been on the market for almost twenty years. For the other fibres, all of very recent origin, broad pictures of the process of production, but not the details of the individual steps, can be obtained.

Therefore, the nature of the technical literature on the subject should be made clear. The published accounts employed are surveys, based on the patent literature and the literature of chemistry, of alternative approaches to each general process step. Since only the general steps are important for the identification of the locationally strategic inputs, no pretense is made that the accounts which follow are of any firm's actual manufacturing operation.

Production of Nylon 66

The monomer from which nylon 66 is made is hexamethylene diammonium adipate, commonly called nylon salt. If fibre manufacture is separate from chemical intermediate manufacture, the salt is made into a water solution to facilitate handling and shipment to yarn plants. In conversion into nylon 66 fibre the salt is first run from storage tanks into evaporator kettles where it is concentrated. Acetic acid is added as a viscosity stabilizer in polymer formation. From the evaporator kettles a 60 per cent solution of the salt is run into autoclaves where polymerization takes place. If a delustered yarn is wanted, an aqueous suspension of titanium dioxide is added to the reaction mass soon after it begins to boil. When polymerization is completed, nitrogen pressure is applied to force the polymer, a sirupy material, to flow out through a slot in the bottom of the autoclave. Nylon 66 is spun by the melt-spinning process. Several steps are required to prepare the polymer for melt spinning.

From the autoclave a ribbon of the polymer flows onto a casting wheel, where it is hardened by sprays of water and cooled with streams of air. Then a mechanical cutter reduces the ribbons of polymer to small flakes; these flakes are conveyed to a blender and then transferred to supply hoppers which ride (via an overhead rail) over the spinning machines. Specially designed spinning machines, consisting of batteries of spinnerets, are used to extrude the polymer. From an overhead supply hopper the flaked polymer is fed into funnel-shaped, nitrogen-filled melt chambers which, in turn, feed one or more spinnerets. In the melt chamber a heated grid melts the polymer, which can then be pumped directly to the spinnerets. The holes of the spinnerets vary in size and number according to the size of the filaments and yarn desired. From the lower face of the spinneret the viscous filaments pass down through the orifices into a cooling chimney and are subjected there to an air blast to aid in solidification.

A converging guide gathers the filaments into a bundle of yarn. To prevent elongation and to keep the yarn from loosening and slipping in a subsequent winding operation, it is passed through a chamber and humidified with steam. After the steam treatment, the yarn is passed over glass finish rollers which apply an oil emulsion as a lubricant for other operations. At the bottom of the spinning unit the yarn is wound up in undrawn and untwisted form on bobbins. At this point the remaining process steps depend on the form in which the yarn is to be sold.

Continuous Filament

To make continuous-filament fibre, bobbins of yarn are conveyed to the textile area of the plant. Most nylon 66 filament is sold as low-twist yarn. In preparing the untreated yarn the filaments are twisted together—on conventional textile machinery-with about one or two twists per inch. Then the yarn is rewound on bobbins for a cold-drawing process on machines which stretch it to about four times the original length by passing it through sets of rollers moving at different speeds. From the viewpoint of chemistry the molecules of the undrawn filaments have random orientation only. After drawing, the linear molecules are aligned parallel to the fibre axis, a change which gives the yarn its particular physical properties. After the yarn has been drawn, it is wound on bobbins, the typical put-up, and is inspected and packaged for shipment.

Staple and Tow

For the manufacture of nylon 66 staple thousands of continuous filament fibres are combined into a rope or tow. The tow is drawn and then crimped by mechanical means to give the yarn the loftiness associated with natural fibres such as wool. The crimped ropes of yarn are set with steam and then cut into the desired length by staple cutters. At the end of this operation the staple is dried and baled for shipment. If tow is to be sold, it is packaged after being drawn.

Production of Orlon

Orlon is made from monomeric acrylonitrile, and Morgan suggests that for a large-scale operation polymerization can best be carried out as a continuous process. A water-soluble catalyst, such as ammonium persulphate, and a reduction agent, such as sodium bisulphite or sodium thiosulphate, can be used. In the suggested continuous process streams of acrylonitrile and aqueous solutions of the catalyst and reducing agent are fed into a reactor.

Low rates of conversion of the monomer to polyacrylo-nitrile make it necessary, after filtration of the slurry issuing from the reactor, to recover unconverted acrylonitrile by fractional distillation so that it may be recycled. As the polyacrylonitrile is formed, it precipitates in fine granular form which can be readily filtered.

Lessing, in writing on the development of Orlon, notes that in the experimental production of filaments both the wet- and dryspinning processes were employed but that the latter method proved best. In either event a spinning solution must be prepared by dissolving the dry polymer in an organic solvent. For Orlon manufacture polyacrylonitrile is dissolved in dimethyl formamide. If a delustered yarn is desired, an agent such as titanium dioxide is introduced into the spinning solution, or dope. Once prepared, the solution for dry spinning must be carefully filtered for the removal of foreign matter and gels of undissolved material. Each batch of polymer solution can be processed through conventional filter presses.

In a typical dry-spinning process sketched by Fremon the filtered dope is stored in tanks which feed the cells of the spinning units. Spinning dope is pumped into the top of each cell by metering pumps. As the polymer is forced through the spinneret, filaments are formed and solidified as the solvent in the solution is evaporated by hot air. The volatized solvent can be drawn off from an aperture in the cell by air flows and conveyed to a solvent recovery unit, a vital operation in the economics of the process. As the new filaments are gathered together in each cell by wind-up devices at the bottom, they are passed over a roller which applies a protective lubricant and wound up on spools for further processing.

Fibres produced by the dry-spinning process are hot-stretched rather than cold-drawn as in the case of nylon. The stretching operation may be preceded by a washing treatment in which the yarn is passed through a bath to remove solvent residue. Frey and Sippel point out three alternative hot-stretching methods.

First, the yarn may be drawn between two or more sets of rollers, moving at different velocities, through a hot bath consisting of a nonsolvent for the fibre or of hot air. In the second approach hot drawing is made possible by passing the yarn through heated rollers or over hot plates. Finally, they note that for polyacrylonitrile fibres the yarn may be stretched in special steam tubes. To stabilize the molecular arrangement brought about by stretching and to relieve the strains in the drawn fibres, the yarn may be "relaxed" by a heat treatment in an annealing oven.

Continuous-filament Orlon is sold with low twist or none at all, so that the final operation is to wind the yarn on bobbins or to give it a low twist and then put it up. In describing the production of Orlon staple Mauersberger cites a seven-step operation. Filaments are gathered together to form a tow which is drawn and then passed through mechanical devices which put a crimp in the fibre. A relaxing step follows, and the yarn is cut into staple fibre. The cut staple is dried and put up into bales for use by the textile industry.

Dynel Output

Dynel, the vinyl fibre of the Union Carbide and Carbon Corp., is made of a copolymer of vinyl chloride (60 per cent) and acrylonitrile (40 per cent). Harris describes an emulsion polymerization procedure for the preparation of constant composition copolymers of vinyl chloride-acrylonitrile containing the monomers in the proportion 60/40. The step is carried out by mixing vinyl chloride and acrylonitrile in water containing a water-soluble catalyst and an emulsifier. At the end of the reaction the polymer is a resin in powder form.

The published literature shows that dynel is made by the wetspinning process. Although none of the wet-spinning methods used in the production of vinyl fibres has been described in detail, the general nature of the operation is known from viscose rayon production. Just as in the case of dry spinning, a solution is prepared by dissolving resinous polymer in a suitable solvent. For dynel production acetone is used.

The solvent, the resin powder, and a viscosity stabilizer are mechanically mixed to form the spinning dope. To prepare it for use in spinning, the solution is filtered. In the wet process the spinning apparatus lies horizontal, submerged in a liquid-bath medium. The spinning solution is extruded into the liquid, and filaments are formed by coagulation. As indicated in the introduction, dynel is manufactured in the form of staple and tow only. Fremon has described a wet-spinning process for making synthetic staple which provides a glimpse of the general steps which follow filament formation. After the fibre in tow form emerges from the coagulating medium it can be taken to a second bath where it is washed and stretched.

In the next step the tow is dried in an oven and is also annealed to stabilize the molecular structure brought about by stretching. From the annealing oven the tow is passed through a mechanical crimper and then can be cut into staple form and baled. In an alternative approach the tow can first be dried to low moisture content, cut, and then passed through a device known as a tunnel dryer. The air temperature of the dryer is maintained as high as possible without injury to the yarn, and the fibre shrinks unevenly, taking on a crimp.

Acrilan Manufacture

Few details have been released by the Chemstrand Corp. on the manufacture of Acrilan, its vinyl fibre. Production so far has consisted of staple and tow only. Like dynel, the fibre is a copolymer. Acrilan is made principally of acrylonitrile (85 to 90 per cent), and the other component has been most frequently identified in the technical accounts as vinyl acetate or vinyl pyridine.

Morgan notes that copolymers of acrylonitrile containing small amounts of a second monomer, such as vinyl pyridine, can be manufactured in a manner similar to the process described in the Orlon section for the polymerization of acrylonitrile. The second monomer can be mixed in with the acrylonitrile stream. In the most recently published account Mauersberger has described the general steps for the manufacture of Acrilan. He reports that the spinning process employed is the wet method. Whatever solvent is used to prepare the spinning solution of the powdered copolymer has not been specifically identified.

Dacron Production

Dacron, Du Pont's polyester fibre, is the condensation product of dimethyl terephthalate and ethylene glycol. A commercial route used in the production of the polymer, polyethylene terephthalate, has been described by Reynolds. Ester interchange between dimethyl terephthalate and excess ethylene glycol yields a condensate which is heated in a vacuum to effect removal of the excess glycol and to give a polymer of high molecular weight. Ester interchange requires a catalyst such as magnesium or zinc borate. As the temperature of the reaction mixture is raised, methanol forms and is removed by distillation. Pressure filtration can be used to purify the initial condensate product before it is transferred to a polymerization reactor. In the polymerization step excess ethylene glycol is evolved which presumably can be recycled.

The molten polymer is released from a heated extrusion valve at the base of the vessel by the application of nitrogen pressure.Dacron is melt-spun, and the manufacturing steps are reported to be very similar to the general procedure followed in the production of nylon. The molten polymer emerges from the polymerization reactor as a solid ribbon, which is run onto a casting wheel, cooled by water sprays, and cut into flakes for the spinning operation. Again, if a delustered yarn is desired, titanium oxide is used and it can be added to the condensate mixture as a dispersion in the ethylene glycol. The fibre is produced as continuous-filament yarn without twist and as staple and tow. Since the remaining steps in a fibre plant with a melt-spinning process were described for nylon, the details are assumed to apply in an approximate way for Dacron.

LOCATION THEORY AND THE SYNTHETIC-FIBRE INDUSTRY

An empirical location study must turn to location theory for a systematic coverage of all factors which might possibly influence the location of an industry and, in general, the geographical distribution of its plants. In this location study of the synthetic-fibre industry an attempt has been made to apply the general substitution approach to the locational equilibrium of the firm. Underlying the location pattern of an industry are decisions made by business firms. The general area of economic theory which embraces this type of decision-making activity is a branch of microeconomics, the theory of the firm, and, in particular, the locational equilibrium of the firm. Traditionally, the theory of the firm has been made synonymous with price theory, the postulates of which are inferred through the implications of profit-maximizing behabiour of the firm in three roles: first, in the process of production, *i.e.*, the transformation of inputs into outputs; second, as a buyer of inputs; and third, as the seller of a product.

Given the familiar data of the problem, including the structure of markets faced by the firm (defined in terms of numbers of buyers and sellers and homogeneity of product), profit-maximizing levels of output can be deduced. Price theory ignores the location problem because buying, production, and selling are looked upon as single-dimension activities, *i.e.*, as occurring at a point.

This assumption is abandoned in the theory of the location equilibrium of the firm. As a buyer of inputs, the firm incurs costs in assembling them over space and is also confronted with regional differences in their availability; the impact of site on the technical possibilities of production is recognized; and the selling activity embraces a market area in which transportation costs are also incurred in delivering the product. Thus the attributes of plant-location theory stem from firm behabiour in a fourth role as well: that of selecting from an array of geographical possibilities, minimum-cost production, and selling sites; and cost minimization is focused on variations which result from the factor of space.

The Weberian Framework

In his pioneering work Weber's signal contribution was the development of the principal categories of industrial location, largely on the basis of technical empirical knowledge. To Weber, the transportation network of a nation is the basic magnetic field of economic attraction which, through its lines of force, determines the position of industrial particles. He regarded transportation orientation as the basic principle of industrial location.

Conceptually, competing magnetic fields are sometimes capable of causing particles to be drawn away from the transportation lines of force to their own. In short, all other types of orientation, such as power or labour, are deviations from minimum-cost transportation sites. Weber also singled out the importance of agglomeration economies in industrial location. Although his types of location are suggestive for empirical studies, Weber's geometrical approach to the locational equilibrium of the firm, which, for example, requires the solution of location triangles, offers only a rough methodological guide.

The General Substitution Approach

That the Weberian location categories could be derived by means of the familiar substitution principle of price theory was suggested by Predohl in 1928, but it was not until 1949 that an operational framework was established for the general substitution approach to location theory by Isard. In his work Isard has extended the bounds of traditional production theory to include the locational equilibrium of the firm. Since cost minimization has a spatial dimension, it is specifically recognized by way of transport inputs. Within the enlarged framework the problem of production is that of choosing the right combination of capital, labour, land, and transport inputs.

From the most significant cost difference which occurs through variation over space the orientation of an industry can be determined. Isard distinguishes two types of substitution. The exploration of alternative outlays from the use of transport inputs can determine whether costs for a transport-oriented industry are minimized at market or at raw-material sites; and the traditional outlay substitutions of production theory between capital, labour, and land can be used to deduce other types of industrial location such as labour or power orientation.

APPLIED GENERAL SUBSTITUTION ANALYSIS AND THE SYNTHETIC-FIBRE INDUSTRY

The general substitution approach to the locational equilibrium of the firm provides an effective operational tool for empirical analysis. This approach can be put in simple terms in the case in which a given market, or set of markets, is to be served by a plant with given physical inputs per unit of output and is able to draw raw materials from alternative supply areas. As this plant is moved from region to region, the effect on its costs and revenue via the substitution of inputs and of diverse outlays can be determined. Consider, for example, the alternative in the synthetic fibre industry of producing the chemicals for fibre synthesis at sites in chemical raw-material regions or of manufacturing them in fibre-market regions in the textile area of the United States.

Output at a raw-material site requires little outlay on movement of raw materials, but outlays are incurred on the transportation of the finished product and on utilities, labour, etc. At a market site a transportation cost on raw materials is substituted for the cost of moving the finished product, and the market-site utility, labour costs, etc., are substituted for those at the raw-material site. In addition, market-site agglomeration economies are substituted for those at the raw-material site. Substitutions such as these can be used to generate regional cost differences so that the most favourable region for plant location can be uncovered.

The Hypothesis

It is clear that the successful use of the general substitution approach in an empirical location study depends in a large measure upon the extent to which production functions, the physical relationship between various inputs (including transportation inputs) and output, can be obtained or estimated. If, for instance, the location of an ammonia plant is in question, it is necessary to know the raw-material inputs per unit of output, the utility and labour inputs, the manner in which these inputs vary with the size of plant, the manner in which the capital investment varies with size of plant, etc.

The hypothesis in this study of the synthetic-fibre industry is that:

- From an examination of the industry's production functions and market areas a spatial framework of regions which are suitable for plant location can be established.
- General substitution analysis can be used to convert the data of the production functions into production and distribution cost differences at selected sites.

In this manner basic regional advantages and disadvantages for plant location can be ascertained. The computation of regional differences in production and distribution costs will also point out the basic type of location orientation for each stage of the synthetic-fibre industry as well as the minimum-cost regions among all those tested.

COSTS AND LOCATION ANALYSIS

Although in the location analysis of the synthetic-fibre industry it might be desirable to employ all the costs which the chemical engineer or the cost accountant would include, from the standpoint of location theory the estimation of costs, hence the estimation of production functions, can be simplified. All inputs which have a locational significance must be identified, but the cost items which do not vary significantly from region to region can be set aside. Thus the application of the general substitution approach to the synthetic-fibre industry begins with prior determination of relevant cost categories to guide the estimation of production functions. The basic cost categories in the theory of the firm are variable and fixed costs, and these are applicable in location analysis. With respect to the locational equilibrium of the firm, variable and fixed costs may vary spatially because of interregional differences in achievable agglomeration economies.

Cost Variation Plants of Equal Size

An examination of specific items in fixed production costs in both the chemical-intermediate and the fibre-producing stages of the synthetic-fibre industry leads to the conclusion that for any given predetermined size of plant regional uniformity in fixed charges can be assumed without doing violence to reality.

The chemical intermediates from which synthetic fibres are produced can be manufactured by more than one process, but the technology of petrochemical production is assumed in almost all cases in this study, since this particular route promises the most locational mobility for the future. In their study of the petrochemical industry Isard and Schooler found the assumption of regional uniformity of fixed costs for plants of given size to be a realistic one, after considering plant construction costs, interest charges, depreciation, plant maintenance, insurance costs, taxes, and land costs. In general, the analysis of the components of fixed charges also applies to the synthetic-fibre stage, even though fibre output is a combined chemical-textile operation.

In the past chemical-plant construction costs in warm-weather regions, such as Texas or California, have tended to be lower than in regions of the North by 10 or 12 per cent because certain structural parts unnecessary in the "outdoor" type of plant were thought to be indispensable in the "indoor" design for northern regions. However, Isard and Schooler note that the recent technical literature of chemical-plant construction points to the feasibility of the cheaper outdoor design for all regions. Plant construction cost inequalities which originate in labour-cost and material-cost variations cannot be systematically forecast and are not considered vital to the location problem in question.

For the large firms engaged in the development of synthetic fibres and their chemical intermediates ready access to the nation's capital market excludes the possibility of interest-cost disparities at different sites; therefore, the United States can be regarded as an area characterized by homogeneous capital availability for the location analysis. In considering the possibility of plant location in Puerto Rico, branch plants of American firms are assumed so that capital-cost uniformity applies to the island as well.

Since depreciation allowances are determined principally by tax legislation at the Federal level, there is little room for regional differences in this item of fixed costs. Plant maintenance has a regional variation and could be expected to be higher for cold weather regions, but the practice of performing major maintenance work in good weather limits the magnitude of the cost difference.

Insurance costs are significant in chemical production because of the risk hazard. Manufacture in a raw-material region with plants concentrated at specific sites raises plant insurance costs over areas in which chemical production is dispersed. Since the insurance cost differences are small and cannot be predicted in their variation with space, they are neglected in the cost analysis.

To account for regional variations in the burden of taxes which enter into fixed charges is a difficult matter. Property taxes are important but are set by legislative bodies and are not amenable to location analysis. Some states offer property-tax exemptions, usually for a five-year period, as an inducement to the location of new industry. The practice is so widespread, however, that its impact on location decisions tends in the long run to become neutralized regionally, although it may be a factor in the selection of a site within a region. For the cost analysis land outlays are assumed to be equal in all regions. The problem of minimizing land outlays in factory construction is not one of choosing the right region but the proper site within a region.

Although interregional equality in fixed charges may be justifiably assumed, in both stages of the synthetic-fibre industry regional differences in variable costs do exist. Computation of the amounts of these differences requires first an estimation of the most important variable physical inputs per unit of output.

Agglomeration Economies

For the synthetic-fibre industry the major agglomeration economies are:

- Large-scale production economies;
- Integration economies from the output of several chemical products in a multiunit plant—savings are realized, for example, by the utilization of the by-product of one unit as a raw material for another;
- localization economies from the geographical concentration of chemical plants in a given area—a single large-scale chlorine plant may serve the needs of several users and make it unnecessary for each to construct a small-scale unit;
- External economies from the purchase of raw materials from large-scale producers-in Texas, for instance, large-scale oil refineries make available to chemical manufacturers huge streams of waste gas at low prices.

It may be valid to assume that for some products plants of the same predetermined size could be built in any region. In many cases this assumption would imply that agglomeration economies are uniform from region to region. However, if plants of the same predetermined size cannot be built in all regions, then location analysis must evaluate regional differences in variable and fixed costs caused by differences in the extent to which large-scale production and other agglomeration economies can be achieved.

All four of the above-mentioned types of agglomeration economy may be significant in the location of plants in the first stage of synthetic-fibre production, the manufacture of chemical intermediates. Since many of the intermediates have markets in other end uses, there may be regional differences in market potential.

A site suitable for fibre manufacture may not draw the chemical stage unless it can serve a market for the excess production over the fibre requirement, for otherwise the chemical unit would suffer a large-scale production economy disadvantage. The other three types of agglomeration economy generally favour raw-material over market regions or, among market regions, those with the most highly developed chemical industries.

In this study agglomeration economies are evaluated for the chemical—intermediate stage through the estimation of the influence of size on fixed and variable costs. The other agglomeration economies-integration, localization, and external—are given qualitative consideration only. In the second stage of the industry, the manufacture of synthetic fibres, the main agglomeration consideration is large-scale production economies. The market for each synthetic fibre is spread out geographically over the entire textile area so that the output of a single plant is not sold entirely in a specific region.

For this reason plant size is largely a function of technological and cost considerations common to all fibre producers and is somewhat independent of the region chosen for the location of a plant. Since the technological and cost considerations permit plants of economic size to be built in any region, agglomeration economies can be eliminated as a factor in the location of synthetic-fibre plants.

Conclusions of Preliminary Cost Analysis

For the first stage of the industry, the output of chemial intermediates, the assumption is made from the foregoing analysis that the location problem can be solved by comparing regional differences in variable costs alone for products for which the assumption is valid that plants of the same predetermined size can be built in any region.

In most cases, however, the comparison of regional differences in variable costs must be augmented by the estimation of the effect on both fixed and variable costs of agglomeration economies. With respect to the second stage, the manufacture of synthetic fibres from chemical intermediates, the assumption is made that the location problem can be solved by comparing regional differences in variable costs for plants of equal size.

7

Techniques of Fabric Construction

WEAVING

In weaving two sets of yarns are interlaced at right angle to one another in an established sequence. *e.g.* 'charpai' being made? On the frame, first one rope is taken and interlooped on two parallel edges and after that a second rope is interlaced perpendicular to it going once over the rope and once under it. This kind of an interlacement gives an even check effect and the weave is quite firm. The weaving of a fabric is also done in a similar way, except for the fact that yarns are used for interlacing and a loom is used to hold the thread instead of a frame.

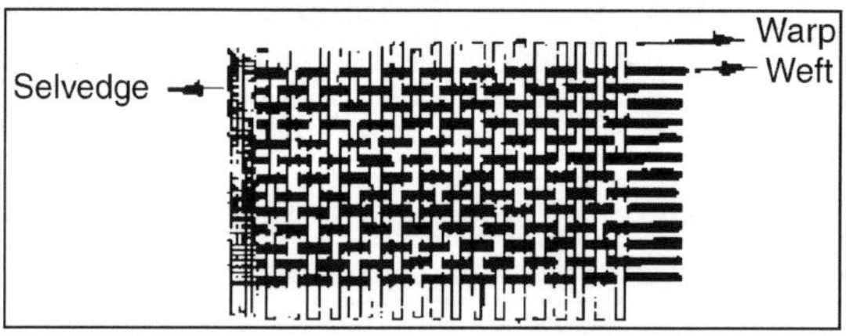

Fig. Common Terms in Textiles

- *Selvedge*: When you examine a fabric, you see two long finished edges, one on each side along the length of the fabric. These edges are called selvedge and give strength to the edges which is important in further processing of the fabric.
- *Warp*: They are the yarns along the length of the fabrics or parallel to the selvedge. They are also called ends.
- *Weft*: Woven fabric, besides warps, another set of yarns move perpendicular to warps. They are called wefts. They are interlaced with warps in a crosswise direction to make a fabric. They are also called picks or fillings.
- *Thread Count*: Some woven fabrics look dense and compact whereas others open. This difference is due to the thread count which refers to the total number of warps and wefts per square inch of a woven fabric. Thread count tells us about the fabric quality and durability.

Fabric with a higher thread count is better than fabric with lower thread count. Also, for good quality fabric, warps and wefts should be more or less equal in number.

- Weaving looms can range from quite simple to very complex. Looms have been used to produce cloth for thousands of years, and while technology has improved the loom, the basic strategies and practices remain much the same. Understanding the parts of the loom can help you learn to weave or simply learn a bit more about how weaving works.
- *Warp Beam*: The warp beam is a beam that holds the warp, or vertical threads, on the loom. Depending on the loom, this can be a simple beam or a more complex arrangement allowing for easier warping of the loom.
- *Heddles*: Heddles are made of cord or wire, and are attached to the shaft of the loom. The warp threads pass through the heddles, separating the warp threads to allow the weft threads to pass between them easily.
- *Harnesses*: The more harnesses or shafts a loom has, the more design possibilities you will have. Most larger looms have four harnesses, and table looms often only one; however, looms of up to 16 harnesses are available.
- *Shuttle*: The shuttle is a bobbin which holds the weft yarns. The shuttle is passed or thrown back and forth to create the weft of the fabric.
- *Beater and Reed*: The beater of the loom presses the newly created weft thread against the already woven fabric using a reed. This process is called battening.
- *Cloth Beam or Cloth Roll*: It is located at the end of the loom on which the fabric is rolled as it is constructed. It also supports the warp yarn at the front side of the loom.

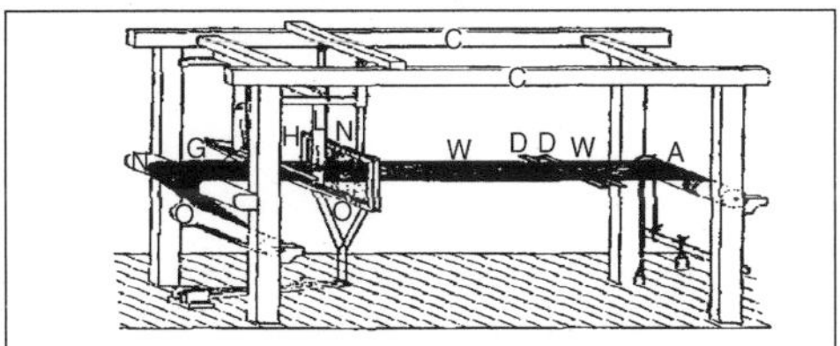

A-Warp Beam; B-Cloth Beam; DD-Less Rods; H-Harness; T-Treddle

STEPS IN WEAVING PROCESS

Weaving process is completed in four steps:

1. *Shedding*: Raising the required wrap yarns by using the harness and forming the shed.
2. *Picking*: Inserting the shuttle containing filling yarns through the shed.

3. *Battening*: Pushing the filling yarns in place with the help of reed to form the compact structure. Its also called beating up.
4. *Taking Up and letting Off*: Winding the finished fabric on the cloth beam and releasing more of the warp from the warp beam.

CONSTRUCTION OF A PLAIN WEAVE

Plain weave is the simplest interlacing pattern which can be produced. It is formed by alternatively lifting and lowering one warp thread across one weft thread. A plain weave fabric in plan view and warp way and weft way cross-sections through the same fabric. The diagrams are idealized because yarns are seldom perfectly regular and the pressure between the ends and picks tends to distort the shape of the yarn cross-sections unless the fabrics are woven from monofilament yarns or strips of film. The yarns also do not lie straight in the fabric because the warp and weft have to bend round each other when they are interlaced. The wave form assumed by the yarn is called crimp and is referred to in greater.

Diversity of Plain Weave Fabrics

The characteristics of the cloths woven will depend on the type of fibre used for producing the yarn and whether it is a monofilament yarn, a flat, twisted or textured (multi-)filament yarn or whether it has been spun from natural or manufactured staple fibres. The stiffness of the fabrics and its weavability will also be affected by the stiffness of the raw materials used and by the twist factor of the yarn, that is the number of turns inserted in relation to its linear density. Very highly twisted yarns are sometimes used to produce special features in plain weave yarns. The resulting fabrics may have high extensibility or can be semiopaque.

The area density of the fabric can be varied by changing the linear density or count of the yarns used and by altering the thread spacing, which affects the area covered by the yarns in relation to the total area. The relation between the thread spacing and the yarn linear density is called the cover factor. Changing the area density and/or the cover factors may affect the strength, thickness, stiffness, stability, porosity, filtering quality and abrasion resistance of fabrics.

Square sett plain fabrics, that is fabrics with roughly the same number of ends and picks per unit area and warp and weft yarns of the same linear densities are produced in the whole range of cloth area densities and cloth cover factors. Low area density fabrics of open construction include bandages and cheese cloths, light area density high cover factor fabrics include typewriter ribbons and medical filter fabrics, heavy open cloths include geotextile stabilization fabrics and heavy closely woven fabrics include cotton awnings.

Warp-faced plain fabrics generally have a much higher warp cover factor than weft cover factor. If warp and weft yarns of similar linear density are used, a typical warp faced plain may have twice as many warp ends as picks. In such fabrics the warp crimp will be high and the weft crimp extremely low. The plan view and cross-sections of such a fabric. By the use of suitable cover factors and choice of yarns most of the abrasion on such a fabric can be concentrated on the warp yarns and the weft will be protected.

Weft-faced plains are produced by using much higher weft cover factors than warp cover factors and will have higher weft than warp crimp. Because of the difference in weaving tension the crimp difference will be slightly lower than in warp-faced plain fabrics. Weft-faced plains are little used because they are more difficult and expensive to weave.

Construct a Point Paper Diagram

To illustrate a weave either in plan view and/or in cross-section, takes a lot of time, especially for more complicated weaves. A type of shorthand for depicting weave structures has therefore been evolved and the paper used for producing designs is referred to as squared paper, design paper or point paper. Generally the spaces between two vertical lines represent one warp end and the spaces between two horizontal lines one pick. If a square is filled in it represents an end passing over a pick whilst a blank square represents a pick passing over an end. If ends and picks have to be numbered to make it easier to describe the weave, ends are counted from left to right and picks from the bottom of the point paper design to the top.

The point paper design is the design for a plain weave fabric. To get a better impression of how a number of repeats would look, four repeats of a design (two vertically and two horizontally) are sometimes shown. When four repeats are shown the first repeat is drawn in the standard way but for the remaining three repeats crossing diagonal lines may be placed into the squares, which in the firsts repeat, are filled in. This method is shown for a plain weave.

Plan View

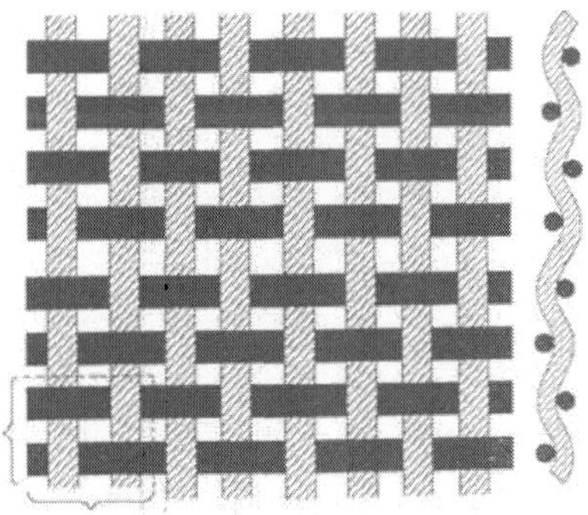

Fig. Fabric woven with plain weave - plan view (4x4 repeats) and warp and weft cross-sections.

RIB FABRICS AND MATT WEAVE FABRICS

These are the simplest modifications of plain weave fabrics. They are produced by lifting two or more adjoining warp threads and/or two or more adjoining picks at the same time. It results in larger warp and/or weft covered surface areas than in a plain weave fabric. As there are fewer yarn intersections it is possible to insert more threads into a given space, that is to obtain a higher cover factor, without jamming the weave.

Matt Fabrics (or hopsack)

Simple matt (or hopsack) fabrics have a similar appearance to plain weave. The simplest of the matt weaves is a 2/2 matt, where two warp ends are lifted over two picks, in other words it is like a plain weave fabric with two ends and two picks weaving in parallel. The number of threads lifting alike can be increased to obtain 3/3 or 4/4 matt structures. Special matt weaves, like a 4/2 matt are produced to obtain special technical effects. Larger matt structures give the appearance of squares but are little used because they tend to become unstable, with long floats and threads in either direction riding over each other. If large matt weaves are wanted to obtain a special effect or appearance they can be stabilized by using fancy matt weaves containing a binding or stitching lift securing a proportion of the floats.

Matt weave fabrics can be woven with higher cover factors and have fewer intersections. In close constructions they may have better abrasion and better filtration properties and greater resistance to water penetration. In more open constructions matt fabrics have a greater tear resistance and bursting strength. Weaving costs may also be reduced if two or more picks can be inserted at the same time.

Rib Fabrics

In warp rib fabrics there are generally more ends than picks per unit length with a high warp crimp and a low weft crimp and vice versa for weft rib fabrics. The simplest rib fabrics are the 2/2 warp rib and the 2/2 weft rib shown respectively. In the 2/2 warp rib one warp end passes over two picks whilst in the 2/2 weft rib one pick passes over two adjoining ends. The length of the floats can be extended to create 4/4,6/6,3/1 or any similar combination in either the warp or weft direction. 3/1 and 4/4 warp ribs.

In rib weaves with long floats it is often difficult to prevent adjoining yarns from overlapping. Weft ribs also tend to be expensive to weave because of their relatively high picks per unit length which reduces the production of the weaving machine unless two picks can be inserted at the same time.

Twill Fabrics

A twill is a weave that repeats on three or more ends and picks and produces diagonal lines on the face of a fabric. Such lines generally run from selvedge to selvedge. The direction of the diagonal lines on the surface of the cloth are generally described as a fabric is viewed along the warp direction. When the diagonal lines are running upwards to the right they are Z twill' or 'twill right' and when they run in the opposite direction they are 'S twill' or 'twill left'. Their angle and definition can be varied by changing the thread spacing and/or the linear density of the warp and weft yarns. For any construction twills will have longer floats, fewer intersections and a more open construction than a plain weave fabric with the same cloth particulars.

Industrial uses of twill fabrics are mainly restricted to simple twills and only simple twills are discussed here. Broken twills, waved twills, herringbone twills and elongated twills are extensively used for suiting and dress fabrics. For details of such twills see Robinson and Marks or Watson. The smallest repeat of a twill weave consists of 3 ends x 3 picks. There is no theoretical upper limit to the size of twill weaves but the need to produce stable fabrics with floats of reasonable length imposes practical limits. The twill is produced by commencing the lift sequence of adjacent ends on one pick higher or one pick lower. The lift is the number of picks which an end passes over and under. In a 2/1 twill, an end will pass over two picks and under one, whilst in a 1/2 twill the end will pass over one pick and then under two picks.

Either weave can be produced as a Z or S twill. There are, therefore, four combinations of this simplest of all twills. To show how pronounced the twill line is even in a 3 x 3 twill, four repeats (2 x 2) lifted ends shown in solid on the point paper. 2/1 twills are warp faced twills, that is fabrics where most of the warp yarn is on the surface, whilst 1/2 twills have a weft face. Weft-faced twills impose less strain on the weaving machine than warp-faced twills because fewer ends have to be lifted to allow picks to pass under them. For this reason, warp-faced twills are sometimes woven upside down, that is as weft-faced twills. The disadvantage of weaving twills upside down is that it is more difficult to inspect the warp yarns during weaving.

Lenos

In lenos adjoining warp ends do not remain parallel when they are interlaced with the weft but are crossed over each other. In the simplest leno one standard end and one crossing end are passed across each other during consecutive picks. Two variations of this structure are shown in plan view and cross-section. Whenever the warp threads cross over each other, with the weft passing between them, they lock the weft into position and prevent weft movement.

Leno weaves are therefore used in very open structures, such as gauzes, to prevent thread movement and fabric distortion. When the selvedge construction of a fabric does not bind its edge threads into position, leno ends are used to prevent the warp threads at each side of a length of cloth from slipping out of the body of the fabric. They are also used in the body of fabrics when empty dents are left in weaving because the fabrics are to be slit into narrower widths at a later stage of processing.

Leno and gauze fabrics may consist of standard and crossing ends only or pairs or multiples of such threads may be introduced according to pattern to obtain the required design. For larger effects standard and crossing ends may also be in pairs or groups of three. Two or more weft threads may be introduced into one shed and areas of plain fabric may be woven in the weft direction between picks where warp ends are crossed over to give the leno effect. Gauze fabrics used for filtration generally use simple leno weaves.

Standard and crossing ends frequently come from separate warp beams. If both the standard and crossing ends are warped on to one beam the same length of warp is available for both and they will have to do the same amount of bending, that is they will have the same crimp. Such a leno fabric is shown in plan view and cross-section. If the two series of ends are brought from separate beams the standard ends and the crossing ends can be tensioned differently and their crimp can be adjusted separately. In such a case it is possible for the standard ends to lie straight and the crossing ends to do all the bending. That crossing threads can be moved either from the right to the left or from the left to the right on the same pick and adjoining leno pairs may either cross in the same or opposite directions. The direction of crossing can affect locking, especially with smooth monofilament yarns. When using two beams it is also possible to use different types or counts of yarn for standard and crossing ends for design or technical applications. The actual method of weaving the leno with doups or similar mechanisms is not discussed in this chapter.

When lenos are used for selvedge construction only one to four pairs of threads are generally used at each side and the leno selvedge is produced by a special leno mechanism that is independent of the shedding mechanism of the loom. The leno threads required for the selvedge then come from cones in a small separate creel rather than from the warp beam. The choice of selvedge yarns and tensions is particularly important to prevent tight or curly selvedges. The crimps selected have to take into account the cloth shrinkage in finishing.

Satins and Sateens

In Britain a satin is a warp-faced weave in which the binding places are arranged to produce a smooth fabric surface free from twill lines. Satins normally have a much greater number of ends than picks per centimetre. To avoid confusion a satin is frequently described as a 'warp satin'. A sateen, frequently referred to as a 'weft sateen', is a weft-faced weave similar to a satin with binding places arranged to produce a smooth fabric free of twill lines.

Sateens are generally woven with a much higher number of picks than ends. Satins tend to be more popular than sateens because it is cheaper to weave a cloth with a lower number of picks than ends. Warp satins may be woven upside down, that is as a sateen but with a satin construction, to reduce the tension on the harness mechanism that has to lift the warp ends.

To avoid twill lines, satins and sateens have to be constructed in a systematic manner. To construct a regular satin or sateen weave (for irregular or special ones see Robinson and Marks or Watson) without a twill effect a number of rules have to be observed. The distribution of interlacing must be as random as possible and there has only to be one interlacing of each warp and weft thread per repeat, that is per weave number.

The intersections must be arranged in an orderly manner, uniformly separated from each other and never adjacent. The weaves are developed from a 1/x twill and the twill intersections are displaced by a fixed number of steps. The steps that must be avoided are: (i) one or one less than the repeat (because this is a twill), (ii) the same number as the repeat or having a common factor with the repeat (because some of the yarns would fail to interlace). The smallest weave number for either weave is 5 and regular satins or sateens also cannot be constructed with weave numbers of 6,9,11,13,14 or 15. The most popular weave numbers are 5 and 8 and weave numbers above 16 are impracticable because of the length of floats. Weave numbers of 2 or 3 are possible for five-end weaves and 3 or 5 for 8-end weaves.

TRIAXIAL WEAVES

Nearly all two-dimensional woven structures have been developed from plain weave fabrics and warp and weft yarns are interlaced at right angles or at nearly right angles. This also applies in principle to leno fabrics and to lappet fabrics where a proportion of the warp yarns, that is the yarns forming the design, are moved across a number of ground warp yarns by the lappet mechanism.

The only exceptions are triaxial fabrics, where two sets of warp yarns are generally inserted at 60° to the weft, and tetra-axial fabrics where four sets of yarns are inclined at 45° to each other. So far only weaving machines for triaxial fabrics are in commercial production. Triaxial weaving machines were first built by the Barber-Colman Co. under licence from Dow Weave and have been further developed by Howa Machinery Ltd., Japan.

Triaxial fabrics are defined as cloths where the three sets of threads form a multitude of equilateral triangles. Two sets of warp yarn are interlaced at 60° with each other and with the weft. In the basic triaxial fabric, the warp travels from selvedge to selvedge at an angle of 30° from the vertical. When a warp yarn reaches the selvedge it is turned through an angle of 120° and then travels to the opposite selvedge thus forming a firm selvedge. Weft yarn is inserted at right angles (90°) to the selvedge. The basic triaxial fabric is fairly open with a diamond-shaped centre. The standard weaves can be modified by having biplane, stuffed or basic basket weaves, the latter being. These modified weaves form closer structures with different characteristics. Interlacing angles of 75° to the selvedge can be produced. At present thread spacing in the basic weave fabric is limited to 3.6 or 7.4 threads per centimetre.

The tear resistance and bursting resistance of triaxial fabrics is greatly superior to that of standard fabrics because strain is always taken in two directions. Their shear resistance is also excellent because intersections are locked. They have a wide range of technical applications including sailcloths, tyre fabrics, balloon fabrics, pressure receptacles and laminated structures.

Selvedge

The selvedges form the longitudinal edges of a fabric and are generally formed during weaving. The weave used to construct the selvedge may be the same, or may differ from, the weave used in the body of the cloth. Most selvedges are fairly narrow but they can be up to 20-mm wide. Descriptions may be woven into the selvedge, using special selvedge jacquards, or coloured or fancy threads may be incorporated for identification purposes. For some end-uses selvedges have to be discarded but, whenever possible, selvedges should be constructed so that they can be incorporated into the final product so as to reduce the cost of waste.

In cloths woven on weaving machines with shuttles selvedges are formed by the weft turning at the edges after the insertion of each pick. The weft passes continuously across the width of the fabric from side to side. In cloths woven on shuttleless weaving machines the weft is cut at the end of each pick or after every second pick. To prevent the outside ends from fraying, various selvedge motions are used to bind the warp into the body of the cloth or edges may be sealed. The essential requirements are that the selvedges should protect the edge of the fabric during weaving and subsequent processing, that they should not detract from the appearance of the cloth and that they should not interfere with finishing or cause waviness, contraction or creasing. Four types of woven selvedges.

HAIRPIN SELVEDGE - SHUTTLE WEAVING MACHINE

A typical hairpin selvedge which is formed when the weft package is carried in a reciprocating media, for example a shuttle. With most weaves it gives a good edge and requires no special mechanism. Frequently strong two-ply yarns are incorporated into the selvedge whilst single yarns are used in the body of the cloth because the edge threads are subjected to special strains during beat-up. To ensure a flat edge a different weave may be used in the selvedge from the body of the cloth. For twills, satins and fancy weaves this may also be necessary to ensure that all warp threads are properly bound into the edge. If special selvedge yarns are used it is important to ensure that they are not, by mistake, woven into the body of the fabric because they are likely to show up after finishing or cause a reduction in the tear and/or bursting resistance of industrial fabrics.

When two or more different weft yarns are used in a fabric only one weft yarn is being inserted at a time and the other yarn(s) are inactive until the weft is changed. In shuttle looms the weft being inserted at any one time will form a normal selvedge whilst the weft yarn(s) not in use will float along the selvedge. If long floats are formed in weaving, because one weft is not required for a considerable period, the floats may have to be trimmed off after weaving to prevent them from causing problems during subsequent processing. Whenever a pirn is changed, or a broken pick is repaired, a short length of yarn will also protrude from the selvedge and this has to be trimmed off after weaving.

Industrial fabrics with coarse weft yarn are sometimes woven with loop selvedges to ensure that the thread spacing and cloth thickness remains the same right to the edge of the cloth. To produce a loop selvedge a wire or coarse monofilament yarn is placed 3 or 4 mm outside the edge warp end and the pick reverses round the wire to form a loop. As the wire is considerably stiffer than a yarn it will prevent the weft pulling the end threads together during beat-up. The wire extends to the fell of the cloth and is automatically withdrawn during weaving. Great care is required to ensure that no broken-off ends of wire or monofilament remain in the fabric because they can cause serious damage to equipment and to the cloth during subsequent processing.

Tucked-in Selvedges

A tuck or tucked-in selvedge. It gives a very neat and strong selvedge and was first developed by Sulzer Brothers (now Sulzer-Textil), Win-terthur, Switzerland, for use on their projectile weaving machines. Its appearance is close to that of a hairpin selvedge and it is particularly useful when cloths with fringe selvedges would have to be hemmed. High-speed tucking motions are now available and it is possible to produce tucked-in selvedges even on the fastest weaving machines.

Tucked-in selvedges, even when produced with a reduced number of warp threads, are generally slightly thicker than the body of the cloth. When large batches of cloth with such selvedges are produced it may be necessary to traverse the cloth on the cloth roller to build a level roll. The extra thickness will also have to be allowed for when fabrics are coated. When tucked-in selvedges can be incorporated into the finished product no yarn waste is made because no dummy selvedges are produced but the reduced cost of waste may be counterbalanced by the relatively high cost of the tucking units.

Sealed Selvedges

When fabrics are produced from yarns with thermoplastic properties the edge of a fabric may be cut and sealed by heat. The edge ends of fringe selvedges are frequently cut off in the loom and the edge threads with the fringe are drawn away into a waste container. Heat cutting may also be used to slit a cloth, in or off the loom, into a number of narrower fabrics or tapes.

For special purposes ultrasonic sealing devices are available. These devices are fairly expensive and can cut more cloth than can be woven on one loom. Whilst they can be mounted on the loom they are, because of their cost, more frequently mounted on a separate cutting or inspection table.

Leno and Helical Selvedges

In most shuttleless weaving machines a length of weft has to be cut for every pick. For looms not fitted with tucking motions the warp threads at the edges of the fabric have to be locked into position to prevent fraying. With some weft insertion systems the weft has to be cut only after every second pick and it is possible to have a hairpin selvedge on one side of the fabric and a locked selvedge on the second side. Leno and helical selvedges are widely used to lock warp yarns into position respectively. When shuttleless weaving machines were first introduced there was considerable customer resistance to fringe selvedges but, in the meantime, it has been found that they meet most requirements.

With leno selvedges a set of threads at the edge of the fabric is interlaced with a gauze weave which locks round the weft thread and prevents ravelling of the warp. As the precut length of the picks always varies slightly, a dummy selvedge is sometimes used at the edge of the cloth. This makes it possible to cut the weft close to the body of the cloth resulting in a narrow fringe which has a better appearance than the selvedge from which weft threads of varying lengths protrude. It also has the advantage in finishing and coating that there are no long lengths of loose weft that can untwist and shed fly.

Because of the tails of weft and because of the warp in dummy selvedges, more waste is generally made in narrow shuttleless woven fabrics than in fabrics woven on shuttle looms. For wide cloth the reverse frequently applies because there is no shuttle waste. Leno selvedges, sometimes referred to as 'centre selvedges', may be introduced into the body of the cloth if it is intended to slit the width of cloth produced on the loom into two or more widths either on the loom or after finishing. Helical selvedges consist of a set of threads which make a half or complete revolution around one another between picks. They can be used instead of leno selvedges and tend to have a neater appearance.

Bibliography

Ainley, Albert: *Woolen and worsted loomfixing; a book for loomfixers and all who are interested in the production of plain and fancy worsteds and woolens*, Lawrence Mass, 1900.

Alford, Marianne Margaret Compton Cust: *Needlework as art*, London: Sampson Low Marston Searle and Rivington, 1886.

Alfred Jenks & Son: *Illustrated Catalogue of Machines Built by Alfred Jenks & Son, Manufacturers of Every Variety of Cotton and Wool Carding, Spinning, and Weaving Machinery, in All Its Departments*, Bridesburg, Pa.: Alfred Jenks & Son, 1853.

Allen, Frederick James: *The shoe industry*, Boston: The Vocation bureau of Boston, 1916.

Allen, Nellie B: *Cotton and other useful fibers*, Boston: Ginn, 1929.

Allington, Sara May: *Practical sewing and dressmaking*, Boston: Page Co., 1913.

Allinson, May: *Dressmaking as a trade for women in Massachusetts*, Washington: Govt. print. off., 1916.

Atkinson, Edward: *Cotton and Cotton Manufactures in the United States*, Boston: Franklin Press, 1880.

Baird, Robert H: *The American Cotton Spinner, and Managers' and Carders' Guide*: A Practical Treatise on Cotton Spinning, Philadelphia: H. C. Baird, 1869.

Batchelder, Samuel: *Introduction and Early Progress of the Cotton Manufacture in the United States*, Boston: Little, Brown and Company, 1863.

Brown Brothers & Co: Brown Brothers & Co., Successors to Butler, Brown & Co.: *Manufacturers of Shaw's U. S. Standard Ring Travelers, Belt Hooks, Wire Goods, Etc., and Dealers in Supplies for Cotton, Woolen, Silk, Jute & Flax Mills, Providence*, R.I.: Brown Brothers & Co., c1881.

E. Jenckes Manufacturing Co: *Illustrated Price List: E. Jenckes Mfg. Co., Sole Manufacturers of Hicks' U. S. Standard Ring Travelers*, Manufacturers' Supplies of Every Description, Pawtucket, R.I.: E. Jenckes Mfg. Co., 1883.

Lowell Machine Shop: *The Lowell Machine Shop, Lowell, Mass.: Builders of Cotton Machinery of Every Description*, Lowell, Mass.: Stone, Bacheller & Livingston, 1882.

Mason Machine Works: *The Mason Machine Works, Taunton Massachusetts, U.S.A.: Inventors and Builders of Cotton Machinery*, Taunton, Mass.: The Company, 1898.

Merrill, Gilbert R: *American Cotton Handbook: A Practical Reference Book for the Entire Cotton Industry*, New York: American Cotton Handbook Co., c1941.

Scott, Robert: *The Practical Cotton Spinner, and Manufacturer: The Managers', Overlookers' and Mechanics' Companion*, Corr. and enl. Philadelphia: H. C. Baird, 1851.

Webber, Samuel: *Manual of Power for Machines, Shafts, and Belts, with the History of Cotton Manufacture in the United States*, New York: D. Appleton and Company, 1879.

Index